GW00982617

THE TOURMALINE GROUP

The Tourmaline Group

R. V. Dietrich
Central Michigan University

VAN NOSTRAND REINHOLD COMPANY
New York

Library of Congress Catalog Card Number: 85-3350
ISBN: 0-30-442-21857-5

Manufactured in the United States of America.

Published by Van Nostrand Reinhold Company Inc.
115 Fifth Avenue
New York, New York 10003

Van Nostrand Reinhold Company Limited
Molly Millars Lane
Wokingham, Berkshire RG11 2PY, England

Van Nostrand Reinhold
480 La Trobe Street
Melbourne, Victoria 3000, Australia

Macmillan of Canada
Division of Canada Publishing Corporation
164 Commander Boulevard
Agincourt, Ontario MIS 3C7, Canada

15 14 13 12 11 10 9 8 7 6 5 4 3 2

Library of Congress Cataloging in Publication Data
Dietrich, Richard V.
The tourmaline group.
Bibliography: p.
Includes index.
1. Tourmaline. I. Title.
QE391.T7D54 1985 553.8′7 85-3350
ISBN 0-442-21857-5

Dedicated to the memory of my parents,
Roy E. and Mida Vincent Dietrich

Contents

Preface

Tourmaline group minerals have graced the cabinets of mineral collectors and museums for untold generations. Their colors and color patterns, along with their diverse shapes and associations, have assured their role as exquisite showpieces. Their workability and durability have, in addition, made them favorites among connoisseurs of colored gemstones. Tourmalines, however, are much more than exquisite showpieces and beautiful gemstones. Their diverse crystal forms and unique structure, their variable chemical compositions, their intriguing physical properties, and their widespread occurrence in nearly all kinds of rocks have long attracted the attention of scientists from several disciplines. Furthermore, they have several potential uses in science and industry.

From an historical standpoint: Tourmaline is possibly the "Lyngurium" —green at one end, light colored at the other—that Theophrastus (ca.315 B.C.) described in *On Stones,* the first known book about minerals. Tourmaline is one of the minerals thought to have been used as a "sunstone" navigation compass by Vikings during the eighth, ninth, and tenth centuries. Tourmaline from South America and the Orient was on the gemstone market as early as the mid 1600s. Tourmaline, primarily because of its electrical properties, occupied the center of the scientific stage during much of the eighteenth century—in fact, Wilcke (1766) thought that tourmaline might provide the key for the grand unification theory that would relate heat, electricity, and magnetism. Tourmaline specimens have been valued by collectors since at least the late 1700s—for example, in 1780, Archduchess Marie-Anne of Austria apparently presented several beautiful specimens of Tyrolean tourmaline from her "extensive cabinet" to Duke Charles of Lorraine for his collection. Tourmaline's varied chemical composition has long been of interest—Arfvedson (1818) discovered the element lithium in tourmaline; Ruskin (1891) characterized the composition of tourmaline as "more like a medieval doctor's prescription, than the making of a respectable mineral"; Nobel laureate W. L. Bragg (1937) called tourmaline "one of nature's catch-all or garbage-can minerals." Tourmaline was first used in tourmaline tongs, a handy polariscope for distinguishing gems, by Karl Michael Marx, in 1827 (duBois-Reymond and Schaefer, 1908). The mineral has subsequently found several additional uses—for example, it has been used in pressure gauges and, because of this use, during World War II, Martin Ehrman was commissioned to find and spirit some high-quality Madagascar tourmaline out of German-occupied France. Tourmaline is currently a leading money-maker in the colored gemstones market, and one predominantly tourmaline specimen from Minas Gerais, Brazil, judged to be one of the world's most valuable mineral specimens, was recently sold for more than $1.25 million.

From the scientific standpoint: Tourmaline has been the subject of more than 2500 articles: articles by chemists, crystallographers, gemologists, geologists, metallurgists, mineralogists, pedologists, and physicists; articles in English, Latin, French, Italian, Portuguese, and Romanian; Dutch, German, Norwegian, and Swedish; Czech, Hungarian, Polish, Russian, and Ukrainian; Turkish; Japanese and Korean.

Despite—or perhaps because of—its long and varied history and its interest to people of so many backgrounds and disciplines, no broad-based summary about tourmaline has been published. Because of this void and because of his long-time love affair with tourmaline—dating back to mineral collecting days in Jefferson and St. Lawrence counties, New York, during the 1930s—the writer decided to prepare this book.

The Tourmaline Group consists largely of a compilation of data and suggestions published through 1980. It also includes information from a few subsequently published and unpublished papers. More than 2500 publications, of which nearly 1000 are included in the bibliography, were reviewed, some only in abstract form. (The complete reference list is available on 3″ × 5″ cards at the U.S. National Museum of Natural History.)

The coverage more or less follows that used by Frondel (1962) in his . . . *Silica Minerals.* The information is treated under the following headings: Nomenclature; Symmetry and Morphological Crystallography; Crystal Structure; Chemistry and Alteration; Inclusions and "Intergrowths"; Color and Optical Properties; Physical Properties; Synthesis; Uses and Recovery; Tourmaline as a Gemstone and in the Decorative Arts; and Occurrences and Geneses. Appendices include lists of additional crystal forms and angles between faces, midpoint analyses for tourmalines of the common solid solution series, and an extensive list of localities from which noteworthy specimens and/or gem materials have been collected.

The text has been written and illustrated so that mineral collectors, students, and professional mineralogists can easily broaden their appreciation of the diverse features, uses, occurrences, and origins of the mineral species belonging to the tourmaline group. The referenced documentation and bibliography should be of particular value to students making literature searches preliminary to, or in conjunction with, research pertaining to tourmaline group minerals.

Acknowledgments

The manuscript for this book was prepared during a coupled research leave (fall semester, 1982) and sabbatical leave (academic year, 1983). Randolph Barton, Jr., David H. Current, Pete J. Dunn, Michael Fleischer, Eugene E. Foord, Carl A. Francis, Clifford Frondel, Richard V. Gaines, Gerald V. Gibbs, Edward J. Gübelin, Raymond W. Grant, Knut S. Heier, Darrell J. Henry, Aphrodite Mamoulides, Dean McCrillis, William C. Metropolis, J. Stewart Monroe, Wayne E. Moore, Joseph A. Nelen, Dusan Němec, Donald R. Peacor, Paul W. Pohwat, Jack Satterly, John Sinkankas, John F. Slack, K. Schmetzer, Dennis J. Thavenet, I. E. Voskresenskaya, John Sampson White, Jr., and Horace Winchell provided specimens, data, suggestions, or other aid. Franklin F. Foit, Jr., Eugene Foord, Clifford Frondel, Raymond Grant, Stephanie M. Mattson, Kurt Nassau, Robert E. Newnham, George R. Rossman, and John White read and criticized part or all of the original manuscript. Andrew V. Bedo, James W. Clarke, Thomas J. Delia, Frances S. Dietrich, Michael Fleischer, Theresa Form, Sven Gavelin, Eric H. Kadler, Rodney C. Kirk, Fumiko Miyazaki, Valerie Peterson, and Karel Topinka translated one or more of the foreign language articles. David D. Ginsburg and Joy Pastucha, librarians at Central Michigan University, and their staffs diligently searched and obtained obscure published materials. I-Ming Aron and William Stanwick of the CMU Computer Center programmed chemical analyses and other data for evaluation and plotting. Dennis A. Pompilius, Coordinator of CMU's Instruction Material Production Unit, supervised preparation of the diagrams. Martha Brian and Brenda Nutzman transcribed the manuscript and the card-file bibliography to word-processor copy and disks. Nancy Dutro, copyeditor, and Caroline Lee, production editor, provided invaluable assistance in converting the manuscript into this book. Grants from the Research Corporation of New York and the Provost's Office of CMU helped defray expenses, including part of the cost of color separations for the plates. Frances S. Dietrich helped in many, many, . . . ways.

I gratefully express my thanks for all of these contributions.

R.V. DIETRICH

THE TOURMALINE GROUP

Chapter 1

Nomenclature

The name "tourmaline" appears to have been derived from the Sinhalese term *turmali*, a designation often applied by ancient Ceylonese merchants to mixed gemstones of unproved identities. The first known use of the name in a Western language publication was in 1707 in *Curiöse Speculasiones bey Schlaf-losen Nächten* (Fig. 1-1). As can be seen, in that book—apparently by Johann Schmidt, using the pseudonym Immer Gern Speculirt (Always Gladly Speculating)—it is stated that "in the year 1703, the Dutch first brought a precious stone called tourmaline, turmale, or trip to Holland from Ceylon, in the East Indies; it had the property of attracting ashes from warm or burning coals, as a magnet does iron, and also of repelling them again . . ." (p.269).

Several authors have outlined interesting aspects of the early historical development of tourmaline nomenclature (e.g., Gerhard, 1777; Romé de l'Isle, 1783; Keferstein, 1849; and E. S. Dana, 1892). Two of the more significant breakthroughs were the recognition that schorl (in part), as well as the gem varieties, is tourmaline (Romé de l'Isle, 1772) and the redefinition of schorl to refer to only black tourmaline (Werner, 1780). Prior to Werner's redefinition, schorl was applied to such diverse materials as rutile needles in quartz, dark amphiboles and pyroxenes, axinite, staurolite, and even the rock basalt (especially that which occurs in columns that resemble crystal forms), as well as to black tourmaline.

Today, tourmaline is used as a group name and is also applied in the generic sense to specimens and occurrences of tourmaline group minerals not identified to species. The word used for tourmaline is essentially the same in all modern languages: tourmaline—English and French; turmalin—German, Russian, and the Scandinavian languages; toermalijn—Dutch; turmalina—Italian, Portugese and Spanish; turmalinie—Polish; turmaliny—Czech; turmala—Arabic; and touramal—Hindi.

A photocopy of the nomenclature section for tourmaline in the sixth

Curiöse
Speculationes
bey
Schlaf-losen
Nächten,
Werden in
Unterschiedlichen Gesprächen
fürgestellet/
Und handeln von allerhand curiösen
sowohl politischen/ theologischen/ medicini-
schen/ physicalischen/ und dergleichen
Dingen;
also
Daß ein ieder curiöser Liebhaber etwas
zu seiner Vergnügung darinnen finden
wird.
Zu eigener nächtlicher Zeit-Verkürtzung aufge-
zeichnet von einem Liebhaber
der
Immer Gern Speculirt.

Chemnitz und Leipzig/
Bey Conrad Stösseln/ Anno 1707.

bey Schlaf-losen Nächten. 

gel-Lac über einwenig Streu-Sand hält/
wenn dieser in einen Augen-Blick sich erhebt/
und begierig in die Höhe an das Siegel-Lac
fähret. Ubrigens habe ich observiret/ daß fast
nichts ist/ das sich nicht solte anziehen lassen
und so man vom Siegel-Lac oder von Agd-
Stein etwas abschabet/ so ziehet es solches
auch wieder an sich als wie Spreu.
Speculator.
Da du nun selbst solche Observationes von
offt genannten Dingen gemacht hast/ so kanst
du mir desto eher Beyfall geben/ daß das Anziehen solcher Dinge nicht magnetisch sey. Daß
du aber vermeynest/ es behielte solchem nach
der Magnet gantz alleine die eigene Krafft zu
ziehen/ das will ich zwar eben nicht bejahen/ weil
ich nicht zweifeln will/ es werden noch mehr
Dinge gefunden/ welche eine magnetische Ei-
genschafft an sich haben/ ob es uns gleich nicht
eben bekannt ist. Wie mir denn ohnlängst nur
der curiöse Herr Daumius, ietziger wohlbestell-
ter Stabs-Medicus bey der Königl. Polni-
schen und Churfl. Sächs. am Rhein stehenden
Militz/ erzehlet hat/ daß Anno 1703. die Hol-
länder einen aus Ost-Indien von Ceylon kom-
menden Edelstein/ Turmalin, oder Turmale,
auch Trip genannt/ zum ersten mahl nach
Holland gebracht hätten/ welcher die Eigen-
schafft

Figure 1-1
First publication known to include the name tourmaline ("Turmalin"); see fourth line from the bottom on right side (page 269). Page to left is title page. (From Schmidt, 1707.)

edition of Dana's *System* . . . (1892) is given as Figure 1-2. Subsequent to its publication, a few additional terms have been introduced for tourmaline species, varieties, and hypothetical end-member compositions. The following alphabetical list includes definitions for pre-1892 terms not recorded by Dana, terms that have been introduced since Dana's listing, and terms on Dana's list that require clarification and/or are still used widely. The terms given in bold face type are the species listed in Fleischer's *Glossary of Mineral Species 1983* plus the species chromdravite, which was subsequently approved by the International Mineralogical Association's Commission on New Minerals and Mineral Names; those in italics are frequently applied designations for color varieties; those in regular type and included in square

TOURMALINE.

426. TOURMALINE. *Early syn. of precious T.* Turamali, Turmalin (fr. Ceylon), ***Ceylon name***, *Garmann*, Curiosæ Speculationes, etc., von einem Liebhaber, der immer gern speculirt, Chemnitz, 1707. Pierre de Ceylan; un petit aiman; *M. Lemery* la fit voir, etc., Hist. Ac. Sci., Paris, p. 8, 1717. Aschentrecker *Holl.;* Aschenzieher *Germ.;* Ash-drawer *Engl.* [alluding to electrical property]. Zeolithus vitreus electricus, Tourmalin, *Rinmann*, Ak. H. Stockh., 1766; *v. Born*, Lithoph., **1**, 47, 1772. Borax electricus *Linn.*, Syst., **96**, 1768. Tourmaline Garnet *Hill*, Foss., 148, 1771. Tourmaline *Kirw.*, Min., 1, 271, 1794.

Early syn. of opaque T. Schurl pt, *Erker*, 1595; Schirl pt. *Brückmann*, 1727 [see p. 206]. Skiörl pt., Corneus crystallisatus pt., *Wall.*, 139, 1747. Basaltes cryst. pt., Skörl-Crystall pt., *Cronst.*, 70, 1758. Schörl, Stangenschörl, *Germ.;* Shorl, Shirl, Cockle, *Engl.* Borax Basaltes *Linn.*, Syst., 95, 1768. Basaltes crystallisatus *v. Born*, Lithoph., **1**, 34, 1772, **2**, 95, 1775. Shorl *Kirw.*, Min., 1, 265, 1794.

Syn. from union of T. and S. in one species. Tourmaline ou Basalte transparent = Schorl, *de Lisle*, Crist., 266, with fig. cryst. (and proofs of ident. of T. & S.), 1772. Schorl transparent rhomboidal dit Tourmaline et Peridot = Schorl. *de Lisle*, Crist., **2**, 344, with figs., 1783. Schörl, Stangenschörl (incl. var. (1) Schwarzer S., (3) Elektrischer S. = Turmalin), *Wern.*, Cronst., **169**, 1780; Bergm. J., **1**, 374, 1789; *Jameson*, Min., 1816. Tourmaline *H.*, Tr., 3, 1801.

Var. introd. as Sp. Rubellite (fr. Siberia) *Kirw.*, Min., **1**, 288, 1794 = Daourite *Delameth.*, T. T., **2**, 303, 1797 = Siberite *l'Hermina*, J. de l'École Polytechn., **1**, 439 = Tourmaline apyre *H.*, **4**. 1801 = Apyrit *Hausm.*, Handb., 642, 1813. Indicolite and Aphrizite (fr. Norway) *d'Andrada*, J. Phys., **51**, 243, 1800, Scherer's J., **4**, 19, 1800. Taltalite *Domeyko*, Min., 139, 1860 = Cobre negro estrellado de Tantal (Atacama).

Var. introd. as Subsp. Achroit (fr. Elba) ***Herm.***, J. pr. Ch., **35**, 232, 1845. **Dravit** *Tschermak*, Min., 472, 1883.

Figure 1-2

Nomenclature section for tourmaline as it appears in Dana's *System* . . . (1892).

brackets have been suggested for end-member compositions not yet verified as existing in nature.

Several of the terms are unacceptable and should be abandoned so far as their application to tourmaline group minerals. The color-based terms, if used at all, should be used only to indicate tourmalines of the appropriate color, as viewed macroscopically; for example, rubellite would apply to red (including pink) tourmalines no matter what their species. Strictly speaking, however, these terms also would be better abandoned—for one reason, several workers, especially in Russia, have used rubellite as a synonym for elbaite; for another, some workers use rubellite to mean red or pink elbaite without realizing that some of their so-designated specimens are, for example, red dravite; and, for still another reason, there is no parallel term for either black or brown tourmalines, many of which could be one or another species on the basis of color alone. All such dilemmas would, of course, be avoided if only color adjectives were used for specimens not identified to species—for example, green tourmaline. Furthermore, this scheme could be logically extended to designate specimens that have been identified to species—for example, green uvite. Chemically based terms such as chromium-tourmaline and lithium-bearing tourmaline that have been used in the literature are not included among the defined terms; thirty such terms are given in Table 1.1.

Table 1-1
Examples of Tourmalines Recorded in the Literature with Chemical Indicators

Designation[a]	*Reference (example only)*
alkali tourmaline	Dana, 1892
alkali-free tourmaline[b]	Rosenberg and Foit, 1979
chromturmalin	Cossa and Arzruni, 1883
chromian dravite	Dunn, 1977*b*
cobaltoan tourmaline[b]	Frondel, 1977, personal communication
cuproan tourmaline[b]	Fairbanks, 1973[c]
ferric-iron tourmaline	Frondel, Biedl, and Ito, 1966
ferroan-vanadian tourmaline	Fairbanks, 1973[c]
gallium tourmaline[b]	Voskresenskaya and Barsukova, 1971
high-iron tourmaline	Slawson, 1936
iron tourmaline	VanHorn, 1926
lead-bearing tourmaline	Lebedev, 1937
lime-dravite	Fairbanks, 1973[c]
lithium tourmaline	Chaudhry and Howie, 1976
magnesia-tourmaline	Becht, 1913
magnesian tourmaline	Bruce, 1916
magnesium-calcian tourmaline	Fairbanks, 1973[c]
magnesium tourmaline	Dana, 1892
magnodravite	Wang and Hsu, 1966
manganese tourmaline	Slivko, 1959*a*
mangano-dravite	Strunz, 1970
nickeloan tourmaline[b]	Frondel, 1977, personal communication
soda-manganesiferous tourmaline	Sandréa, 1949
stannian tourmaline	Fairbanks, 1973[c]
titan (titano, titanian) tourmaline	Kunitz, 1936
vanadian tourmaline	Fairbanks, 1973[c]
vanadium-bearing tourmaline	Foit and Rosenberg, 1979
zincian tourmaline[b]	Fairbanks, 1973[c]
zinciferous tourmaline	Jedwab, 1962

[a]Designations are given in the form used in the cited reference.
[b]To the present known only as a synthetic material.
[c]Names not supported with data or references.

Achroite: colorless tourmaline; name based on Greek *a* (= without) and *chroma* (= color) in allusion to its lack of color (Hermann, 1845).

[Aluminobuergerite]: a suggested end-member, $Na_{1-x}Al_3Al_6B_3Si_6O_{27}O_{3-x}(OH)_{1-x}$ (Foit and Rosenberg, 1975).

Alumoelbaite: name applied by Povarennykh (1972) to the elbaite part of his elbaite-schorl series.

Andalusite: according to Aitkens (1932), this name—the accepted name of one of the Al_2SiO_5 trimorphic mineral species—has been used as a "trade name for brown tourmaline." Application of the term andalusite to a tourmaline is unacceptable.

[Belbaite]: in Vernadsky's (1913) scheme, one of the hypothetical tourmaline molecules ($M'_{14}Al_2B_2Si_4O_{21}$); the name appears to be elbaite with a prefixed B, Vernadsky's designator for this "molecule."

Brazilian chrysolite: unacceptable designation; used by some members of the gem trade to refer to yellow-green tourmaline that resembles chrysolite.

Brazilian emerald: unacceptable designation; used by some members of the gem trade to refer to green and transparent tourmaline that resembles emerald. (This term and some of the similar terms date back to at least the 1700s—see Romé de l'Isle, 1783.)

Brazilian peridot: unacceptable designation; used by some members of the gem trade to refer to honey-yellow to green tourmaline that resembles peridot.

Brazilian ruby: unacceptable designation; used by some members of the gem trade to refer to transparent red tourmaline that closely resembles ruby (cf. Siberite).

Brazilian sapphire: unacceptable designation; used by some members of the gem trade to refer to transparent blue tourmaline that closely resembles sapphire.

Buergerite: tourmaline (bronzy brown) made up predominantly of $NaFe^{3+}_3Al_6B_3Si_6O_{30}F$; type specimen from Mexquitic, San Luis Potosi, Mexico; named for American crystallographer, M. J. Buerger (Donnay, Ingamells, and Mason, 1966).

Ceylon chrysolite: unacceptable designation; used by some members of the gem trade to refer to yellow or yellow-green tourmaline that resembles chrysolite.

Ceylon(ese) peridot: unacceptable designation(s); used by some members of the gem trade to refer to honey-yellow tourmaline that resembles peridot.

Chameleonite: unacceptable designation; according to Shipley (1971), this designation has been applied to alexandritelike tourmaline (cf. Deuterolite).

Chromdravite: tourmaline (dark green, nearly black) made up predominantly of $NaMg_3(Cr_5Al)B_3Si_6O_{27}(OH)_4$; type specimen from the Onezhkii depression, central Karelia, U.S.S.R.; named for composition (Rumantseva, 1983).

Coronite: name suggested for dravite from Crown Point, New York; name based on *corona*, in reference to the locality (Hunt, 1886).

Cromolite: unacceptable designation; used by some members of the gem trade to refer to medium dark green tourmalines with good transparency along *c;* although the term was originally applied only to Brazilian Cr-rich tourmalines, it has subsequently been applied rather widely to similarly appearing green tourmalines regardless of their place of origin or their Cr-content (D. Sauer, 1983, personal communication).

Deuterolite: name suggested for alexandritelike (i.e., green in daylight or fluorescent tube light and red in incandescent light) tourmaline; specimen described was a chromium-bearing dravite from Šabry (= Shabrovsk?) in the Urals (Cossa and Arzruni, 1883).

Donnayite: term applied to an Fe^{3+} tourmaline with Al_2O_3 20-26 weight percent by Kornetova (1975); this use of the term was not approved by the IMA Commission on New Minerals and Mineral Names (J. A. Mandarino, 1983, personal communication). (Donnayite is the accepted name of triclinic $Sr_3NaCaY(CO_3)_6 \cdot 3H_2O$.)

Dravite: tourmaline (typically brown, greenish or brownish black) made up predominantly of $NaMg_3Al_6B_3Si_6O_{27}(OH,F)_4$; type material from Dobrowa on the Drau (now Drava) River in Carinthia, Austria; named for type locality; (Tschermak, 1885).

Eicotourmaline (= eicoturmalin; also spelled eikotourmaline): name suggested by Lodochnikov (1933) for a mineral described as being like tourmaline but containing no boron and being optically biaxial, with 2V up to 40°.

Elbaite: tourmaline (several diverse colors—especially green, pink, and colorless—commonly having zone relations within individual crystals) made up predominantly of $Na(Al,Li)_3Al_6B_3Si_6O_{27}(O,OH)_3(OH,F)$; named for Island of Elba in the Mediterranean Sea; in Vernadsky's (1913) scheme, elbaite was one of the hypothetical tourmaline molecules ($M'_2Al_6B_2Si_4O_{21}$)—compare with kalbaite and belbaite. Also applied by Povarennykh (1972) to the elbaite-schorl series.

Emeralite: unacceptable designation; formerly used by some members of the gem trade and a few geologists (e.g., Jahns and Wright, 1951) to refer to emerald-green tourmaline (Aitkens, 1932).

Eukotourmaline: apparently an erroneous recording of eicotourmaline by English (1939) (see Spencer, 1934).

Ferridravite: tourmaline (black) made up predominantly of (Na,K) $(Mg,Fe)_3Fe_6B_3Si_6O_{27}(O,OH)_3(OH,F)$; type from the San Francisco mine near Villa Tunari (Alto Chapare), Bolivia; named for dravite plus ferric iron content (Walenta and Dunn, 1979).

Ferroelbaite: name applied by Povarennykh (1972) to schorl part of elbaite-schorl series.

Ferroschorlite: name applied by Povarennykh (1972) to schorl part of dravite-schorl series.

Gouverneurite: name suggested for brown, magnesium-rich tourmalines from Gouverneur, Saint Lawrence County, New York; name based on locality (VanHorn, 1926).

Indicolite (also indigolite): blue tourmaline; name based on Latin *indicum* (= indigo) in allusion to its color (d'Andrada, 1800).

Iochroite (originally Jochroite): the specimen so named by K. Pipping (as recorded—with no B-content!—by Nordenskiöld, 1863) was shown to be violet tourmaline by LaCroix (1918).

Jochroite: see Iochroite.

[Kalbaite]: in Vernadsky's (1913) scheme, one of the hypothetical tourmaline molecules ($M'_8Al_4B_2Si_4O_{21}$); name is for DeKalb, Saint Lawrence County, New York, a locality from which the tourmaline appeared to match this "molecule."

Lapis Lyncurius: the contention that Theophrastus' (ca. 315 B.C.) description of Lyngurium applied to tourmaline may be correct (see, for example, Watson, 1759). On the other hand, Pliny's (77) use of the term appears to refer to amber or possibly yellow zircon (see, for example, Hamlin, 1873).

Liddicoatite: tourmaline (several deverse colors, like elbaite, commonly with zoned relations within individual crystals) made up predominantly of $Ca(Li,Al)_3Al_6B_3Si_6O_{27}(O,OH)_3(OH,F)$; type specimen from detrital overburden at Antsirabe, Madagascar; named for American gemologist, R. T. Liddicoat, Jr. (Dunn, Appleman, and Nelen, 1977).

Magnesioschorlite: name applied by Povarennykh (1972) to the dravite part of the dravite-schorl series.

Mineral H: designation applied to a titanium-rich tourmaline in the pegmatite of the Commercial quarry, Crestmore, California, and also recorded from the North Hill and Victoria City quarries in the Riverside area of California (Woodford, Crippen, and Garner, 1940).

Oxytourmaline: name suggested by Frondel, Biedl, and Ito (1966) for application to tourmalines with noteworthy replacement of (OH,F) by O. (The prefix is used in the same way, so-to-speak, as it is in the names oxyhornblende, oxyannite, etc.).

Pierrepontite: name suggested for black, iron-rich tourmaline from Pierrepont, Saint Lawrence County, New York; name based on locality; VanHorn (1926). (Dunn et al., 1977, recommend that all black tourmaline from this locality be called uvite ((as redefined)).)

Rubellite: pink or red transparent to nearly opaque tourmaline; name based on Latin *rubellos* (= reddish) in allusion to its color (Kirwan, 1794).

Schorl: tourmaline (black) made up predominantly of $NaFe_3Al_6B_3Si_6O_{27}$-$(OH)_4$; derivation of name in doubt (see Dana, 1892, xliv-xlv); term first(?) recorded by Mathesius (1564) (see Fig. 1-3).

Die IX. Predig/Vom zin/bley/

Da bergkleut ein schubwand finden die quartzigt ist/vnd derquartz weit fürsticht/haben sie jhre rechnung sie sey weit gangen/schlagen dem gang deste weyter für/vnd treiben jhre röschen biß sie auffs gefert kommen. Da ein fahl oder tumpel in einem fliessenden wasser inn der sicherung viel stein hat/ist ein starcke vermutung es sey ein gang oder fletz in der nehe.

Von den bergkarten so in vnd neben den zwittern pflegen zu brechen.

Wenn nu die genge außgericht vnd beritten sein/da gehören bergkverstendige leute zu/die sichern/probieren/vnnd guten zwitter von falschen vnterscheiden können/den der zwitter bricht selten rein vnd allein.

Schürl.

Es ist ein schwartz bergkart/schürl genant/sihet dem zwitter ehnlich/die ist auch schwartz/biß weilen mult vnd leicht/vn im wasser fluchtig/offtmals ist sie auch schwer vnnd feste/das sie vnterm zinstein sitzen bleibet/das man sie im brennofen mit schaden des zwitters teuben muß/denn sie reubet im fewr/vnd gibt viel schlacken/vnd macht das zin hart vnd weyßflecket.

Wolfrumb.

Wolfrumb/welches die Lateiner Wolffsschaum/etliche wolffshar heyssen/darumb das es schwartz vnd lenglicht ist/bricht auch neben dem zwitter/wie glantz neben dem silber ertz/welcher lind vnnd mild ist/der ist schidig vnnd fluchtig im wasser/den groben vnnd spissigen muß das fewer wegreumen/damit die zin nicht vnscheinlich werden.

Es bricht auch kobelt in zwitter (wie sich im Buchholtz silber vnd zwittergenge mit einander schleppen) der macht die zin auch hart vnd weyßflecket /

Spießglaß oder giffiiger kiß.

Spießglaß oder wie andere fürgeben / der giffiige kiß/so dem zwitter bricht/sihet schier dem Wolfrumb ehnlich/ist sehr ein schedlich vnnd giffiig metal/ welches durchs fewer vberweltiget/vnnd vom zwitter abgebrent wird/der rauch vnd stanck daruon/verderbet laub/graß/hoppen vnnd getreyde/vnnd das wasser so von den lautertrögen vnd henden fellet/ist sehr vergifftet/wie viel leut vnnd viehe daruon zu Erbarfdorff gestorben sein.

Eysenmal.

Eysenmal oder eysenschüssig art ist gilblicht vnd rötlicht/wenn es schwertzlich ist/betreugt es viel newe Zinherrn/so es gerieben vnd gesichert wird/so rötet es/vnnd weil das wasser rot wird im sichertrog/ist noch eisenmal drunter/Im brennofen mattet man die art/sonst werden die zin vnartig vnd flecket daruon.

Weysser vnnd grawer kiß.

Weysser vnd grawer kiß/raubet den stein/vnd macht die zin mürbe vnd grießlicht/wenn er zumal kupfferig ist/darumb muß man jn im ofen zu todte brennen/vnd offt reissen oder wenden/so lang die zwitter noch ein rauch von sich geben.

Spat.

Roter/weysser/gelber spat/ist er leicht/so ist er flüchtig/der schwere setzt sich / den kan man one schaden nicht verbrennen/ob er aber wol reubet vnd viel schlacken gibt/macht er doch die zin nicht vnschmeydig vnd dönnig.

Wismut

Figure 1-3
Schorl ("Schürl") as recorded by Mathesius (1564); see first marginal note.

Schorlite: name sometimes used for schorl (see, for example, Hunt, 1886, and Winchell and Winchell, 1951); also applied by Povarennykh (1972) to the dravite-schorl series.

Schorlomite: name used, apparently by mistake, for schorl by Leckebusch (1978). (Schorlomite is the accepted name for $Ca_3(Fe^{3+},Ti)_2(Si,Ti)_3O_{12}$ of the garnet group.)

Siberian ruby: unacceptable designation; used by some members of the gem trade to refer to ruby-red tourmaline (specifically that from the vicinity of Murzinka, Siberia); compare with Siberite.

Siberite: unacceptable designation; used by some members of the gem trade to refer to a violet-red to purplish rubellite from Shaitanka, near Murzinka, Siberia; named for locality (l'Hermina, 1799).

TOURMALINE: group name for all species given in bold face on this list; derivation of name is noted in preceding text.

[Tsilaisite]: a hypothetical tourmaline group molecule ($H_8Na_2Mn_6Al_{12}Si_{12}$-$B_6O_{62}$) (or, in line with formulae used herein, $NaMn_3Al_6B_3Si_6O_{27}(OH)_4$; originally used to account for a Mn-rich tourmaline of the elbaite-schorl series from Tsilaisina, Madagascar; named for the locality (Kunitz, 1929). (Slivko, 1959*a*, has used the modified formula $Na(MnAl_2)Al_6$-$(BO_3)_3Si_6O_{18}(O,OH)_4$.)

Uvite: Tourmaline (typically dark brown to black) made up predominantly of $CaMg_3(Al_5Mg)B_3Si_6O_{27}(OH,F)_4$; originally used to designate an hypothetical molecule (Kunitz, 1929); used herein as redefined by Dunn, Appleman, and Nelen (1977), who designated a gemstone specimen from Ceylon (U.S. National Museum of Natural History specimen #C5212) as the neotype; named for province of Uva, Sri Lanka (formerly Ceylon), by Kunitz (1929).

Verdelite: green tourmaline; name based on Latin *viridis* (or French *verd*) in allusion to its color (Quensel and Gabrielson, 1939). [Some jewelers call dark green tourmaline "African tourmaline" or "Transvaal tourmaline," no matter what its source (G.I.A., notes for Colored Stones course); cf. Cromolite.]

Watermelon: adjective applied widely to tourmaline—in most, if not all, cases elbaite—that is color-zoned concentrically around *c* with a pink central zone surrounded by a green outer zone.

Xeuxite: variant of zeuxite given by Chester (1896).

Zeuxite: acicular tourmaline from Huel Unity, Cornwall, England; named by Thomson (1836); later shown to be a ferriferous tourmaline by Greg (1855).

Chapter 2

Symmetry and Morphological Crystallography

SYMMETRY

Four different sets of axes have been used in descriptions of crystals in the trigonal subsystem of the hexagonal crystal system. The axial setups for these schemes—generally referenced to Weiss (1816-1817), Miller (1839), Bravais (1851), and Schrauf (1861)—are described and illustrated by, for example, Frondel (1962).

Bravais indices, which are based on three *a* axes with their positive ends at 120° to each other and a *c* axis perpendicular to the plane of the *a* axes (see Fig. 2-1), are used throughout this book, except on Table 3-6. However,

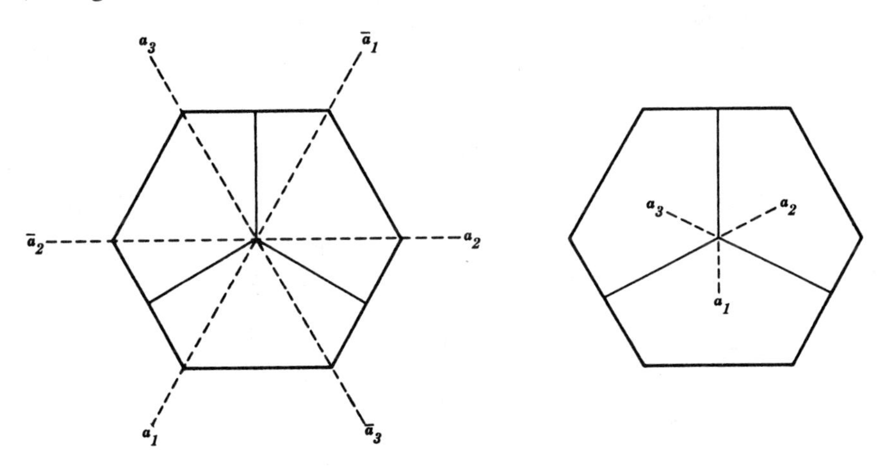

Figure 2-1
Axes for hexagonal system. *Left,* Bravais (1851) four-axis setup; *right,* Miller (1839) three-axis setup. In both diagrams, *c*, which is not shown, is perpendicular to the plane of the *a* axes.

However, Miller indices, which are based on three equivalent axes that are parallel to the polar edges of unit rhombohedra faces (the interaxial angle thereby becoming the definitive constant), are also given in Table 2-3.

Minerals of the tourmaline group crystallize in the ditrigonal pyramidal ($3m$) class of the trigonal subsystem of the hexagonal crystal system. The symmetry of this class is illustrated by the stereogram, Figure 2-2. The threefold axis of symmetry is c; the mirror planes are normal to each a axis. The class is hemimorphic and thus has no axis or planes of symmetry perpendicular to c and no center of symmetry. Consequently, {0001} is a pedion and all forms $\{h\,o\,\bar{h}\,l\}$, $\{h\,h\,\overline{2h}\,l\}$, and $\{h\,k\,\bar{i}\,l\}$, are pyramids rather than dipyramids.

The general forms of the ditrigonal pyramidal class are given in Table 2-1.

Polar Asymmetry—Antilogous versus Analogous Poles

The c crystallographic axis is the polar axis. By definition, the positive end of the c axis is called the antilogous pole and the negative end ($\bar{c}$) is called the analogous pole (Riess and Rose, 1843).

The poles can be distinguished by structural analysis (Barton, 1969), etching (Wooster, 1976), and by tests that measure pyroelectric or piezoelectric effects (see Chap. 7). The structure is such that the SiO_4 tetrahedra point toward the analogous pole (see Fig. 3-3). Etch figures on prism faces have acute-angle ends that also point toward the analogous pole (see Fig. 4-8). Charges developed in response to temperature or pressure differences are given in Table 2-2.

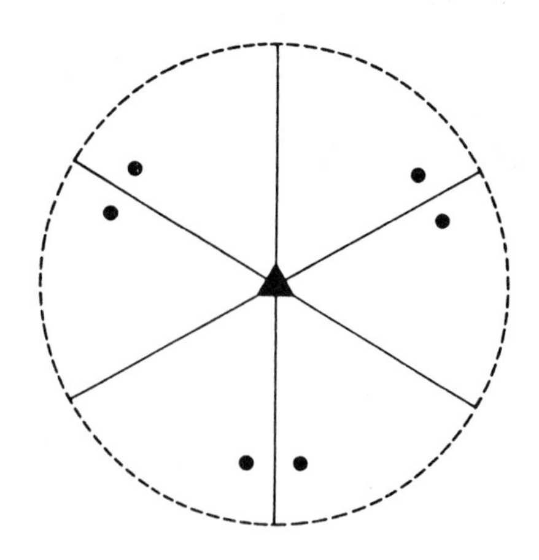

Figure 2-2
Stereogram showing symmetry and general form $\{21\bar{3}1\}$ of the $3m$ class.

Table 2-1
General Forms: Ditrigonal Pyramidal Crystal Class

Name	*Form*			*Faces*
	Top		*Bottom*	
pedion	{0001}		{0001}	1 each
trigonal prism		$\{10\bar{1}0\}$		3
trigonal prism		$\{01\bar{1}0\}$		3
hexagonal prism		$\{11\bar{2}0\}$		6
ditrigonal prism		$\{hk\bar{i}0\}$		6
ditrigonal prism		$\{kh\bar{i}0\}$		6
trigonal pyramid	$\{h\,0\,\bar{h}\,l\}$		$\{h\,0\,\bar{h}\,\bar{l}\}$	3 each
tirgonal pyramid	$\{0\,h\,\bar{h}\,l\}$		$\{0\,h\,\bar{h}\,\bar{l}\}$	3 each
hexagonal pyramid	$\{h\,h\,\overline{2h}\,l\}$		$\{h\,h\,\overline{2h}\,\bar{l}\}$	6 each
ditrigonal pyramid	$\{h\,k\,\bar{i}\,l\}$		$\{h\,k\,\bar{i}\,\bar{l}\}$	6 each
ditrigonal pyramid	$\{k\,h\,\bar{i}\,l\}$		$\{k\,h\,\bar{i}\,\bar{l}\}$	6 each

Note: These forms except for the pedions, are consistent with the setting in which the *a* axes have twofold symmetry and/or the planes $\{11\bar{2}0\}$ are reflection planes; if the symmetry axes are the lattice rows [[21.0]] and/or the planes $\{10\bar{1}0\}$ are reflection planes, the symmetry symbol becomes *31m,* and the nature of the {*h*0— —} and {*hh*— —} forms are transposed.

Table 2-2
Characteristics of Antilogous Versus Analogous Ends of Tourmaline Crystals

Antilogous end (c)	*Analogous end* ($\bar{c}$)
(−) electrical charge upon heating (+) electrical charge upon cooling	(+) electrical charge upon heating (−) electrical charge upon cooling
(−) electrical charge upon decompression along *c* (+) electrical charge upon compression along *c*	(+) electrical charge upon decompression along *c* (−) electrical charge upon compression along *c*
pyramidal faces: typically more acute (*i.e.*, steeper)[a] (Fig. 2-3, *left*)	pyramidal faces: typically less acute (*i.e.*, less steep) (Fig. 2-3, *left*)
forms: more numerous on doubly terminated crystals	forms: less numerous on doubly terminated crystals
crystal faces: typically present on singly terminated crystals (e.g., Johnsen, 1907; LaCroix, 1922; Nemec, 1954; Čech, 1958; Barton, 1968[a]; Dunn, 1975*b*)	crystal faces: commonly absent on singly terminated crystals

(continues)

[a]According to a census made of specimens in the U.S. National Museum of Natural History in the 1960s, G. Donnay found about 80 percent of the singly terminated crystals to exhibit the steeper antilogous end; the observation was recorded incorrectly (i.e., in the opposite sense) in Donnay and Barton (1967), Barton and Donnay (1968), and Barton (1969)—G. Donnay, 1982, personal communication.

Table 2-2 *(continued)*
Characteristics of Antilogous Versus Analogous Ends of Tourmaline Crystals

Antilogous end (c)	*Analogous end ($\bar{c}$)*
edges between *r* {$10\bar{1}1$} terminations and *m* {$10\bar{1}0$} prisms: parallel to the (0001) plane (Fig. 2-3, *right*) (Deer, Howie, and Zussman, 1962, state that exceptions exist in crystals that also exhibit the supplementary prism {$01\bar{1}0$}.)	edges between *r′* {$1\bar{1}0\bar{1}$} terminations and *m* {$10\bar{1}0$} prisms: at an angle to the (0001) plane (Fig. 2-3, *right*)
(0001): high luster (J. D. H. Donnay, 1947 personal communication to C. Frondel)	($000\bar{1}$): commonly low luster (Donnay, 1947 pers. commun. to C. Frondel)
striations on ($10\bar{1}1$): parallel to the opposing edge (Donnay, 1947 pers. commun. to C. Frondel)	striations on $10\bar{1}\bar{1}$: parallel to adjacent edges (Donnay, 1947 pers. commun. to C. Frondel)
growth, in nature: more rapid and even preferential toward antilogous end on crystals (e.g., Tertsch, 1917; Frondel, 1936; Němec, 1954) and during authigenic growth on detrital grains (Alty, 1933)	however, Němec (1954) found tourmaline crystals from 16 European and American localities to have their analogous ends directed away from their substrates.
growth, during synthesis: dominant growth on verdelite seeds is toward the antilogous end (Voskresenkaya and Barsukova, 1971)[b]	growth, during synthesis: dominant growth of both black and light green tourmaline on schorl seeds and of both colorless and greenish blue tourmaline on elbaite seeds is toward the analogous end (see Fig. 8-1) (Voskresenskaya and Barsukova, 1971)
etching: deeper on faces at the antilogous end; that is, the rate of solution is greater toward the analogous end (Kulaszewski, 1921; Frondel, 1935 and 1948)	etching: less marked (in fact, commonly absent) at the analogous end; that is, the rate of solution is lesser toward the antilogous end (Kulaszewski, 1921; Frondel, 1935 and 1948) (Fig. 2-4)

[b]The question arises: Do elevated temperatures and/or pressures during synthesis (and natural growth) cause electric field differences that are responsible for the preferential growth?

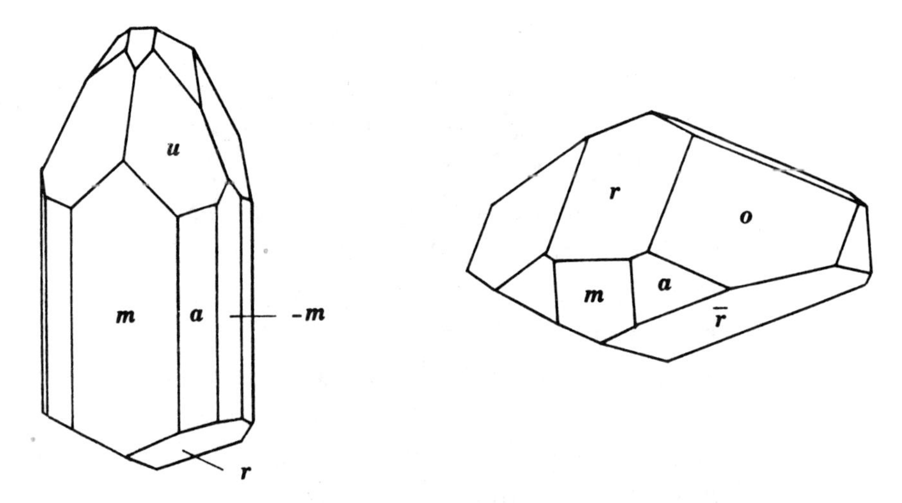

Figure 2-3
Morphological indications of antilogous *(top)* versus analogous *(bottom)* ends of tourmaline crystals. For explanation, see text.

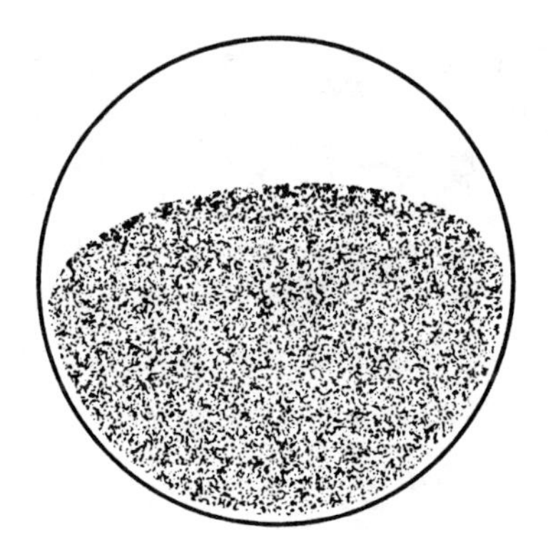

Figure 2-4
The remains (indicated by stippled pattern) after etching of a tourmaline sphere with *c* axis vertical and antilogous end at top. (After Kulaszewski, 1921).

MORPHOLOGICAL CRYSTALLOGRAPHY

The first known illustration of a tourmaline crystal is in Gesner's book *De Omne Rerum Fossilium,* dated 1565 (Fig. 2-5). As can be seen, striae parallel to *c*, which are so typical of tourmaline, are indicated. The probability that the label, "Smaragdus Bresilicus," which is usually translated as Brazilian emerald, refers to tourmaline rather than beryl is permissively supported by

De figuris lapidum, &c.

minatur. Amianti veri è Cypro quem Franciſcus Calceolarius, pharmacopola Veronenſis vt artis ſuæ peritiſſimus, i omnium ſimplicium medicamentorum indagator acerrimus, ad me miſit, fruſtulum delineatũ hîc exhibeo: quod fila molliuſcula neri texiq́; apta remittit, & ab alumine ſciſſili tum vero & aluminis vim ſaporemq́; referente, tum falſo & inſipido, quod alumen plumę hactenus vocare ſoliti ſunt Medici, non parum differt.

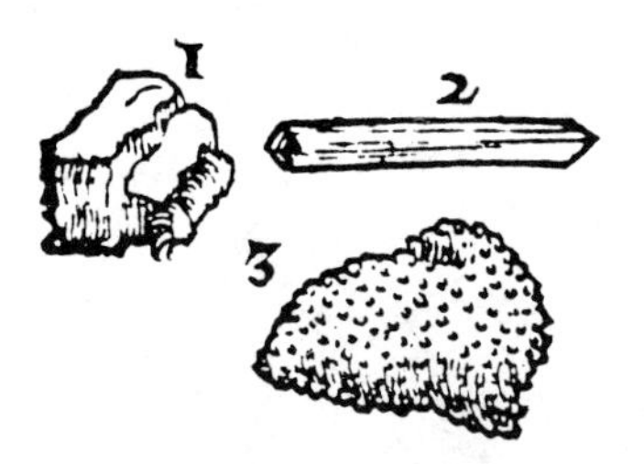

1 *Amiantus è Cypro.*
2 *Smaragdus Breſilicus, cylindri ſpecie.*
3 *Hammites vel Ammonites, minor, minimis piſcium ouis vel araneorum ſimilis, velutiq́; ex arenulis coagmentatus.*

Hic & alijs aliquot in locis, optime Lector, non vnam ſolùm, vel plures, ſed inſtituto ordini conuenien-

Figure 2-5
Earliest known illustration of a tourmaline crystal *(2)*, erroneously labeled as Brazilian emerald. (From Conrad Gesner's essay *De Omne Rerum Fossilium*, published in Zurich, in 1565.)

early descriptions of, for example, deLaet (1647). In any case, indisputable diagrams of tourmaline became relatively common in the late 1700s and early 1800s. Examples are those of Romé de l'Isle (1772), Born (1790), and Haüy (1801 and 1822) (Fig. 2-6).

Victor Goldschmidt's *Atlas der Kristallformen* (1922) lists 222 forms and includes 403 diagrams for tourmaline. The forms are given in Table 2-3 and Appendix A. Among the major contributors referenced in Goldschmidt's compilation were Haüy (1801 and 1822), Rose (1836), Erofeyev (1871), D'Achiardi (e.g., 1896), von Worobieff (1900), Reimann (1907), and LaCroix (1893, 1910, and 1914). Especially noteworthy, so far as updating Goldschmidt's compilation, are the papers of Frondel (1936), Tarnovskii (1961), and Slivko (1963).

Table 2-3 gives 16 principal forms, along with the appropriate angles for each. Appendix A gives 313 additional forms, some that are rare, others questionable, and still others descredited.

In Table 2-3, most of the letter designations are those used by Goldschmidt (1922); the *h k i l* indices are G_1 (not the G_2) values of Goldschmidt, a few of which have been revised to agree with the Miller indices that he also recorded: Miller indices are given in the next column. The tabulated angles are as follows:

ϕ = the azimuthal angle to $(11\bar{2}0)$,
ρ = the interfacial angle to (0001),
A_1 = the interfacial angle to $(2\bar{1}\bar{1}0)$,
A_2 = the interfacial angle to $(\bar{1}2\bar{1}0)$.

They are, as noted, based on the nominal value of 0.4465 for the axial ratio *c*:*a*. As might be expected because of the variation in composition within the tourmaline group, measured values of *c* and *a* have a broad range (see Table 3-1). Different methods of measurement, calculation, and refinement may, however, also account for some of the recorded differences.

Frequency of occurrence of forms is indicated as follows:

VC—very common (i.e., those forms that characterize the morphology of most crystals);
C—common (i.e., those forms that may or may not be present and, where present, commonly occur as relatively small modifying faces);
LC—less common (i.e., those forms that have been recorded relatively few times and in nearly all cases as modifying faces).

Habit

The following statements relating to crystal size, crystal shape, and crystal groups are based on descriptions in the literature plus the writer's examination

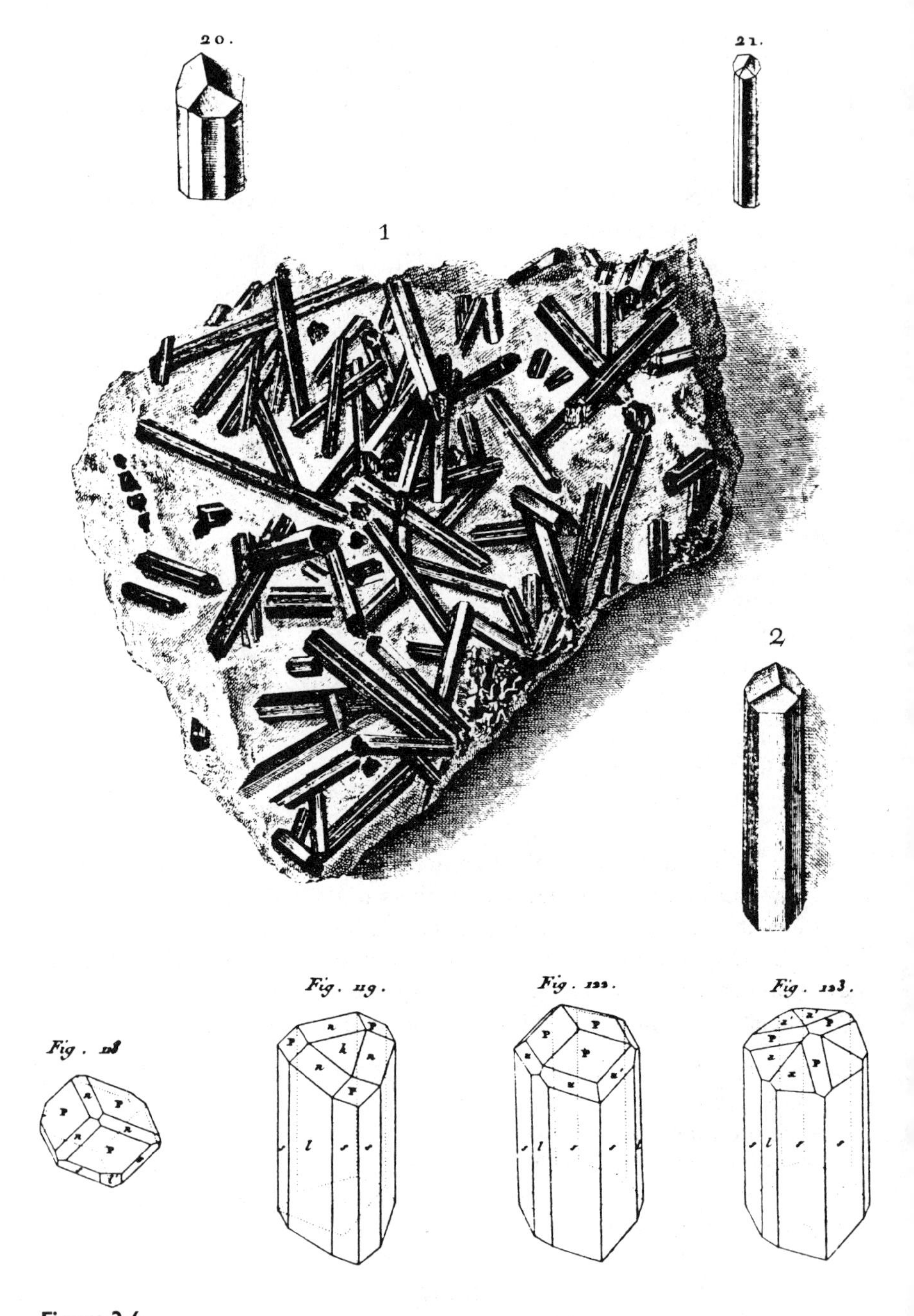

Figure 2-6
Early illustrations of tourmaline. (*20* and *21* from Romé de l'Isle, 1772; *1* and *2* from Born, 1790; *118, 119, 122,* and *123* from Haüy, 1801.)

Table 2-3
Principal Forms

Tourmaline: Hexagonal—R; ditrigonal pyramidal—$3m$ [C^5_{3v}]. Z = 3
a:c ≃ 1:0.4465; α ≃ 114°; p_o:r_o ≃ 0.52; λ ≃ 41°45′

Lower	*Upper*	hkil	*Miller*	ϕ	ρ	A_1	A_2	*Frequency*
c′	*c*	0001	111	— — —	00°00′00″	90°00′00″	90°00′00″	VC
	m	$10\bar{1}0$	$2\bar{1}\bar{1}$	30°00′00″	90°00′00″	30°00′00″	90°00′00″	VC
	a	$11\bar{2}0$	$10\bar{1}$	00°00′00″	90°00′00″	60°00′00″	60°00′00″	VC
	σ	$21\bar{3}0$	$5\bar{1}\bar{4}$	10°53′36″	90°00′00″	49°06′24″	70°53′36″	LC
	l	$52\bar{7}0$	$4\bar{1}\bar{3}$	13°53′52″	90°00′00″	46°06′08″	73°53′52″	LC
	h	$41\bar{5}0$	$3\bar{1}\bar{2}$	19°06′24″	90°00′00″	40°53′36″	79°06′24″	LC
g′	*g*	$10\bar{1}2$	411	30°00′00″	14°56′27″	77°30′54″	90°00′00″	LC
r′	*r*	$10\bar{1}1$	100	30°00′00″	27°16′29″	69°08′36″	90°00′00″	VC
y′	*y*	$40\bar{4}1$	$3\bar{1}\bar{1}$	30°00′00″	64°16′38″	38°48′33″	90°00′00″	LC
e′	*e*	$01\bar{1}2$	110	−30°00′00″	14°56′27″	90°00′00″	77°30′54″	C
z′	*z*	$01\bar{1}1$	$22\bar{1}$	−30°00′00″	27°16′29″	90°00′00″	69°08′36″	LC
o′	*o*	$02\bar{2}1$	$11\bar{1}$	−30°00′00″	45°53′45″	90°00′00″	51°33′34″	VC
v′	*v*	$13\bar{4}1$	$21\bar{2}$	−16°06′08″	61°42′55″	50°36′38″	77°47′17″	LC
x′	*x*	$12\bar{3}2$	$21\bar{1}$	−10°53′36″	34°17′45″	68°21′13″	79°22′15″	C
u′	*u*	$32\bar{5}1$	$30\bar{2}$	6°35′12″	66°00′43″	57°00′02″	68°42′30″	C
t′	*t*	$21\bar{3}1$	$20\bar{1}$	10°53′36″	53°45′11″	58°07′55″	74°41′00″	C

of specimens in several collections—for example, those at Harvard University, the U.S. National Museum of Natural History, and those of E. E. Foord and of William Larsen. The citations in this section are less inclusive than in most other sections because scores of persons have described many of these features and to cite all references would serve little purpose.

Crystal Size

Crystals range from microscopic "needles" and "thin films" up to individuals that are more than a meter in greatest dimension. "Needles," actually small prisms elongated parallel to *c*, have been recorded from many localities; those from the Rutherford pegmatite of the Amelia district of Virginia measure 0.005 × <1 mm (Mitchell, 1964). Discrete crystals that comprise films "so dünn, dass sie auf dem Glimmer kaum einen Abdruck hinterlassen" (p. 492) occur on cleavage surfaces of muscovite from a pegmatite near Brissago, Switzerland (Stuker, 1961). At the other end of the scale, several large crystals and crystal groups have been recorded; a few of the more notable ones are:

A cross-section of an elbaite crystal, probably from Mawi, in a Kabul (Afghanistan) merchant's home is 12 inches (~30 cm) in diameter (Bariand and Poullen, 1979).

Doubly terminated elbaite "floaters" in pockets in the Virgem da Lapa pegmatites of Minas Gerais, Brazil, range up to 3¼ feet (~1 m) long (Cassedanne and Lowell, 1982). The "Foguete" ("Rocket") from the Jonas mine, Itatiaia district in Minas Gerais, is 30 inches (~75 cm) long, 10 inches (~25 cm) across, and weighs ~700 pounds (320 kg) (Lallemont, 1978)—see Figure 2-7.

The "Jolly Green Giant," an elbaite "log" from Newry, Maine, is ~10 × 4 inches (~27 × 10 cm) and appears to have been much longer (Dunn, 1975*b*).

Termier (1908) described a crystal, apparently elbaite, from the Ankaratra district, Madagascar, as about 8 × 5 inches (20 × 13 cm); LaCroix (1908) described another zoned elbaite from Madagascar as ~15 × 3½ inches (38 × 9 cm); and Frondel (1983, personal communication) reported some Madagascar elbaite crystals to weigh "30-50 pounds [~13-23 kg] apiece."

The "huge tourmaline crystal" from the vicinity of Overlook, New York, (Rowley, 1942) is apparently a group of at least three black (schorl?) crystals weighing some 18 pounds (~ 8 kg) and measuring 9 × 11 inches (~21 × 29 cm).

Figure 2-7
Foguete ("Rocket") from the Itatiaia district, Minas Gerais, Brazil. (Photograph courtesy of A. Barbosa and R. V. Gaines.)

"Crude rosettes" of schorl up to 7 feet (~2.2 m) in diameter, at the Stewart pegmatite mass in San Diego County, California, and individual crystals, also termed schorl, up to more than 5 feet (~1.5 m) long but only 3 inches (~7.5 cm) across, from pegmatites near Sage in Riverside County, California, are recorded by Jahns and Wright (1951) and Jahns (1953).

A dravite crystal from Yinnietharra, Australia, was ~6 inches (15 cm) in greatest dimension and weighed ~25 pounds (11.5 kg) (Bridge, Daniels, and Pyrce, 1977).

In a report with no references, Metz (1964) recorded the following large crystals, none of which the writer has been able to verify: deep red gem-quality tourmaline up to approximately 16 × 4 inches (40 × 10 cm) from pegmatites of Alto Ligonha, Mozambique; prisms of an otherwise undefined tourmaline up to nearly 3 yards (2.7 m) long from Nuevo, Riverside County, California; and schorl up to about 3¼ yards (3 m) long and a yard (1 m) across from Skrumpetorp, eastern Götland, Sweden.

Crystal Shape

Individual tourmaline crystals have been described as predominantly prismatic, predominantly pyramidal, tabular, waferlike, or equant (Figs. 2-8–2-10). As can be seen, doubly terminated crystals are hemimorphic, and terminations on both doubly and singly terminated crystals range from relatively simple (especially on analogous ends) to extremely complex. Also, doubly terminated crystals with as few as 3 forms and 8 total faces (Fig. 2-9) and as many as 59 forms and more than 170 faces (Fig. 2-10) have been reported. Prism faces range from well defined to so highly modified that individual faces are unrecognizable. The latter, which give the characteristic, rounded triangular cross-sections, result where a trigonal prism is modified by hexagonal or ditrigonal prisms. The predominantly prismatic crystals range from short and stubby with length:cross-section ratios of less than 1 to long and slender with the same ratio being well over 500.

A few tourmalines—for example, some of the schorls from the Himalaya and Little Three mines of San Diego County, California, and greenish blue elbaite from Barra do Salines, Minas Gerais, Brazil—have prismatic forms with equilateral triangular cross-sections. Their faces are vitreous (on the schorl, mirrorlike), but with sporadic steplike features that resemble negative crystals of tetrahedral character; that is, each step has a rectilinear shape. A specimen from the Little Three mine, now in the collection of E. E. Foord, has two of these triangular tourmalines, each of which appears to represent a corrosion remnant of an originally prismatic crystal having the more typical, rounded triangular cross-section.

Surface Features

In general, prism faces have subadamantine to vitreous lusters whereas terminations have lusters that range from subvitreous to dull.

The highly modified prism faces mentioned in the preceding section are commonly grooved and/or striated parallel to *c* (Fig. 2-11). Both features result from the multiplicity of faces responsible for the common, just mentioned, rounded triangular cross-sectional shapes.

Terminal faces on some crystals from a few localities are also grooved or striated. For example, some Madagascar specimens (e.g., U.S. National Museum of Natural History specimen #140880) have striations parallel to their opposing trigonal prism $\{11\bar{2}0\}$ faces on their antilogous and/or analogous end pyramids; and some of the tourmalines from Newry, Maine, have striations perpendicular to *c* on $\{h0\bar{h}l\}$ pyramids. Dunn (1975*b*) attributes the latter to the competition in growth between pyramid and $\{11\bar{2}0\}$ trigonal prism faces.

Some tourmaline crystals have some or all of their faces partly coated with small, typically colorless to white or light pink elbaite prisms, the *c* axes of which lie more or less flat on the coated surfaces. Foord (1976) has referred to this relationship as "snow on the roof."

Some crystal faces, especially those that transect *c*, have sculptured and/or etched appearance (Fig. 2-12). Interpreting some of these features as resulting from growth, including abutment-controlled growth, versus natural etching is hazardous. However, the fact that some synthetic tourmalines have similar surfaces seems to support a growth origin for at least many of the features. The following are examples of such features:

Rounded triangular growth layers and vicinal hillocks occur on pedions of tourmalines from Arassuai, Minas Gerais, Brazil (Seager, 1952).

Growth features, some of which are said to have been controlled by dislocations, are characterized as follows by Slivko (1966*a*): steps, conical protuberances, truncated triangular prisms, discoidal shapes, splinter-like forms, and combinations of these forms (cf. Breskovska and Eskenazi, 1959-1960). Most of these features are also on pedion faces.

Small markings described as "chicken tracks" occur on some terminal faces of some Newry, Maine, elbaites (Dunn, 1975*a*). A cause was not suggested.

Surfaces "indented and pockmarked by phlogopite and plagioclase" are characteristic of the brown dravite crystals from Yinnietharra, Western Australia (Bridge, Daniels, and Pyrce, 1977).

On Fe- and Li-tourmalines synthesized by Voskresenskaya and Barsukova (1971) there are, respectively, "growth steps" and "growth hillocks." Some of the latter have well-defined triangular shapes quite distinct from typical ditrigonal etch figures.

Other Features

"Individual crystals . . . [that] bifurcate in offset to form a single crystal at one end and two at the other" occur at one of the pyrophyllite deposits of the North Carolina Piedmont (Furbish, 1968).

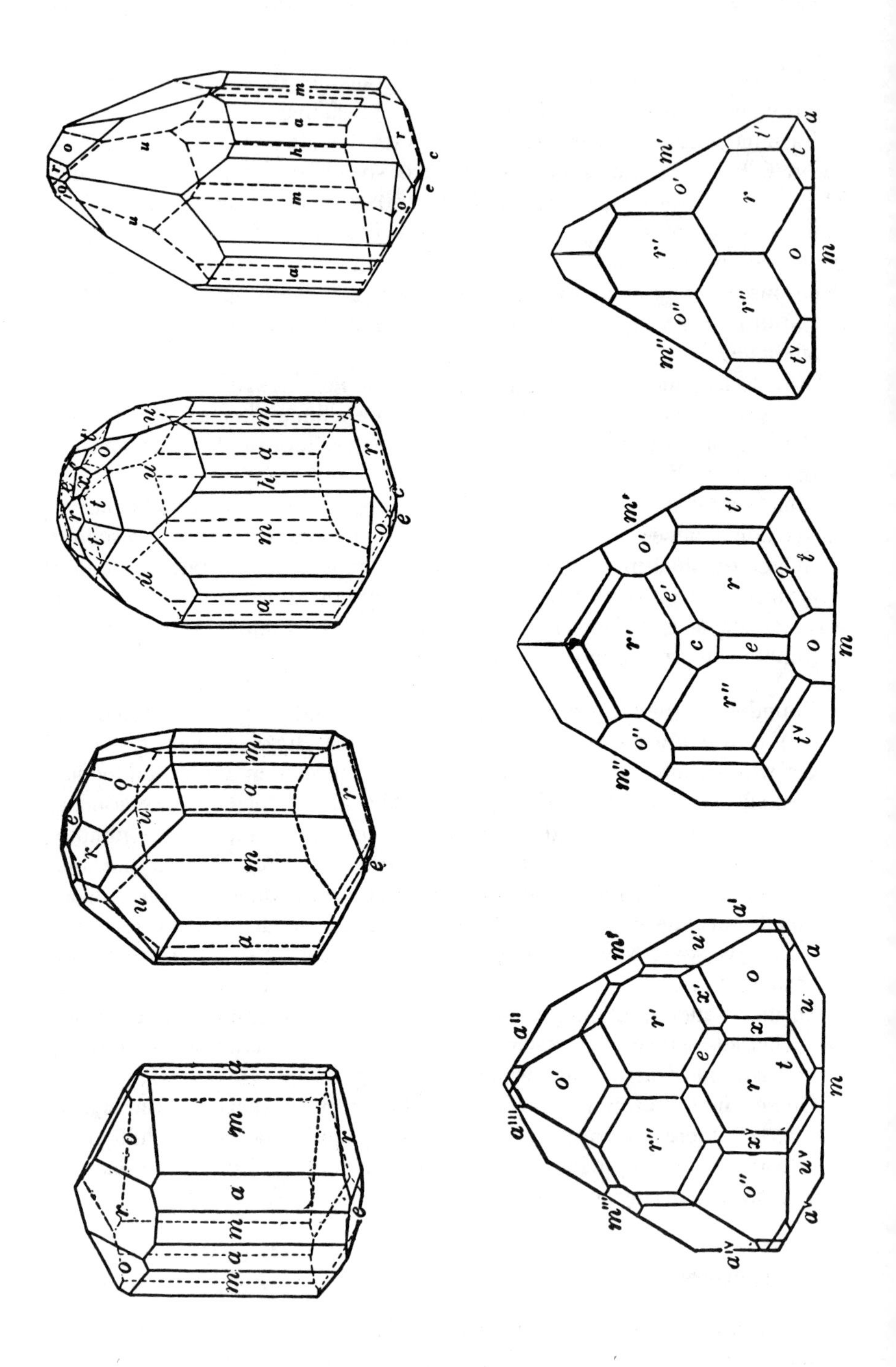

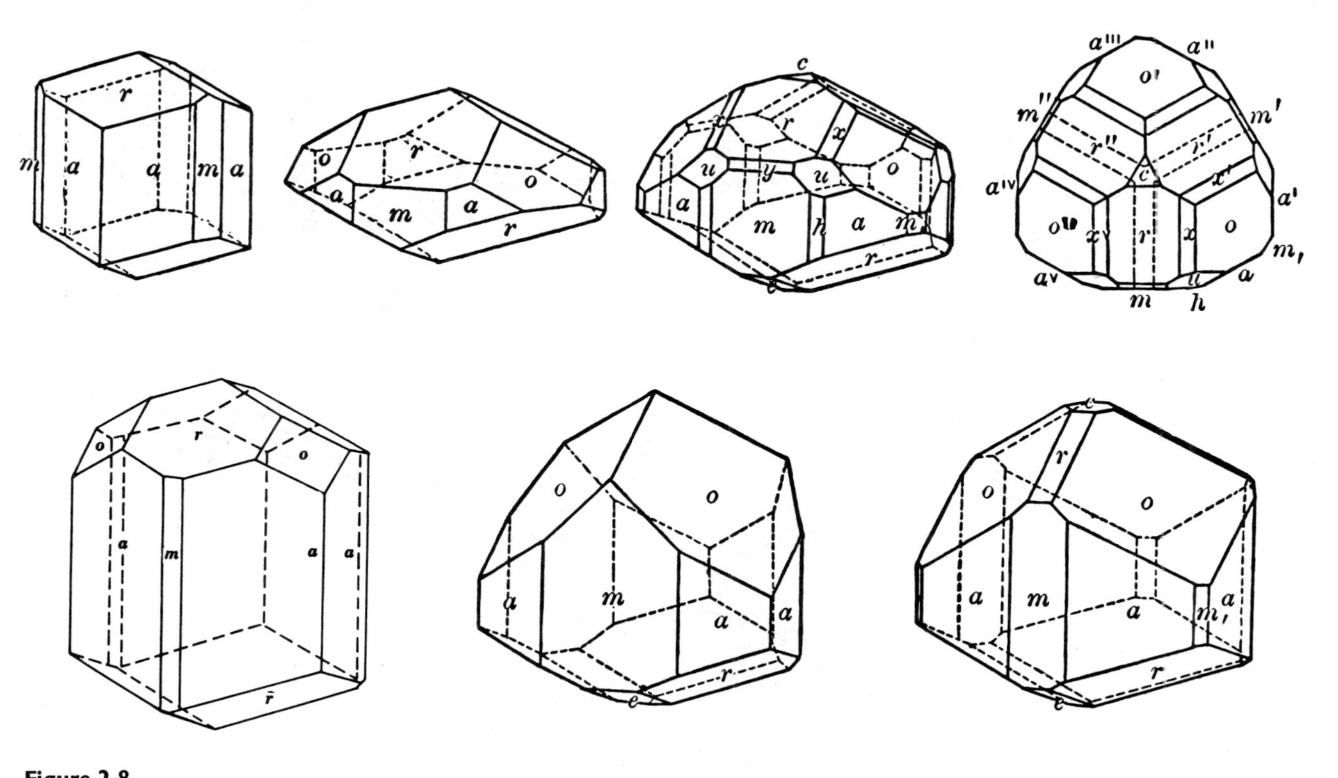

Figure 2-8
Common forms (after Dana, 1892).

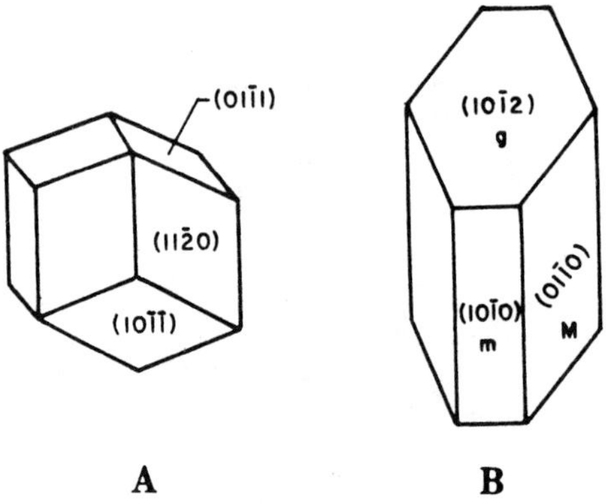

A B

Figure 2-9
A, dravite crystal that resembles a rhombic dodecahedron; *B*, dravite crystal that appears monoclinic. (*B* is after de Camargo and Souza, 1970.)

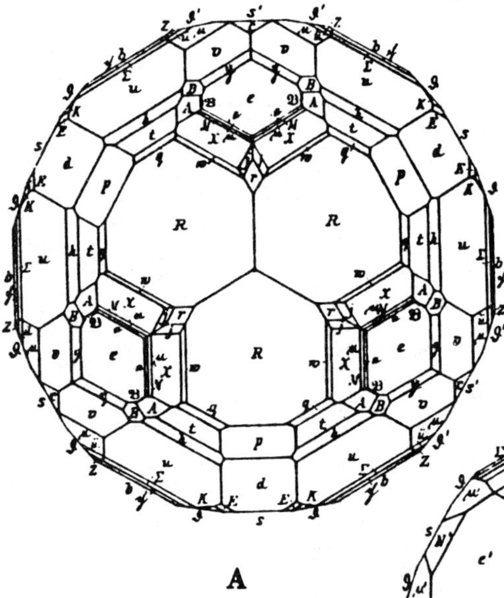

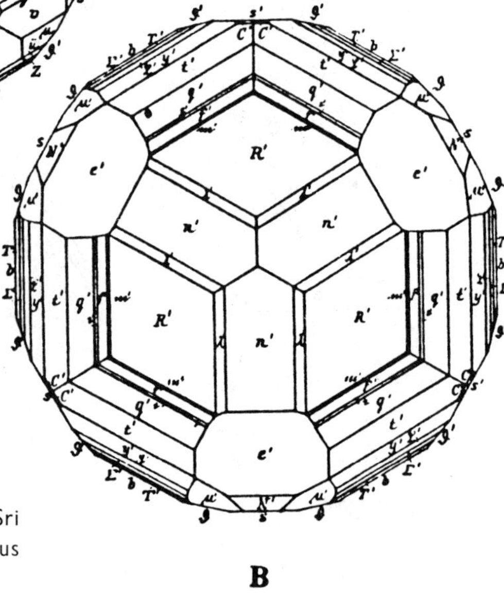

A B

Figure 2-10
Tourmaline crystal from Ceylon (now Sri Lanka): *A*, antilogous end; *B*, analogous end (from von Worobieff, 1900).

Figure 2-11
Elbaite from Lower Alamo, Baja California, Mexico, that exhibits the typical grooved (or striated) prism faces; height of specimen 4½ inches (~11.5 cm); NMNH #C6477. (Photograph courtesy of the Smithsonian Institution.)

Flaw-free zones, called "nodules" in otherwise highly fractured crystals are described in a succeeding section, "Cleavage and Fracture".

Bent and fractured crystals are common, especially in some pegmatites (see Figs. 2-13–2-15). Typically, the "bent" crystals appear to have bent (but not twisted) *c* axes, and the fracturing, where present, is more or less perpendicular to *c*. In some cases, the apparent bending can be seen to represent closely spaced, healed fractures; the "healing" material is typically fibrous tourmaline (Fig. 2-15) but may be another mineral such as quartz. Segments of fractured crystals are typically "in line" within their host, commonly quartz. Some, however, are twisted around *c* with respect to their originally juxtaposed pieces (Jahns, 1953).

Tourmaline-bearing pockets in the pegmatites of, for example, Oxford County, Maine, and San Diego County, California, commonly contain fragments of crystals that are separated up to several centimeters from their

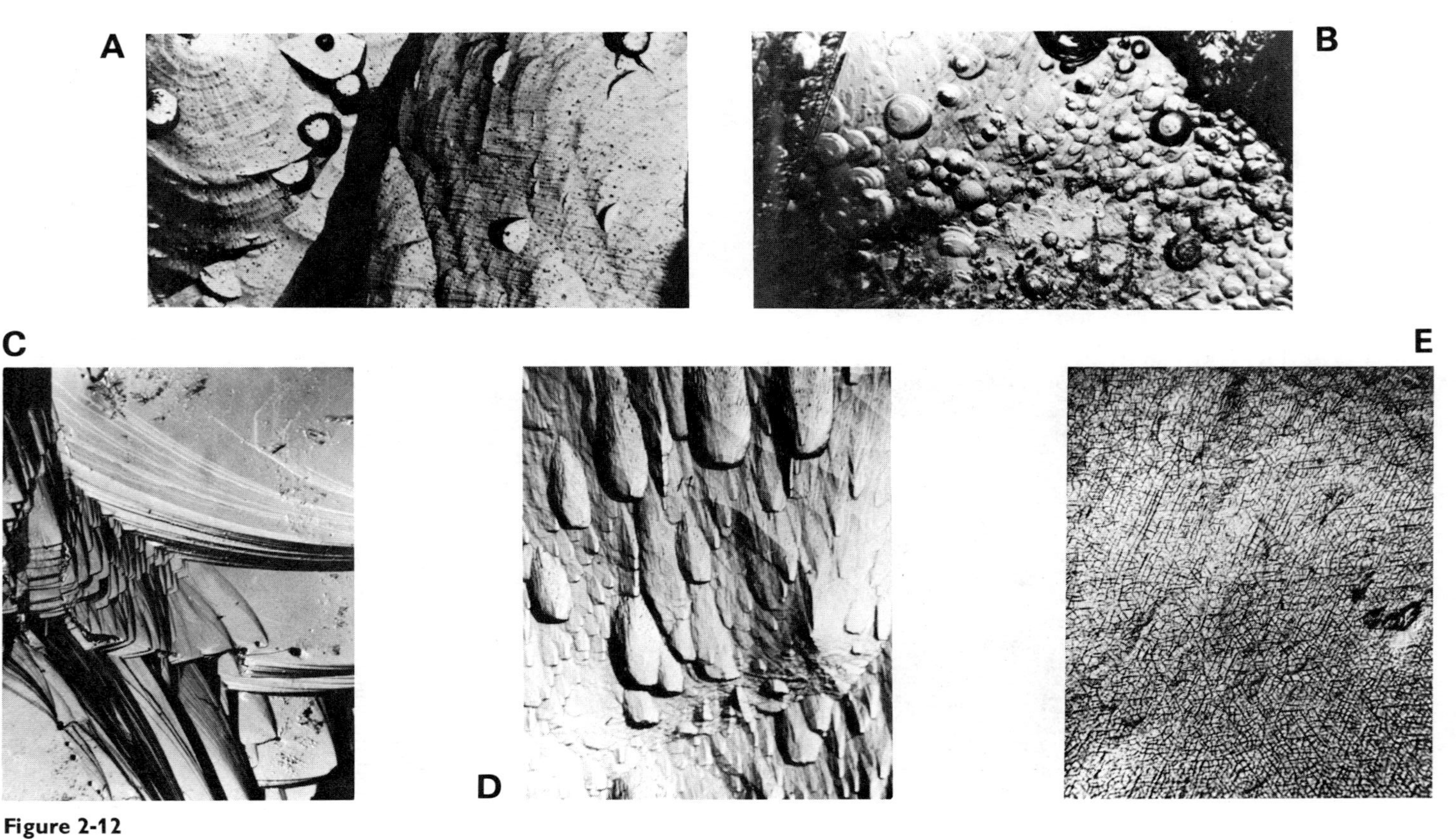

Figure 2-12
Sculptured growth and etched surfaces on elbaites from San Diego County, California. From upper left, clockwise, magnifications are as follows: *A* and *B*, 5x; *C*, 7x; *D* and *E*, 12x.

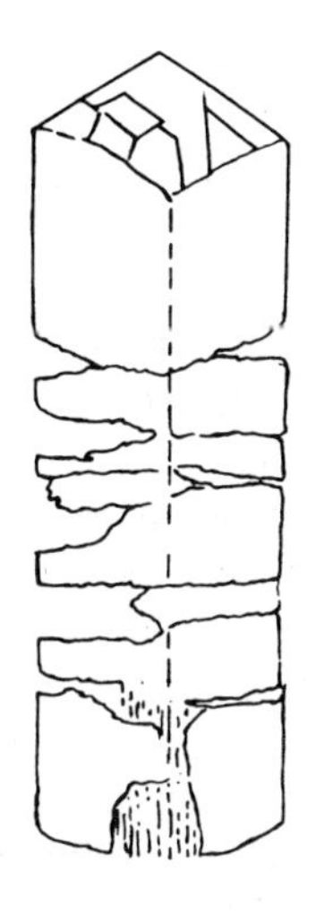

Figure 2-13
Fractured crystal from Mt. Mica, Maine (after Hamlin, 1895).

Figure 2-14
Bent crystal of schorl(?) from an unrecorded locality; length of crystal 3 inches (~7.6 cm); Harvard Museum #44381. (Photograph courtesy of Harvard Museum.)

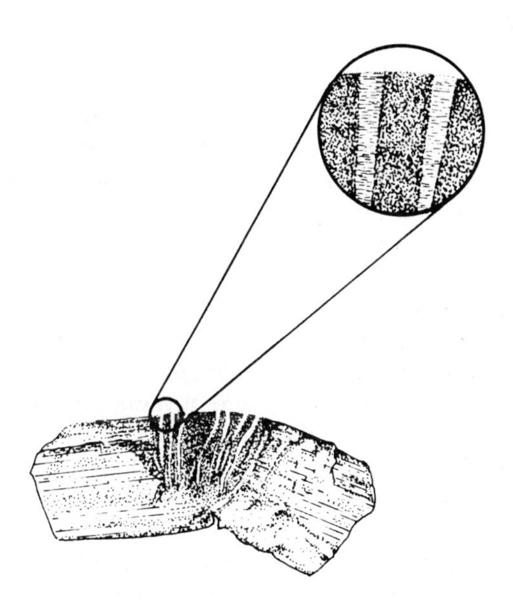

Figure 2-15
Bent and healed elbaite crystal from San Diego County, California. Specimen is ~.71x; magnified section shows "healing," which is by fine cross-fibers of elbaite. (Sketch of specimen in collection of E. E. Foord.)

originally adjacent parts; they are known to be such because they can be fitted together like the pieces of jigsaw puzzles. A few pieces exhibit scratches, apparently formed by natural abrasion. In fact, in a few pockets in San Diego County, California, fragments that resemble tumbled pieces of tourmaline (and/or quartz) crystals have been found. At least three hypotheses have been suggested to account for the abrasion: (1) movement as a result of turbulence during escape of late (pegmatite) stage gases (John Sinkankas, 1983, personal communication); (2) movement in through-flowing aqueous—either hydrothermal or groundwater—solutions after pegmatite formation (Foord, 1976); and (3) repeated jostling within, for example, clays within open pockets, as a consequence of the continual earthquake activities of the region (Foord, 1976). Corrosion, rather than abrasion, seems to be precluded. The presence of what appear to be abraded channels along the floors of a few of these pockets (E. E. Foord, 1984, personal communication) may support origin (2), so far as accounting for their contained "tumbled" crystal fragments. The lack of such channels in other host pockets, however, appears better to support origin (3), at least for their contained "tumbled" crystal fragments.

Crystal Groups

Tourmaline occurs rather commonly in parallel, divergent, and radial groups of crystals and less commonly as fibrous masses and as "jackstrawlike" and massive aggregates (see Figs. 2-16–2-18). Some of the subparallel, divergent groups appear macroscopically to extend outwards from single crystal bases; that is, individual crystals appear to change into groups of crystals along their lengths. A noteworthy occurrence is the schorl of Santa Cruz, Sonora, Mexico (e.g., NMNH #R15501-6). Erofeyev (1871) and Mafveyev (1946) have formulated hypotheses whereby multinucleation and essentially simultaneous unidirectional growth is responsible for such groups, including some of those that yield tourmaline cat's-eyes.

The so-called parallel groups of many investigators are columnar composites of prismatic crystals. Some groups tend to flare (Fig. 2-16); others are typical "sunbursts" (Fig. 2-17). The parallel to slightly divergent groups are fairly common in pegmatites and veins; the more highly divergent flare groups and particularly the "sunbursts" are relatively common in granular lepidolite-rich pegmatites and in tourmaline-rich gneisses and schists, and also occur in other rocks such as granites and rhyolites. An especially noteworthy specimen in the Harvard Museum consists of a "sunburst" on a joint surface in granite from Portland, Maine. Subspherical masses up to ~2½ inches (6.5 cm) in diameter, which are apparently three-dimensional "sunbursts," have been collected from Cabeza de Vaca, Chile.

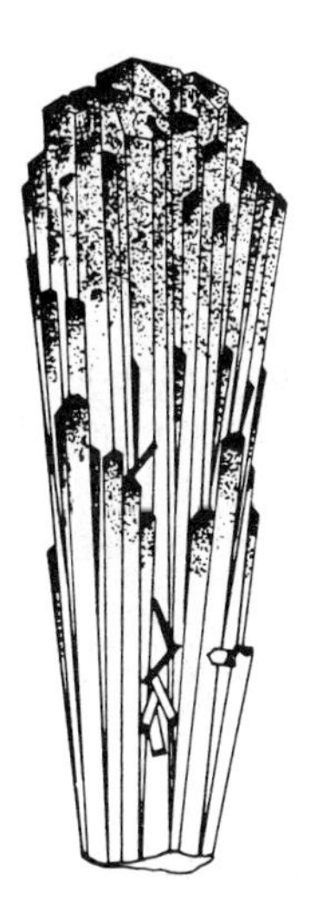

Figure 2-16
Crystal cluster of green (stippled) and pink tourmalines (after Tarnovskii, 1961).

Figure 2-17
Radial groups of pink elbaite crystals in light lavender lepidolite mass from Pala, San Diego County, California; specimen is ~10¾ inches (~27 cm) across mid-section; NMNH #82267. (Photograph courtesy of the Smithsonian Institution.)

Figure 2-18
Fibrous mass of crystals extending outwards from a single crystal substrate from Minas Gerais, Brazil. Specimen, black (single crystal) and light gray (fibrous part), is ~7 inches (~18 cm); Harvard Museum #101955. (Photograph courtesy of Harvard Museum.)

Masses up to ~3 × 4 inches (7 × 10 cm), termed "colloform schorls," occur in metasomatized silica-rich rocks at an ore deposit in the Azerbaydzhan district of Caucasia, U.S.S.R. (Martirosyan, 1962).

Acicular tourmaline in sheaflike and radiating groups from, for example, the fluorite-tourmaline deposits in Siberia (Getmanskaya et al., 1970) are perhaps transitional between typical sunbursts and certain kinds of fibrous tourmaline. The individual units of each of these kinds of composite masses tend to be elongate parallel to *c* (e.g., Jansen, 1933).

Fibrous tourmalines (Fig. 2-18), however, are of diverse character and occurrence. Descriptions in the literature include the following:

Flexible matted, filiform white dravite fibers parallel to *c* occur in the Rutherford pegmatite of the Amelia, Virginia, district (Mitchell, 1964).

Extremely brittle, gray fibers parallel to *c* surround single black tourmaline crystal cores in specimens from Mysore State, India (Ramaswamy and Iyengar, 1937).

"Velvety druses of acicular crystals" in vugs at the tungsten mine of Panasqueira, Portugal, "can be a real hazard to the miners, as the little crystals are sharp and when handled penetrate the skin of the hands and body like nettles" (Gaines and Thadeu, 1971).

In pegmatites, there are: "whiskers" in cleavelandite masses at Strickland's quarry, near Portland, Connecticut (Bantly, 1964)—a few masses at the locality comprise "mountain leather"; asbestiform masses in Minas Gerais, Brazil (e.g., Coelho, 1948); feltlike masses in the Pala district of California (Jahns and Wright, 1951); and the previously mentioned fibrous masses that "heal" fractured crystals in, for example, California (e.g., Sterrett, 1904) and Java (Koomans, 1938).

In veins and along open joints, tourmaline occurs: in quartz veins of the southern Urals (Boriskov, 1969; Slivko, Boriskov, and Kornilov 1972); in veins of Morbehan, Brittany, France (LaCroix, 1910); (dravite) along joints in the Swiss Alps (Dietrich, de Quervain, and Nissen, 1966); and (white dravite) in quartz veins near Chvaletice, eastern Bohemia (Novak and Zak, 1970).

In cavities, there is: filiform tourmaline, some of which is matted, within vugs in tourmalinized barite-rich rocks in the pyrophyllite deposits of the Piedmont of North Carolina (Furbish, 1968); and similarly described tourmaline in miarolitic cavities in granite in the vicinity of Alzo, northern Italy (Cocco, 1952).

"Chalcedony-like" cryptocrystalline fibrous blue tourmaline occurs as the chief constituent of a fault gouge and in a nearby contact zone between the country rocks and an andesite body near Barstow, California (Kramer and Allen, 1954).

Most of the tourmaline described as massive comprises fine- to coarse-granular masses of nearly equant grains, typically black schorl or green dravite. Some "massive" tourmaline, however, is described as sooty, and that observed by the writer consists of extremely fine black fibers rather than of nearly equant grains (cf. Hitchen, 1935; Bayramgil, 1945). In addition, Petrovskaya and Andreeva (1969) have described "cryptocrystalline" tourmaline within fractures in an explosion breccia in the Klyuchevskoe deposit of eastern Transbaikal.

Crystal Shape-to-Species Correlations

Correlations given in Table 2-4 are, as indicated, generalizations to which exceptions are known. In addition, it should be kept in mind that some of the samples, though well described morphologically, have not been identified to species, and even most of those that have been both well described and so identified are collectors' specimens and, therefore, may be far from representative. Furthermore, a few investigators have directed attention to the absence of such correlations; for example, Dunn et al. (1977) state that "Crystals of uvite do not exhibit any peculiar or diagnostic morphology which might aid in their identification" (p. 100).

The tabulated generalizations are based on examination of several descriptions such as the following:

Dravite crystals from Yinnietharra, Australia, are "so similar to dodecahedral garnets that they initially confused collectors" (Bridge, Daniels, and Pyrce, 1977); dravite porphyroblasts in chlorite schist from Goias State, Brazil, have only $\{10\bar{1}0\}$ and $\{01\bar{1}0\}$ plus $\{10\bar{1}2\}$ terminations and closely resemble monoclinic crystals (de Camargo and Souza, 1970); and the Madagascar tourmaline reported as intermediate between dravite and ferridravite "has a distinctly monoclinic aspect simulating class *2/m* . . . on hexagonal axes, the crystal is elongated along the direction of the edge between (10.1) and (02.1)" (Frondel, Biedl, and Ito, 1966, 1504).

Elbaite crystals described and illustrated from the type locality (e.g., Grill, 1922) and most other localities are prismatic.

Most of the crystals identified as schorl in the collections examined by the writer are prismatic. Some of those crystals, however, may not be schorl; adequate descriptions are rare. It remains to be shown whether any regular relation exists whereby, for example, schorls of the elbaite-schorl series tend to be prismatic whereas schorls of the dravite-schorl series tend to be more nearly equant.

Table 2-4
Common Crystal Shape-to-Species Correlations

Species	*Common morphology*	*Remarks*
Buergerite	prismatic	a few nearly equant crystals occur with the typical prismatic crystals
Dravite	equant	including the red dravite from Kenya (Dunn, Arem, and Saul, 1975)
		brown prismatic crystals occur (e.g., at Macomb, New York, and Rumford, Maine)
Elbaite	prismatic	stubby crystals occur (e.g., on Ceylon [Sri Lanka] [von Worobieff, 1900] and at Bom Jesus das Meiras, Bahia [Brazil])
Ferridravite	equant	
Liddicoatite	prismatic	designation is made on basis of close relationship to elbaite
Schorl	prismatic/equant	prismatic crystals are most common in collections;
		nearly equant crystals occur (e.g., on Madagascar [NMNH 123338, etc.], at Shaitanka in the Urals [NMNH #134300], and in a pegmatite, southwestern Montana [Heinrich, 1963]);
		waferlike crystals occur at Mitchells, Virginia (NMNH #1594641) and on Ceylon (Rose, 1836 and Nündel, 1973)
Uvite	equant	cf. Dunn et al. (1977*b*); also crystals tabular on (0001), for example, from Pierrepont, New York (Solly, 1884), and prismatic crystals occur from, for example, Horicon, New York (NMNH #134441)
Chromdravite	pyramidal	

Note: With exceptions, prisms tend to be striated on elbaites, liddicoatites, and prismatic schorls but not on dravites and uvites (even the prismatic ones) and nearly equant schorls.

Crystal Shape-to-Genesis Correlations

Too few data are currently available to correlate definitely between crystal shape and genesis. It has, nonetheless, been recorded widely that crystals from a given locality commonly exhibit similar forms and similar development of forms (and also colors and arrangements of colors where the crystals are zoned), and this may suggest that intensive study of conditions of formation might lead to the establishment of noteworthy correlations. Also, it is rather widely accepted that, for example, most elbaites occur in pegmatites whereas most dravites with good crystal shapes are in metamorphosed carbonate-rich rocks.

The following relations reported for certain kinds of occurrences and/or individual districts may be indicative of the kinds of relationships to be discovered.

As the result of a statistical study, Slivko (1952) found that, in general, tourmaline crystals of Na-Li pegmatites have relatively few terminal faces, and that in some individual pegmatites and pegmatite districts (e.g., those of Borshchevoch Range, Transbaikal) crystallographic differences correlate with color hue and/or color saturation differences.

A predominance of the prism $\{11\bar{2}0\}$ over the prism $\{10\bar{1}0\}$ in Mg-rich tourmalines from Hörlberg, Bavaria, and the opposite relation in Fe-rich tourmalines from the Drachselsrieder mine, near Arnbruck, was found by Pfaffl and Niggemann (1967). Also, Dunn (1975*b*) reported that "Newry [Maine] elbaite crystals are predominantly prismatic with . . . $\{10\bar{1}0\}$ dominant and . . . $\{11\bar{2}0\}$ subordinate" (p. 23).

Twinning

Twinning is rare in tourmaline. In an early record of twinning, Madelung (1883) described twinning in tourmaline specimens from Brazil, Campo Longo (Switzerland), and Bavaria as "schwach zweiaxig." More recently, simple twinning with twin planes $\{10\bar{1}1\}$ and $\{40\bar{4}1\}$ has been recorded from several localities, for example, from the pegmatites of Hörlberg, Bavaria (Pfaffl and Niggemann, 1967) and from pegmatites of the Mesa Grande deposits of San Diego County, California (Foord and Mills, 1978). Also, Frondel (1948) reported that in observing about 50 slabs of KOH-etched Madagascar and Brazil tourmalines, a few small isolated areas were found to be twinned on $\{10\bar{1}1\}$. Twinning with twin axis *c* and contact plane $\{10\bar{1}0\}$ has been reported for a brownish green, "nearly black" tourmaline from the nepheline-bearing pegmatites of Seiland, Norway (Hoel and Schetelig, 1916).

Other kinds of "twinning," now doubted, have been reported by, for example, Ramsay (1886). Also, penetration twinning with parallel axes has been recorded (Dana, 1892) but noted as "not common," and cruciform twinning with $\{10\bar{1}1\}$ as the twinning plane has been recorded by Shepard (1830) and Bauer (1890) (Fig. 2-19). Shepard stated that "two prisms crossing each other (with mutual penetration) at right angles" were found among specimens from Paris, Maine, an observation not subsequently repeated.

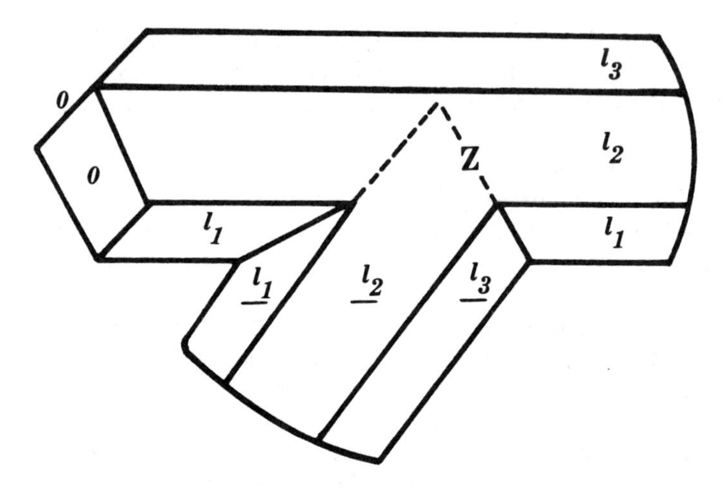

Figure 2-19
Tourmaline twin recorded by Bauer (1890) (after Goldschmidt, 1922).

Figure 2-20
Possible twinning in tourmaline from Minas Gerais, Brazil. Specimen is 2 inches (~6 cm) across; Harvard Museum #112557. (Photograph courtesy of Harvard Museum.)

Figure 2-21
Cleavage pattern in tourmaline (black) inclusions in muscovite from pegmatite of Brissaga; Ӕ indicates optic axes of mica (after Stuker, 1961).

Among hundreds of specimens examined by the writer, only a few possibly twinned crystals were seen (see, for example, Fig. 2-20). In any case, tourmaline crystals, like those of several other minerals, commonly occur as intergrowths that can simulate twins whenever the angular relations are nearly (or exactly) rational. Consequently, statistics may be needed to help establish twinning in tourmaline.

Cleavage and Fracture

Cleavage has frequently been reported as lacking in tourmaline and, indeed, well-developed, readily observable cleavage is typically absent. Actually, however, tourmaline has long been reported to have two difficult (poor, very poor, indistinct, etc.) cleavages: one parallel to $\{11\bar{2}0\}$ (see, for example, Hoel and Schetelig, 1916; Stuker, 1961; Mukherjee, 1968; and Fig. 2-21); the other parallel to $\{10\bar{1}1\}$ (see, for example, Dana, 1892). In addition, Thakur (1972) and Dunn, Appleman, and Nelen (1977) recorded a $\{0001\}$ cleavage for a black tourmaline, otherwise unidentified, and liddicoatite, respectively, and Donnay, Ingamells, and Mason (1966) described buergerite

Figure 2-22
Thin-section of tourmaline, with *c* axis as indicated, exhibiting fracture roughly parallel to (0001); magnification ~90x.

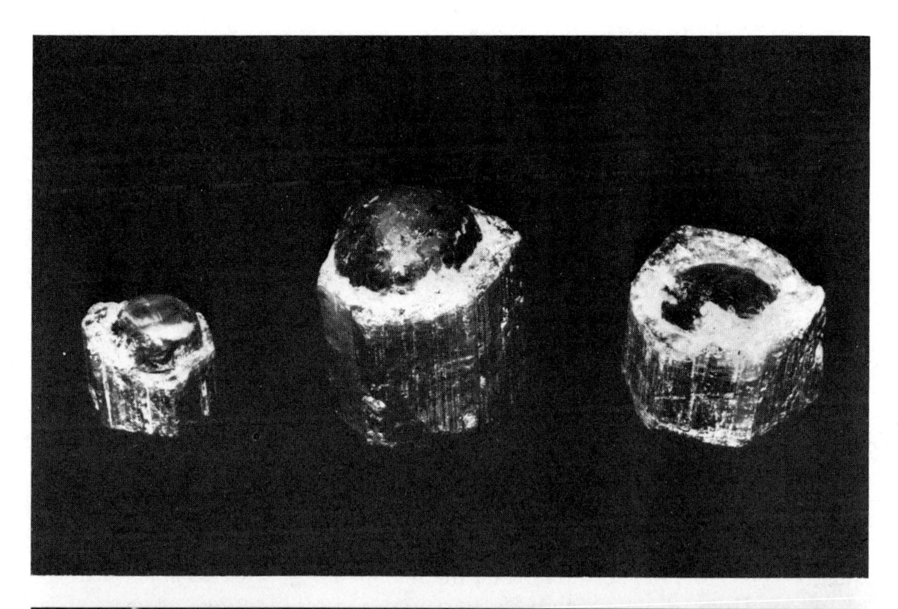

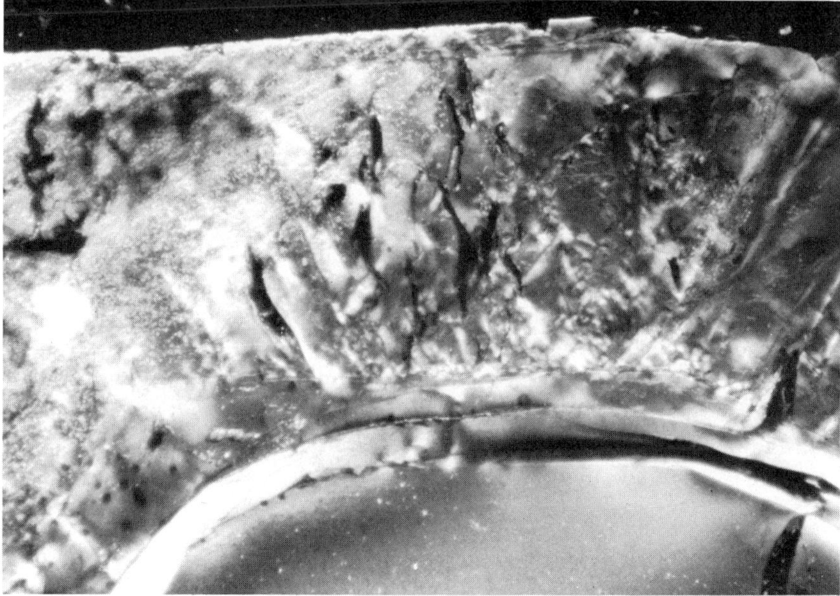

Figure 2-23
Top, nodules from Alto Ligonda, Mozambique; largest specimen is 1¼ inches (~3.2 cm) across. (Photograph courtesy of Harvard Museum.) *Bottom*, photomicrograph (~2.8x) of nodule and surrounding zone from San Diego County, California. (Photograph courtesy of E. E. Foord.)

as having "a distinct prismatic cleavage, in contrast to the very poor cleavage usually reported for tourmaline."

Fracture is usually recorded as conchoidal to uneven, with all species being brittle. Although this is true, the predominant fractures in tourmaline have the crystallographic relationship shown in Figure 2-22. Both optical mineralogists and lapidaries have recognized for many years that these roughly parallel, uneven fractures more or less perpendicular to *c* are frequently developed during grinding of thin-sections or gems. In fact, the quality of these fractures as compared to the perfect cleavage of biotite and the relation whereby the extinction and greatest absorption in tourmaline are perpendicular, rather than parallel (as in biotite), to the "cracks" serve to distinguish readily between the two minerals in thin-section.

Recognition of the fracture tendencies by lapidaries depends largely on the fact that there are globular to subspherical zones bounded by such fractures, and these zones are essentially flawless. These zones, which are called "nodules" (Fig. 2-23), have long been recognized to constitute especially valuable gemstone material; for example, Shepard (1830) described their quality as " 'of the finest water' " and their shape as "*gouttes de suif*" (= drops of tallow). John Sinkankas (1983, personal communication) notes that, at some Brazilian deposits, "miners have found long crystals of tourmaline cracked in sections that can be broken apart with the fingers and then 'peeled' like an onion to obtain the small clear nodules present in the core of each section."

It has been suggested that the fractures that surround the nodules represent discontinuous growth. However, this hypothesis has generally been supported only by the fallacious and, in fact, inconsequential notion that individual nodules consist of only one color of tourmaline, even though some nodules are multicolored. Actually, the critical feature—whether the fractures do or do not coincide with growth zones—has not been determined so, in reality, the suggestion has not been tested.

Liddicoat (1975), in comments about tourmaline gemstones, writes that red tourmaline is commonly fractured roughly parallel to *c*, whereas green tourmaline is not. This generalization has not been found to obtain for specimens examined by the writer.

Chapter 3

Crystal Structure

[Rhombohedral: $R3m$ ($\equiv C_{3v}^5$); Z = 3]

SPACE GROUP

In an early attempt to establish the space group for tourmaline, Kulaszewski (1921) indexed 763 reflections on Laue X-ray patterns made from slices of a clear rose-red tourmaline from Pening, Saxony. The slices were oriented parallel to (0001), $(0\bar{1}11)$, $(01\bar{1}0)$, $(11\bar{2}0)$, $(10\bar{1}1)$, $(50\bar{5}2)$, and $(10\bar{1}0)$. She also studied oscillation X-ray patterns made from all of the slices except the one parallel to $(50\bar{5}2)$. As a result, she gave the cell dimensions of $a = 16.23$Å and $c = 7.26$Å and concluded—as is now known, incorrectly—that the space group is either C_{3v}^1 or C_{3v}^2, between which she said it was not possible to distinguish on the basis of her data. Eight years later, Machatschki (1929) studied three other tourmalines—a schorl from Sondelag, Norway, and two pink elbaites from San Diego County, California—by X-ray methods, and concluded (also incorrectly) that, although the indexing of some reflections was uncertain because of the length of a, any rhombohedral lattice possibility could be eliminated; thus, he implicitly agreed with Kulaszewski's assignment of either C_{3v}^1 or C_{3v}^2 as the space group. Still later, on the basis of X-ray investigations of a "bottle green" tourmaline of unknown source, Barnes and Wendling (1934) concluded (again incorrectly) that "The space group has been unequivocally established as C_{3v}^1" (p. 174).

Fortunately, after J. D. H. Donnay directed Buerger's attention to the fact that the morphology of tourmaline crystals appears to reflect a rhombohedral, rather than a hexagonal lattice, Buerger and Parrish (1937) reinvestigated the space group. They studied a tourmaline from the Etta mine, South Dakota, using the equi-inclination Weissenberg method. (It is of at least passing interest that they wrote "Since hexagonal crystals have not been previously treated by the inspective method, this investigation constitutes a type case for the determination of the cell characteristics of crystals of this system without indexing procedure," p. 1141.) They recorded the cell dimensions to be a = 15.928Å and c = 7.15Å (or, on a rhombohedral basis, a = 9.500Å and $\alpha = 66° 05'$), the lattice to be rhombohedral, and the space group to be C_{3v}^5 (≡*R3m*).

The *R3m* assignment is now accepted. Nonetheless, in the first application of high-resolution electron microscopy to a natural mineral, Iijima, Cowley, and Donnay (1973) obtained direct images of the atomic arrangements in buergerite and elbaite-schorl, and found the images for the elbaite-schorl to show the unexpected plane-group symmetry ($h3\bar{1}l$). As a possible resolution of this apparent anomaly, they suggested that perhaps the electron irradiation caused structural changes that gave rise to the ($h3\bar{1}l$) symmetry image. It seems that this kind of investigation needs to be extended.

CELL DIMENSIONS

Some of the c:a ratios that are recorded for tourmalines have been based on morphological measurements; most, however, have been calculated from X-ray data. Cell dimensions, c:a ratios, and cell volumes ($V = a^2c \sin 60°$) for representative specimens of the diverse tourmaline species are given in Table 3-1. Essentially all of the differences can be resolved on the basis of considerations of the diverse occupancies of the *X*-, *Y*-, and/or *Z*-sites, which are described in the next section of this chapter.

Recorded values for c range from 6.86Å for dravite from Moravia (with a = 15.75Å; Kurylenko, 1949) to 7.47Å for the type specimen of ferridravite from Bolivia (with a = 16.20Å; Walenta and Dunn, 1979). Recorded values for a range from 15.676Å for another Moravian dravite (with c = 7.034; Kurylenko, 1951) to 16.23Å for a pink elbaite from Saxony (with c = 7.26Å; Kulaszewski, 1921). It is noteworthy, however, that measurements by Epprecht (1953) indicate that Kurylenko's measurements may be incorrect.

Values for c:a, based on morphological measurements, range from 0.4423 (Lévy, 1838) to 0.4534 (Niggli, 1926), and values based on X-ray data range from 0.4356 for the just mentioned Moravian dravite (Kurylenko, 1949) to 0.4611 for the type specimen ferridravite (Walenta and Dunn, 1979).

The range in cell dimensions of the tourmalines reflects the variation in their compositions. In fact, as one might suspect, some individual crystals have different cell dimensions for different zones; for example, Donnay (1969) reports differences of up to about 0.7 percent for a and 0.3 percent for c within some Brazilian elbaite crystals.

Correlations between cell dimensions and chemical compositions have been recorded by several workers (e.g., Schaller, 1913; Epprecht, 1953; Slivko and Iorysh, 1964; Donnay, 1969; Voskresenskaya and Barsukova, 1971; Afonina et al., 1974; Schmetzer, Nuber, and Abraham, 1979; and

Table 3-1
Representative Cell Dimensions of Tourmaline Species

Species (description)	*a,c* (Å)	*c:a*	*Volume* (Å^3)	*Reference*
Buergerite (type)	15.869, 7.188	0.4530	1567.617(5)	Barton, 1969
Dravite ("ideal")	15.910, 7.210	0.4532	1580.551(1)	Epprecht, 1953
Elbaite ("ideal")	15.810, 7.085	0.4481	1533.686(2)	Epprecht, 1953
Ferridravite (type)	16.20, 7.47	0.4611	1697.788(4)	Walenta and Dunn, 1979
Liddicoatite (type)	15.867, 7.1354	0.4497	1555.753(7)	Dunn, Appleman, and Nelen, 1977
Schorl ("ideal")	16.000, 7.135	0.4459	1581.855(7)	Epprecht, 1953
Uvite (neotype)	15.981, 7.207	0.4510	1594.025(8)	Dunn et al., 1977
Chromdravite (type)	16.11, 7.27	0.4513	1634.023(9)	Rumantseva, 1983

Sahama, Knorring, and Tornröos, 1979). In the last cited paper, Sahama, Knorring, and Tornröos plotted cell volumes versus total Fe, Mn, Mg, and Ti contents, and showed that the data they considered comprise two rather distinct clusters of points, one representing tourmalines of the elbaite-schorl series, the other, tourmalines of the schorl-dravite (+ uvite) series.

The most frequently cited correlations are those of Epprecht (1953) whereby cell dimensions were used to differentiate tourmalines both between and within the elbaite-schorl and the schorl-dravite series (see Fig. 3.1). According to plots made by the writer (e.g., Fig. 3-2) the scheme does not work for most tourmalines, only for those with compositions along or near the joins between the "ideal" compositions. This apparently depends on the effects of the several diverse substitutions in the *X*-, *Y*-, and *Z*-sites on the *c* and *a* cell dimensions.

For the record: Wilkins, Farrell, and Naiman (1969) have recorded heat-promoted changes in cell dimensions of tourmaline, and Kovyzhenko (1974) has discussed differences in cell dimensions between natural tourmalines and their synthetic analogs.

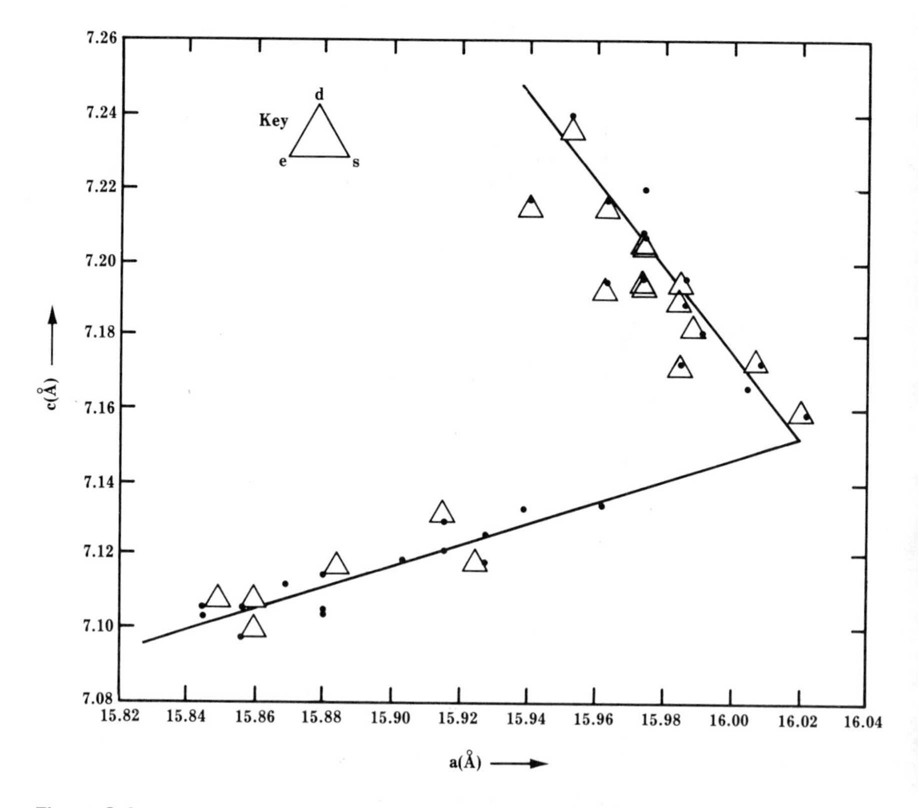

Figure 3-1
Cell dimensions of tourmalines with composition triangles given for specimens having known compositions. (Data from Epprecht, 1953 and diagram after Donnay, 1963.)

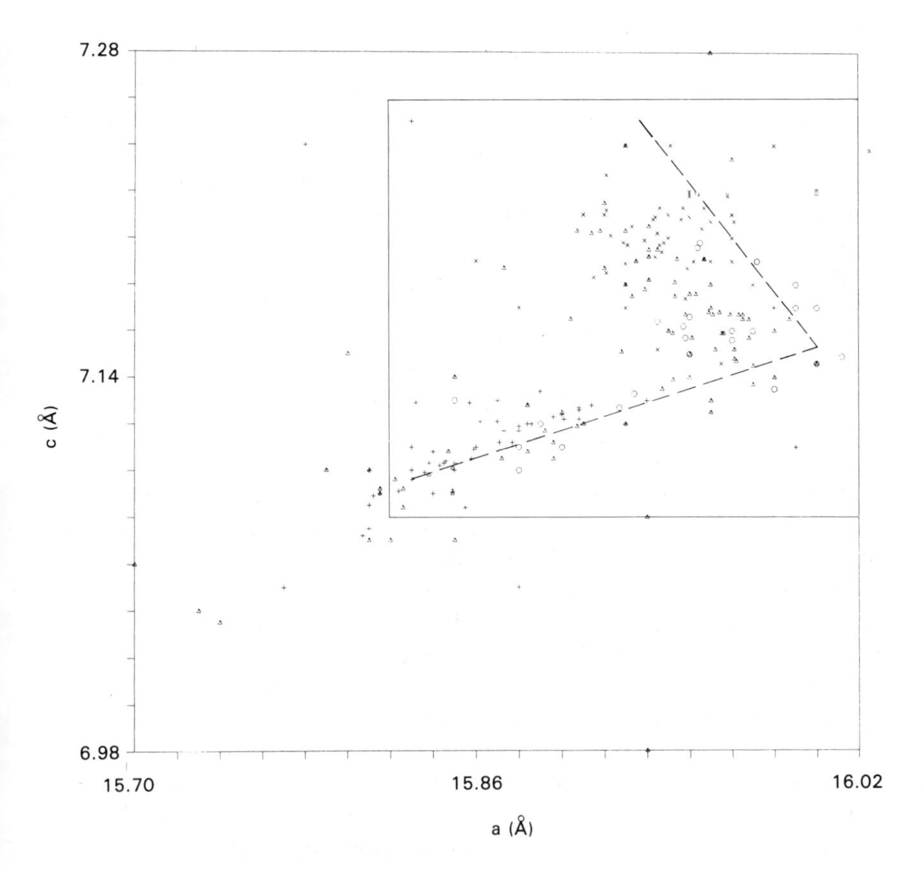

Figure 3-2
Cell dimensions of tourmalines recorded as elbaites, schorls, or dravites in the literature. Dashed line is drawn in same position as solid line on Figure 3-1.

STRUCTURE

Three different structure models were proposed in the late 1940s and early 1950s (Sadanaga, 1947, and Ito and Sadanaga, 1951; Hamburger and Buerger, 1948, and Donnay and Buerger, 1950; and Belov and Belova, 1949 and 1950). The specimens used for those structural analyses were as follows:

Ito and Sadanaga—rose colored "rubellite", assumed to be elbaite, from Brazil (a = 16.0Å and c = 7.17Å);
Donnay and Buerger—nearly colorless uvite/dravite from DeKalb, New York (a = 15.95Å and c = 7.24Å) and black "Fe-tourmaline" from Andreasburg, Harz (a = 16.01Å and c = 7.18Å);
Belov and Belova—dravite from an unrecorded Moravian locality (a = 16.00Å and c = 7.24Å).

In 1953, Epprecht noted that his measurements of the variations of lattice constants could not be "brought into line with the structure proposed by Belov and Belova . . . [but] tend to confirm the tourmaline structures worked out by Donnay/Buerger and Ito/Sadanaga" (p. 504). He also suggested that the Donnay/Buerger structure might obtain for dravite and the Ito/Sadanaga structure for elbaite, a suggestion more or less echoed by Donnay, Wyart, and Sabatier (1959).

Apparently considering such polymorphism to be less than likely, Buerger, Burnham, and Peacor (1962) refined the atomic coordinates given for the three models by least squares and found that the Donnay/Buerger and Ito/Sadanaga coordinates converge to identical values [with $R(hkl)$] = 6.9 percent whereas the Belov/Belova coordinates cannot be so refined without redistribution. This, of course, suggested, albeit only permissively, that the Belov/Belova model is not correct.

Subsequently completed structural analyses have corroborated the refined structure of Buerger, Burnham, and Peacor (1962), which was for a uvite-dravite ($Uvite_{60}$ $Dravite_{40}$) from DeKalb, New York (see Figs. 3-3 and 3-4). Examples of these later refined structures are: buergerite—Barton (1969) and Tippe and Hamilton (1971); elbaite—Donnay and Barton (1972); liddicoatite—Nuber and Schmetzer (1981); schorl—Fortier and Donnay (1975); "chromium-rich Mg-Al tourmaline"—Nuber and Schmetzer (1979); and vanadium-bearing dravite—Foit and Rosenberg (1979).

The elbaite refinement was made by Donnay and Barton because they suspected that the specimen investigated by Ito and Sadanaga (1951) may have been dravite. They felt that if their conjecture was correct, a structural analysis of a true elbaite was needed before any acceptable resolution could be made of the apparent no-solid-solution relation between elbaite and dravite.

Probably the most important later modification relating to the structure of tourmaline is the c-to-$\bar{c}$ transposition, which is based on the fact that Barton (e.g., 1969) found the "absolute orientation" of the structure to have the O_6 apices of the Si-tetrahedra directed toward the analogous, rather than toward the antilogous, pole. (See Figs. 3-3, 3-4, and 3-5.)

Several investigators have studied certain aspects of the tourmaline structure by utilitizing infrared spectroscopy, nuclear magnetic resonance, Mössbauer spectroscopy, etc. Results of these investigations are described in a succeeding part of this chapter and were considered while writing the section that deals with solid solution, in Chapter 4, "Chemistry. . . ."

The Belov/Belova model is based on 30 oxygens (etc.) rather than on the 31 oxygens (etc.) of the Donnay/Buerger and Ito-Sadanaga models. The "missing oxygen" is the one designated O_1 in Figure 3-3; that is, it is the O (or OH) in the central part of the structure and common to three Mg-octahedra.

The Ito/Sadanaga and Donnay/Buerger models, which, as already noted, are similar, are both based on 31 oxygens. The main difference between the structures is in the configuration of the ring of Si-tetrahedra: hexagonal in the Ito-Sadanaga model, trigonal in the Donnay-Buerger model (see Fig. 3-6).[1] In addition, however, the two structures differ in suggested atomic coordinates, especially for *X*-site cations. Nonetheless, both are

characterized by a hexagonal ring of Si-tetrahedra which has the general shape of a hexagonal fragment isolated from a phyllosilicate sheet. The Mg [*Y*-site] atoms are in octahedral coordination and the octahedra share edges to form a trigonal fragment of a brucite-like layer. Six oxygen atoms of the lower part of this layer are also the apices of the hexagonal ring of Si-tetrahedra. The trigonal nucleus of Mg- [*Y*-site] octahedra is extended by six Al- [*Z*-site] octahedra through edge sharing. [And,] Each of the B atoms is in triangular coordination, linked to the vertices of the octahedra of this layer (Buerger, Burnham, and Peacor, 1962, p.583).

Table 3-2 gives the correlation of *X, Y,* and *Z*-sites with Wyckoff's notation and the point symmetry for each kind of site. Table 3-3 gives the sites and multiplicity for each for ideal elbaite, schorl, and dravite. Table 3-4 gives a comparison of atomic fractional coordinates of elbaite, schorl, and dravite.

The following paragraphs consider the structure in more detail:

1. The *X*-site cations (predominantly Na and/or Ca plus or minus K) are located along the threefold axis of symmetry, between the levels of the BO_3 triangles of one "structural island" (as they are termed by Donnay and Buerger, 1950) and the SiO_4 tetrahedra of the underlying "structural island." Each *X*-site cation has nine- or possibly tenfold coordination. The bonds are with three oxygens of the BO_3 triangles (e.g., O_2, $O_{2'}$, and $O_{2''}$ in Fig. 3-3) and six oxygens of SiO_4 tetrahedra (three each of the O_4s and O_5s), and possibly with the unsaturated O ($O_1 = OH$), which is also located on the threefold axis of symmetry (Ito and Sadanaga, 1951, among others, have not included O_1 as partially bonded to *X,* apparently because the distance between the *X*-site cation and O_1 seems to be too great to constitute a bond; it is, for example, 4.52Å in the Ito/Sadanaga model.)

Wang and Hsu (1966) suggested the existence of a magnodravite in which the *X*-site was interpreted to have been occupied to the extent of nearly 50 percent by Mg. Rosenberg and Foit (1979) found their vanadium-bearing dravite to have about 18 percent of its *X*-sites occupied by "Mg^{2+} (or, less likely, Fe^{2+})."

In addition, Foit and Rosenberg (1977) directed attention to the fact that many analyses show substantial *X*-site deficiencies if only Na, Ca, and K are considered as occupants and if the generally accepted formula [$XY_3Z_6B_3Si_6O_{27}(O,OH,F)_4$] is used. Povondra (1981) interpreted his chemical data also to support the conclusion that vacancies are common in *X*-sites. Taylor and Terrell's (1967) synthesis of an alkali-deficient Fe-tourmaline, with only 0.12 atoms in the *X*-site, is permissively corroborative.

2. The *Y*-site cations [predominantly Fe^{2+} and/or Mg^{2+}, ($Al \pm Li$), and/or Fe^{3+}], of which three are shown on Figure 3-3, are equidistant from the threefold axis of symmetry at a level between the layer of BO_3 triangles and the layer of O_6 nadir apices[2] of the SiO_4 tetrahedra. Each *Y* cation is shown to

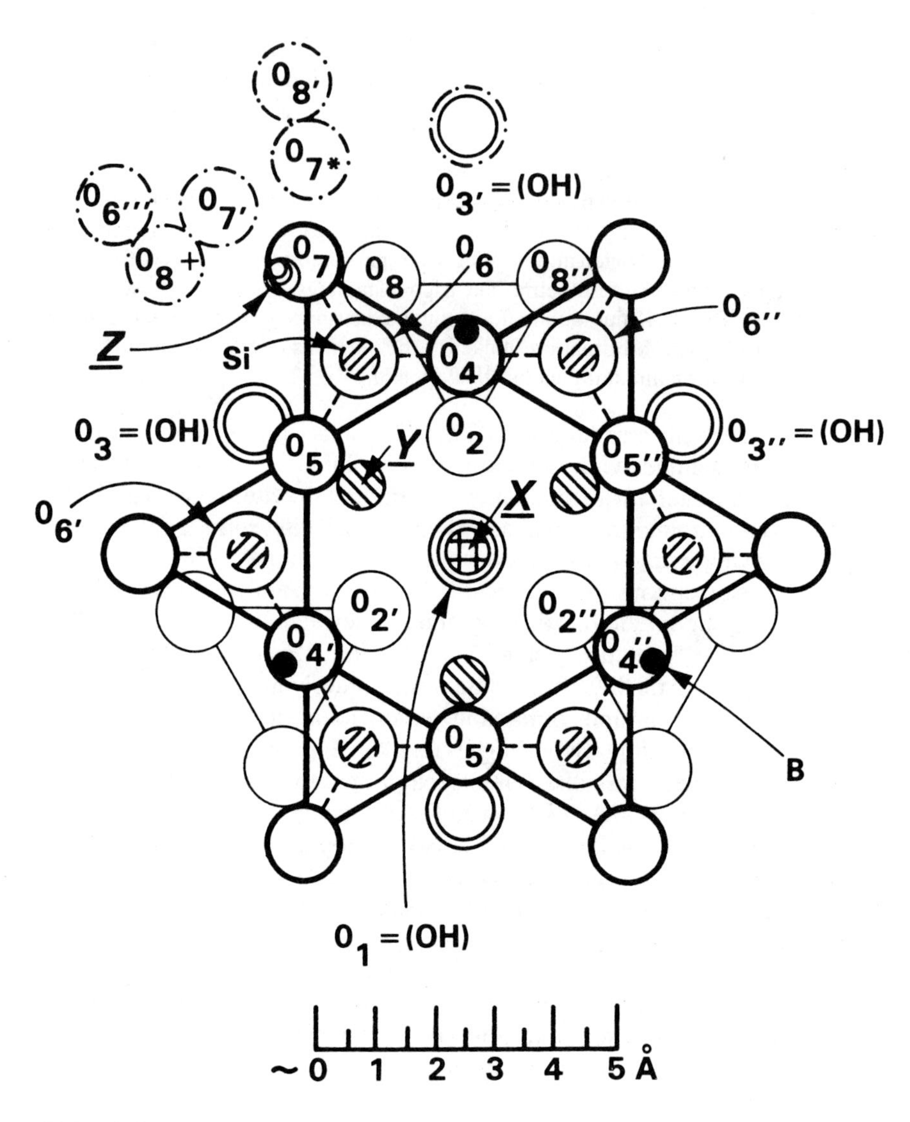

Figure 3-3
Structure of tourmaline projected onto (0001) (after Ito and Sadanaga, 1951).

Figure 3-4
Components of tourmaline structure projected onto (0001). Figures on left give view from the top (i.e., down from antilogous end); figures on right give view of each layer from bottom. Top to bottom: *A* and *A'*, *X*-site cation, located along central axis; *B* and *B'*, Si_6O_{18} ring of six SiO_4-tetrahedra, with apices pointing toward analogous end; *C* and *C'*, three *Y*-site and six *Z*-site octahedra; *D* and *D'*, three BO_3 triangles.

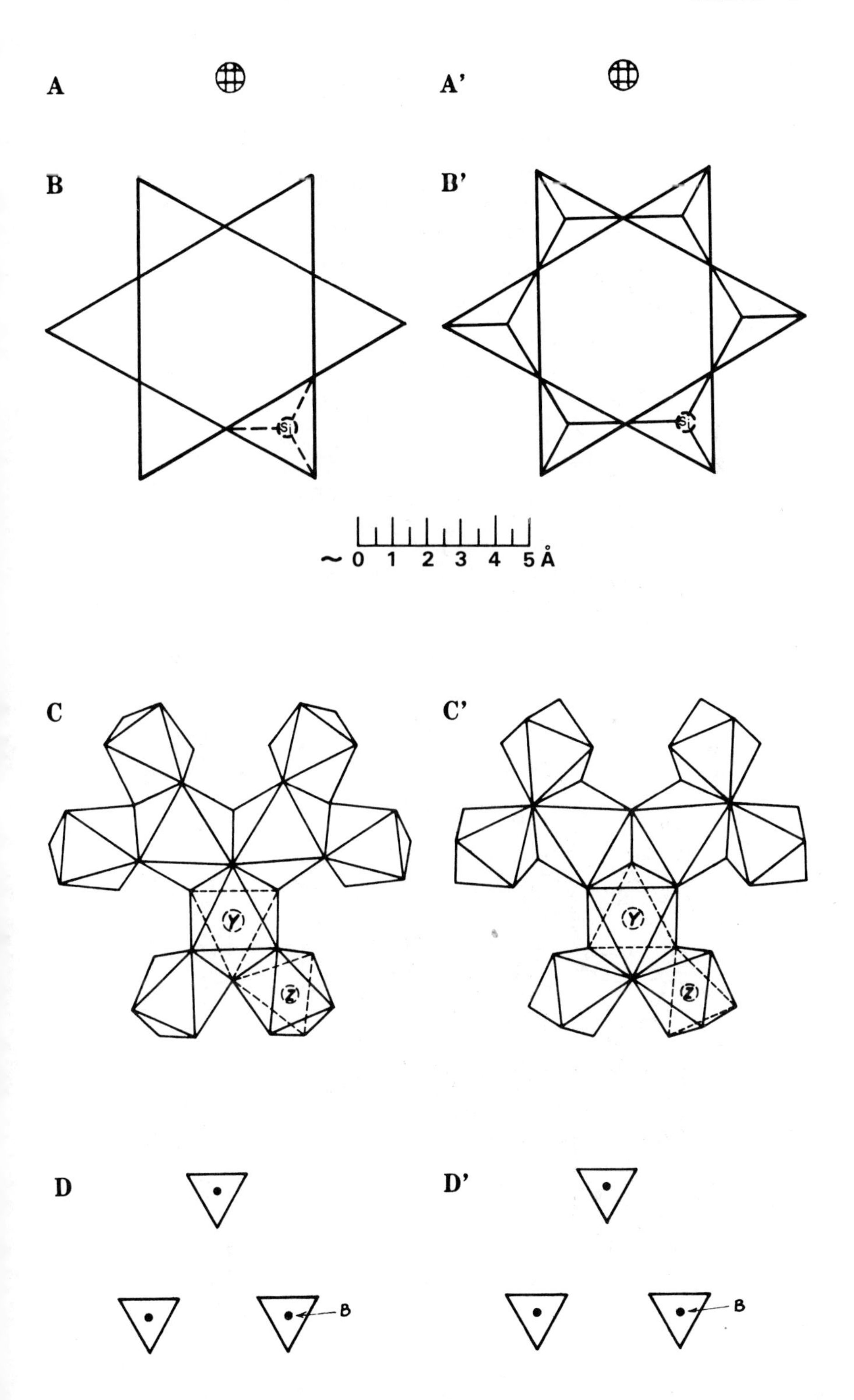
A
A'
B
B'
Si
Si
~0 1 2 3 4 5 Å
C
C'
Y
Y
Z
Z
D
D'
B
B

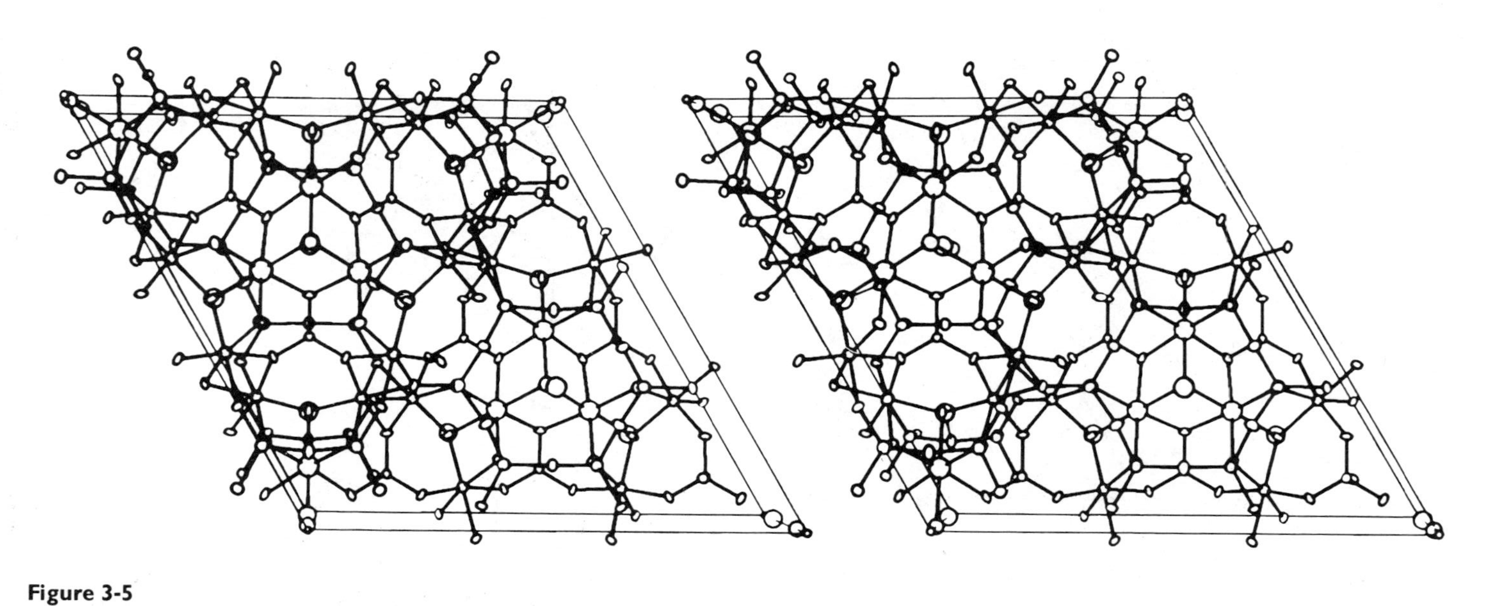

Figure 3-5
Stereographic pair showing tourmaline structure bonds (rather than coordination polyhedra). The small ellipses represent anions whereas the circles represent cations. The circles, in small-to-large order, represent B, *Z*-site cations, Si, *X*-site cations, and *Y*-site cations. (Figure, prepared by W. C. Hamilton, first appeared in Donnay and Barton, 1972; republished by permission of R. Barton, Jr.)

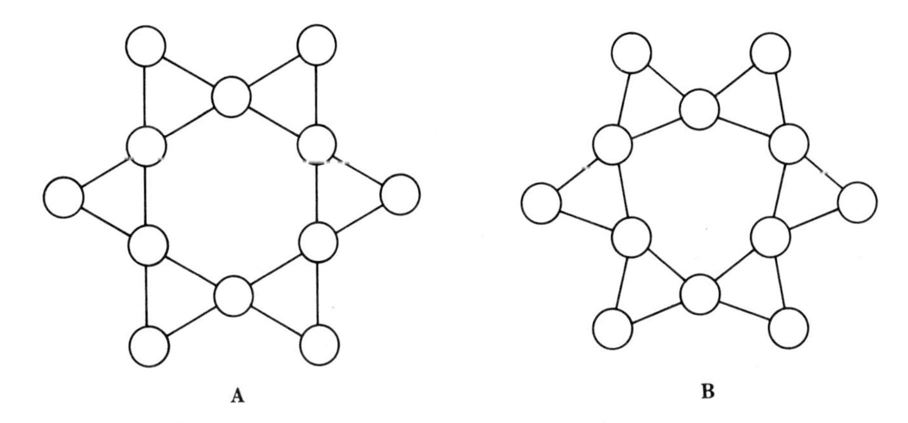

Figure 3-6
Silica ring of tourmaline structure. *A*, as interpreted by Ito and Sadanaga (1951); *B* after Donnay and Buerger, (1950).

Table 3-2
Correlation of the *X*-, *Y*-, and *Z*-sites with Wyckoff's Notation and the Point Symmetry for Each Site

Site	*Wyckoff Notation*	*Point Symmetry*
X	3a	*3m* (C_{3v})
Y	9b	*m* (C_s)
Z	18c	1 (C_1)

Table 3-3
Sites and Multiplicity for Each for Ideal Elbaite, Schorl, and Dravite

Site and Multiplicity	*Elbaite*[a]	*Schorl*[b]	*Dravite*[c]
18c	Si	Si	Si
9b	B	B	B
18c (*Z*)	Al	Al, Fe^{3+}, . . .	Al
9b (*Y*)	(Al,Li)	Fe^{2+}, Fe^{3+}, Al . . .	Mg
3a (*X*)	Na	Na . . .	Na
3a	O_1(F,OH)	OH,O	(F,OH)
9b	O_2	O_2	O_2
9b	O_3(O,OH)	OH,O	(O,OH)
9b	O_4	O_4	O_4
9b	O_5	O_5	O_5
18c	O_6	O_6	O_6
18c	O_7	O_7	O_7
18c	O_8	O_8	O_8

[a]Donnay and Barton, 1972. [b]Fortier and Donnay, 1975. [c]Donnay and Buerger, 1950.

Table 3-4
Coordinates of Refined Tourmaline Structures of Elbaite (El), Schorl (Sh), and Uvite (Uv)

	x	*y*	*z*
3a: *X*, El	O	O	.2347
Sh	O	O	.2235
Uv	O	O	.2324
O_1, El	O	O	.7794
Sh	O	O	.7837
Uv	O	O	.7783
9b: *Y*, El	.1234	x/2	.6348
Sh	.1257	x/2	.6279
Uv	.1270	x/2	.6282
B, El	.1092	2x	.4548
Sh	.1103	2x	.4546
Uv	.1103	2x	.4538
O_2, El	.0614	2x	.4868
Sh	.0615	2x	.4830
Uv	.0609	2x	.4787
O_3, El	.2676	x/2	.5069
Sh	.2706	x/2	.5115
Uv	.2680	x/2	.5145
O_4, El	.0937	2x	.0729
Sh	.0928	2x	.0691
Uv	.0935	2x	.0724
O_5, El	.1858	x/2	.0937
Sh	.1857	x/2	.0904
Uv	.1812	x/2	.0888
18c: *Z*, El	.2964	.2598	.6105
Sh	.2988	.2617	.6116
Uv	.2976	.2615	.6138
Si, El	.1917	.1896	0.0
Sh	.1918	.1899	0.0
Uv	.1922	.1898	0.0
O_6, El	.1966	.1879	.7746
Sh	.1977	.1859	.7774
Uv	.1952	.1866	.7789
O_7, El	.2863	.2859	.0783
Sh	.2846	.2852	.0797
Uv	.2844	.2851	.0810
O_8, El	.2092	.2693	.4397
Sh	.2099	.2705	.4425
Uv	.2085	.2698	.4445

Sources: data from Donnay and Barton (1972), Fortier and Donnay (1975), and Buerger, Burnham, and Peacor (1962); summary modified after Fortier and Donnay (1975).

have octahedral coordination with two oxygens of the BO_3 triangles (e.g., O_2 and $O_{2'}$), two oxygens (of OHs) labelled O_1 and O_3, and two of the O_6 nadir apices of overlying SiO_4 tetrahedra (e.g., O_6 and $O_{6'}$). Each *Y* octahedron shares one edge with two adjacent *Y* octahedra, the three thus having a common apex on the threefold symmetry axis ($= O_1 =$ OH), and each *Y* octahedron also shares two edges with the peripheral *Z* octahedra. The *Y* octahedra are, as shown, generally larger than the *Z* octahedra (e.g., for the DeKalb uvite studied by Buerger, Burnham, and Peacor, 1962, the metal-oxygen distances are for $Y \simeq 2.05$Å and for $Z \simeq 1.93$Å). However, if both are occupied chiefly by Al, they may be more nearly equal in volume.

The *Y*-site may also be occupied by noteworthy amounts of Mn^{2+}, Cr^{3+} (e.g., Dunn, 1977*b*), V^{3+} (e.g., Foit and Rosenberg, 1979), and possibly Ti^{4+} (Povondra, 1981). Also, Voskresenskaya and Plyusnina (1975) have shown by infrared spectroscopy that the Co and Ni of synthetic Co- and Ni-rich tourmalines of the elbaite-schorl series are in the *Y*-site.

3. The *Z*-site cations, predominantly Al^{3+}, Fe^{3+}, Cr^{3+}, or [Al,Mg—($\simeq$5:1)], of which only one is shown on Figure 3-4*C*, are located at nearly the same level as the *Y*-site cations. Each is interpreted to be in octahedral coordination within a somewhat distorted octahedron, the corners of which for the example shown, are one O_3 oxygen (or OH), two O_8 oxygens of BO_3 triangles (one each from two different "structural islands") and three oxygens—O_6, O_7, and O_{7*} for the example shown—of SiO_4 tetrahedra (one each from three different "structural islands"). These relations mean, of course, that the *Z*-site cations serve to link the "structural islands" together in a so-to-speak screw-axis mode (Fig. 3-7).

As already noted, the typically smaller *Z* octahedra are located peripherally around the typically larger *Y* octahedra in such a manner that each *Y* octahedron shares an edge with two *Z* octahedra. As an example of the distortion of these octahedra: for the DeKalb uvite studied by Buerger, Burnham, and Peacor (1962), the edges shared between the *Z* [Al] octahedra and *Y* [Mg] octahedra were found to be 2.58Å as compared to 2.43Å for shared edges of Al-Al octahedra and 2.75Å for shared edges between adjacent Mg-Mg octahedra. Optical spectra studies of Fe^{2+} have also shown noteworthy distortions in both kinds of octahedra (Wilkins, Farrell, and Naiman, 1969).

4. The Si ions are in SiO_4 tetrahedra that constitute six-membered (i.e., Si_6O_{18}) rings. All tetrahedra of these rings point in the same direction; they are shown on the diagram as pointing toward the analogous ($\bar{c}$) pole, in accordance with Barton's (1969) determinations.

Although some mineralogists (e.g., Deer, Howie, and Zussman, 1962) have concluded that Al does not substitute for Si in tourmaline, others (e.g., Buerger, Burnham, and Peacor, 1962) have indicated that some Al can occupy the predominantly SiO_4 tetrahedra. In fact, Buerger, Burnham, and Peacor gave Al:Si = 0.16:5.84 for the tetrahedral site in the DeKalb uvite, and Foit and Rosenberg (1979) recorded a value of 0.37:5.63 for their

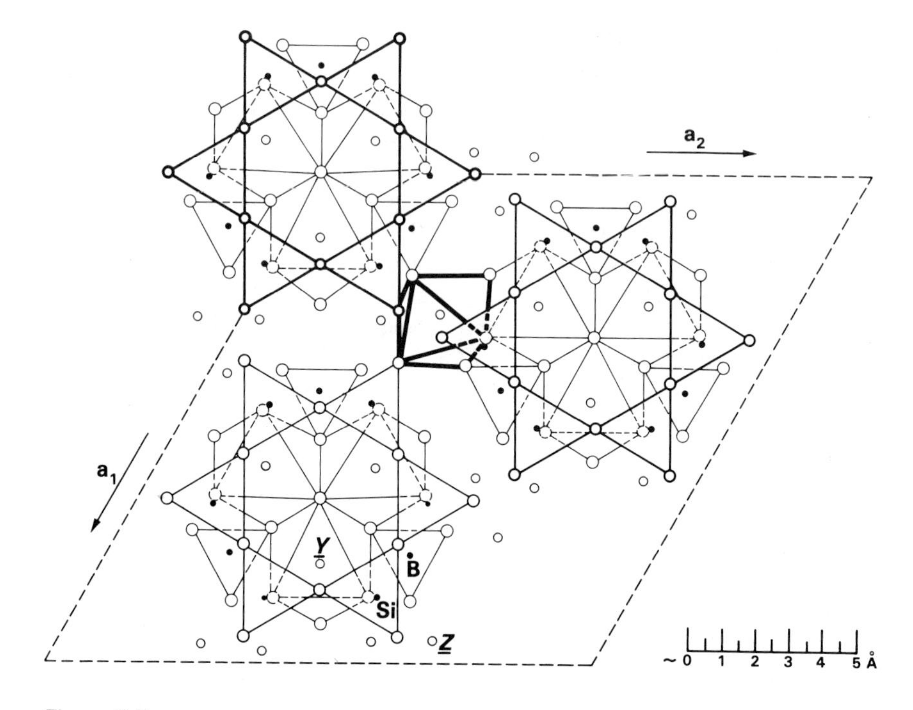

Figure 3-7
Projection of tourmaline structure on (0001). Si-tetrahedra, *Y* and *Z* octahedra, and B-triangle positions are marked in lower "structural island." Heavy lined *Z*-octahedron shows linkage of the "structural islands." (After Barton, 1969.)

vanadium-bearing tourmaline. Plyusnina and Voskresenskaya (1974) concluded from their infrared study of tourmalines that the amount of Al for Si substitution appears to vary directly with the total Al-content.

Still others (e.g., Barton, 1969) have suggested that B in excess of that required for the BO_3 triangles may replace Si in the tetrahedra, and, on the basis of an EPR study of elbaites, Novozhilov, Voskresenskaya, and Samoilovich (1969) reached the same conclusion.

In addition, in a study of the "diadochy of silicon and phosphorus . . .", Pehrman (1962) found <0.07 percent P_2O_5 in the tourmalines included in his investigation.

5. The B ions are shown to have three fold triangular coordination with the planes of the triangles essentially parallel to (0001).

Loewenstein (1956) directed attention to the proximity of each boron ion to an O_4 oxygen of a nearby SiO_4 tetrahedron and suggested that boron should be considered to have tetrahedral rather than triangular coordination. It is true that the BO_3 triangles are distorted, in that the B is off-center, and that the nearest oxygen of the nearest SiO_4 group may be responsible for this distortion. However, both infrared spectroscopic and nuclear magnetic resonance studies appear to militate against Loewenstein's interpretation and, also, against Barton's just mentioned conclusion that B may replace Si in some SiO_4 tetrahedra. Moenke (1962), using infrared spectroscopy, has shown that all of the B in schorl specimens from Penig, Sachsen, and San Diego County, California, is in BO_3 groups—that is, no BO_4 groups were detected; and, Bray et al. (1961), as a result of NMR investigations of transparent tourmaline crystals, not otherwise identified or characterized, have drawn similar conclusions (but, cf. the Novozhilov, Voskresenskaya, and Samoilovich, 1969, conclusion alluded to in paragraph **4**.

Also noteworthy: Ja (1972) concluded from EPR studies that Fe^{3+} replaces B^{3+} in a tourmaline that he studied; and Serdyuchenko (1980) suggested that BO_3 triangles may replace OH groups in some tourmalines. (The latter suggestion seems, among other things, to be unlikely in that it would appear to violate crystal chemistry principles.)

6. The anions (predominantly O and OH + F) occupy the apices of the coordination polyhedra as shown on Figure 3-3.

Although chlorine has not been shown in tourmaline formulae, Fuge and Power (1969) found 0.024-1.102 weight percent Cl in 31 schorls from high-chlorine granites of southeastern England (cf. Kuroda and Sandell, 1953).

As noted by Foit and Rosenberg (1977), some schorl/dravite and elbaite/schorl analyses may be interpreted to indicate that dehydroxylation may extend well beyond their $R^+ R_3^{3+} R_6^{3+} (BO_3)_3 Si_6 O_{18} O_3 (OH)$ end-member (see Fig. 4-6) in the direction of hydroxyl-free tourmaline.

On the basis of their study of the near infrared spectrum of a tourmaline (originally reported incorrectly as beryl), Wickersheim and Buchanan (1968) concluded: (1) OH occupies three different kinds of sites; (2) there is no detectable H_2O present; (3) OH vibration directions are not quite parallel to the symmetry axis of the crystal structure; (4) OH ions are coupled rather strongly to the rest of the structure; and (5)—which is incorrect—the O of each OH group is bonded directly to a silicon ion.

7. The absolute orientation, first established by Barton (e.g., 1969), is based on anomalous dispersion (i.e., $I(hkil) \neq I(\overline{hkil})$), apparently attributable to iron content, that is detectable under copper radiation. Etch figures also indicate this orientation (Wooster, 1976).

INFORMATION FROM AUXILIARY STUDIES

Results of investigations utilizing methods other than X-ray diffraction to study certain aspects of the structure of tourmalines have served to amend site assignments and/or other aspects of the structure. These results are outlined in the remainder of this section, under the following subheadings: Infrared Spectroscopy (IR), Nuclear Magnetic Resonance (NMR), Mössbauer Spectroscopy (NGR), Raman Spectroscopy, Electron Paramagnetic Resonance (EPR), and Other Structure-related Auxiliary Studies.

Infrared Spectroscopy

As noted in the review of the history of the use of infrared absorption in mineral investigations by Alexanian, Morel, and LeBouffant (1966), tourmaline was included by Merritt (1895) in his early study of infrared dichroism of minerals. Subsequently, several investigators have recorded and/or studied infrared spectra of diverse tourmalines (e.g., Reinkober, 1911; Omori, 1961; Plyusnina, 1961; Gebert and Zemann, 1965; Alexanian, Morel, and LeBouffant, 1966; Plyusnina, Granadchikova, and Voskresenskaya, 1969; Novak and Zak, 1970; Vierne and Brunel, 1970; Hunt, Salisbury, and Lenhoff, 1973; Moenke, 1974; Plyusnina and Voskresenskaya, 1974; Liese, 1975; and Rosenberg and Foit, 1979); see Figure 3-8.

In one of the more extensive investigations, Plyusnina, Granadchikova, and Voskresenskaya (1969) presented infrared spectra of several natural tourmalines—nine schorls, six dravites, and nine elbaites—and four of their synthetic analogs. Later, in a follow-up study. "A more exact definition of certain crystal structure positions in tourmaline through infrared spectroscopy," Plyusnina and Voskresenskaya (1974) confirmed (again!) the refined structure given by Buerger, Burnham, and Peacor (1962), as distinct from the structure proposed by Belov and Belova (1949 and 1950), and concluded that:

1. Al replacement for Si takes place in the tourmaline group, the amount of replacement evidently varying directly with the total Al-content.;
2. Infrared spectra afford a means of distinguishing elbaites from schorls from dravites and perhaps even schorls of the elbaite-schorl series from schorls of the dravite-schorl series;
3. The tourmaline structure contains at least two nonequivalent BO_3 groups;
4. Infrared spectra for synthetic tourmalines, though similar to spectra of natural tourmalines, show less distinct bands.

With regard to item 2, the differences are: (a) the shape and position of the main band, which is in the range 1100-1000 cm^{-1} (for schorl, 1077-983 cm^{-1}; for dravite, 1095-995 cm^{-1}; for elbaite, 1100-1015 cm^{-1}); (b) the position of the B^{III} (−0) band (for schorl, 1347-1236 cm^{-1}; for dravite, 1350-1270 cm^{-1}; for elbaite 1350-1306 cm^{-1}) (this band can apparently also be used to distinguish the schorls of the two series in that elbaite-schorls yield a narrow band with position closer to that of elbaite whereas dravite-schorls give a broad band closer in position to that of dravite); and (c) the band(s) due to the stretching OH vibrations (e.g., for schorl, a single band at ~3585 cm^{-1}; for dravites, a single band at ~3562 cm^{-1}; and for elbaite, two distinct bands at ~3595 and 3482 cm^{-1}).

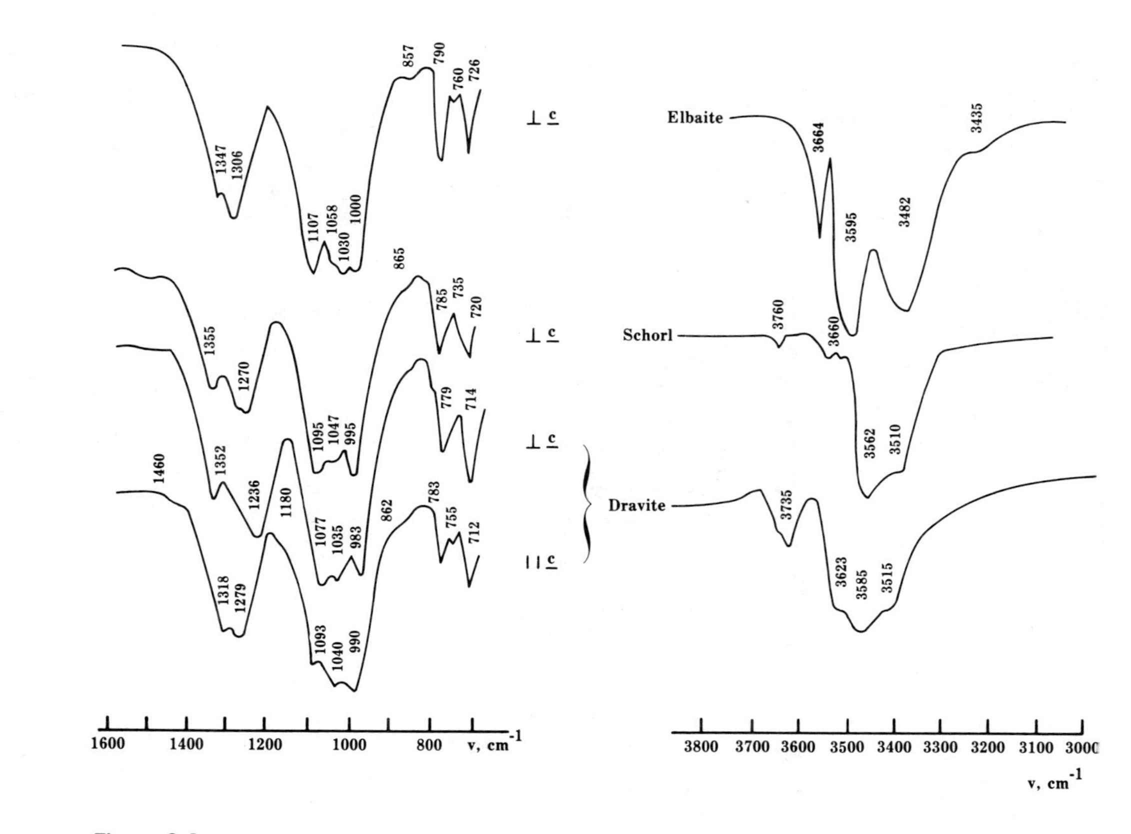

Figure 3-8
Infrared spectra (after Plyusnina, Granadchikova, and Voskrensenskaya, 1969).

With regard to item 3, Vierne and Brunel (1970), as a result of their infrared studies of a dravite and an elbaite, concluded that there are BO_3 triangles in the structure but no BO_4 tetrahedra (cf. Loewenstein, 1956).

With regard to items 2 and 4, Rosenberg and Foit (1979) found the infrared spectra of their "alkali-free tourmalines" to be similar to the spectra reported by Plyusnina, Granadchikova, and Voskesenskaya (1969) and Plyusnina and Voskresenskaya (1974) and by Moenke (1974) except for differences in the OH-stretching region. In that region, the "alkali-free" synthetic tourmalines give spectra more like those of elbaite than of dravite.

Gebert and Zemann (1965) concluded that the strong pleochroism with $\epsilon > \omega$ in the OH-stretching frequency of infrared spectra (i.e., $\lambda \simeq 2.9\mu$) is best accounted for by OH dipoles approximately parallel to {0001}.

Nuclear Magnetic Resonance

As a result of an NMR study of ^{11}B in a transparent tourmaline not otherwise characterized, Bray et al. (1961) concluded that essentially all B is in triangular coordination in tourmaline. Later, on the basis of their NMR investigation of an elbaite from Elba (American Museum of Natural History #12210), Tsang and Ghose (1973) stated that the tourmaline yielded no sharp resonance lines; B is present in BO_3 triangles with the B-O bond highly covalent; and OH groups (but no H_2Os) are present.

Mössbauer Spectroscopy

The first suggestion that Mössbauer spectroscopy might be applied to minerals was by Pollak et al. (1962). Subsequently, several such studies have been made of tourmaline. Examples are reported in papers by deCoster, Pollak, and Amelinckx (1963), Zheludev and Belov (1967), Marfunin et al. (1970), Burns (1972), Burns and Simon (1973), V. F. Belov et al. (1973*a*, 1973*b*), Hermon, Simpkin, and Donnay (1973), Faye, Manning, and Gosselin (1974), A. F. Belov et al. (1975), Alvarez et al. (1975), Bhandari and Varma (1975), Pollak and Bruyneel (1974), Dambly et al. (1976), Scorzelli, Baggio-Saitovitch, and Danon (1976), Gorelikova, Perfil'yev, and Bubeshkin (1978), Korovushkin, Kuz'min, and Belov (1979), Pollak et al. (1979), and Saegusa, Price, and Smith (1979).

Most of these studies have, of course, been directed toward determining the site distribution of Fe^{2+} versus Fe^{3+} in different iron-bearing tourmalines. Two spectra are shown in Figure 3-9. Additional data, including the parameters used by the investigators for their diverse doublet fitting schemes, are given in the referenced papers.

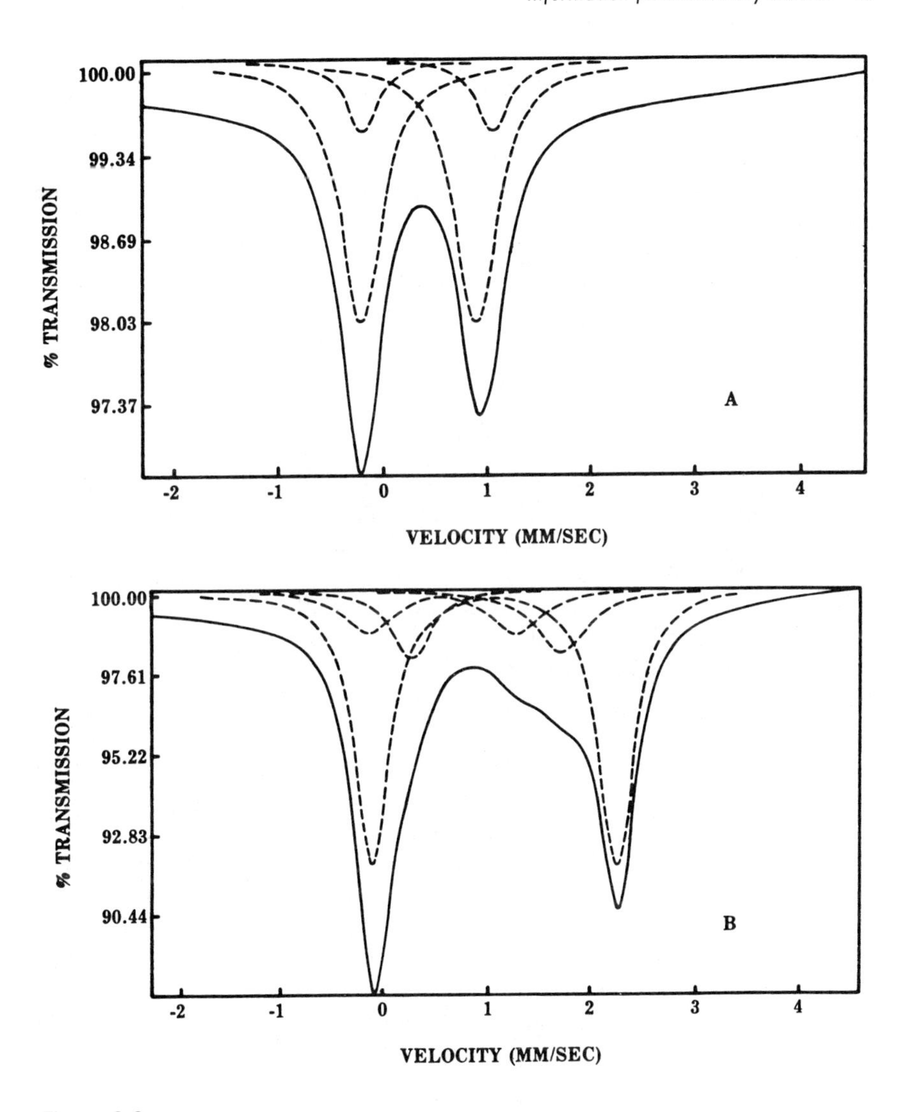

Figure 3-9
Mössbauer spectra (room temperature): *A*, buergerite from Mexquitic, Mexico; *B*, elbaite from San Diego County, California. (After Hermon, Simkin, and Donnay, 1973).

In general, it can be said that Mössbauer spectra corroborate the presence of substantial quantities of Fe in both *Y*- and *Z*-octahedral sites of at least some tourmalines. More specifically, the spectra indicate that:

1. Fe fills up to several percent of the *Z*-sites in some iron-bearing tourmalines, even those that contain sufficient Al to fill those sites com-

pletely (e.g., Burns, 1972, and Hermon, Simpkin, and Donnay, 1973). (This, of course, supports the interpretations based on X-ray [e.g., Barton, 1969] and neutron [Tippe and Hamilton, 1971] diffraction and on optical spectra studies [Wilkins, Farrell, and Naiman, 1969], and is consistent with magnetic susceptibility measurements [Tsang and Thorpe, 1971].)

2. Significant amounts of Fe^{2+}, as well as of Fe^{3+}, can occupy *Z*- as well as *Y*-sites, and the existence of nonequivalent Fe sites with statistical cation distribution in *Y* and *Z* positions has been confirmed (Korovushkin, Kuz'min, and Belov, 1979). Hermon, Simpkin, and Donnay (1973) have postulated this site occupancy phenomenon to represent a chemical response, during formation, to size requirements associated with edge-sharing of *Y* and *Z* octahedra, and have suggested that the lack of such flexibility in iron-poor and iron-free tourmalines (e.g., tourmalines along the dravite/uvite-elbaite/liddicoatite tie line) may account for the nonexistence of solid solution between these species.
3. Fe ions in *Y*-sites have a lower ionization potential than Fe ions in *Z*-sites (Korovushkin, Kuz'min, and Belov, 1979).
4. Some tourmaline specimens showing Si deficiencies and no Fe^{3+} in their chemical analyses may have Fe^{2+} in the cation site having tetrahedral coordination (Saegusa, Price, and Smith, 1979).
5. Mössbauer spectroscopy may be used to determine Fe^{2+}:Fe^{3+} ratios in tourmalines (and other minerals that frequently resist such analyses by other means); see, for example Perfil'eva, Gorelikova, and Bubeshkin (1975) and Gorelikova, Perfil'eva, and Bubeshkin (1978). However, some extraordinary results have led to some questionable interpretations involving such things as tentatively postulated electron hopping between neighboring Fe^{2+}-Fe^{3+} pairs (e.g., Saegusa, Price, and Smith, 1979). And, Pollak and Bruyneel (1974) showed how the determined differences, although they may be real, may be misleading in that the method itself can cause changes in the Fe-valence state and thus yield Fe^{2+}:Fe^{3+} ratios that differ from those originally present in the specimens checked!

In addition, there have been investigations of the effects of Mössbauer spectra of tourmalines subjected to conditions of, for example, high temperatures (e.g., A. F. Belov et al., 1975), low temperatures (Scorzelli, Baggio-Saitovitch, and Danon, 1976, and Saegusa, Price, and Smith, 1979), and under infrared irradiation (Dambly et al., 1976). These studies have indicated, among other things, that:

1. The valence state of Fe^{2+} is increased to Fe^{3+} upon heating in, for example, air between 300°-500° and 900°C (A. F. Belov et al., 1975,

and Gorelikova, Perfil'yev, and Bubeshkin, 1978). (The number of Fe^{2+}-to-Fe^{3+} changes depends upon the duration of heating; Fe^{2+} in *Y*-sites changes more readily than Fe^{2+} in *Z*-sites; and the changes do not appear to disrupt the tourmaline structure.)

2. At relatively low temperatures (e.g., down to 4.2 K) reversible changes, as yet unexplained, take place. As a result of one investigation of the temperature dependence of the magnitude of quadrupole splitting for Fe^{2+} in different nonequivalent positions in tourmaline, Belov et al. (1973) found a linear relationship between −196° and 200°C.
3. During infrared irradiation, tourmaline undergoes effects tentatively explained as arising from *cis-trans* transitions in response to proton hopping between two oxyanions around the same Fe site and/or from electron relaxation between Fe^{2+} and Fe^{3+} in two neighboring sites (Dambly et al., 1976).

Furthermore, it has been suggested that the oxidation coefficient, $Fe^{2+}/(Fe^{2+} + Fe^{3+})$, and other characteristics determined by Mössbauer studies may reflect genetic environments (Gorelikova, Perfil'yev, and Bubeshkin, 1978, and Korovushkin, Kuz'min, and Belov, 1979, respectively).

Raman Spectroscopy

Early Raman spectra for tourmaline were reported by Nisi (1929, 1932) and Hibben (1939). Griffith (1970) presented the data given in Table 3-5. He also remarked that, as with other cyclosilicates, there are noteworthy "ring bands" in the 530-680 cm^{-s} region and the terminal SiO_2 stretches [v^s and v^{as} (SiO_2)] are higher than for orthosilicates (apparently because of the general lack of Si-O π-bonding in the silicate rings).

Contrariwise, on the basis of their study of Raman spectra of two red and two green elbaite-schorl crystals, Alvarez and Coy-Yll (1978) concluded that "From a dynamical point of view it appears that the tourmaline structure cannot be subdivided into separate vibrational units . . . [and thus it] must be considered as a framework lattice" (cf. Foit and Rosenberg, 1979, p.899).

Electron Paramagnetic Resonance

As a result of the EPR study (frequency ~9400 MHz), Novozhilov, Voskresenskaya, and Samoilovich (1969) concluded that: (1) ~0.01 percent of the Si^{4+} is characteristically replaced by B^{3+} (cf., for example, Bray et al., 1961); (2) when irradiated, paramagnetic O^- hole centers are formed in the BO_4 tetrahedra; and (3) when irradiated at 77 K, H^+ ions, which are relatively common in rubellite, are converted to atomic hydrogen (H^0) centers.

Table 3-5
Raman Frequencies (cm^{-1}) and Infrared Data for Tourmaline

Description Mode[a]	*Raman*	*Infrared*
$v^s(MX_2)$	1040 s	1098 m
		1116 m
		1059 m
$v^{as}(MX_2)$		794 s
		1031 s
		995 s
		1004 s
Ring stretches^s	569 s	
	464 vs	
Ring stretches^{as}	929 w	
		760
	682 m	714 s
Ring deformations	353 s	
	340 w	

Note: Raman data and assignments are from Griffith (1970); infrared data are from Plyusnina and Bokiĭ (1958).
[a]Modes for BO_3 are not included.

Also, Ja (1972) found a g = 4.302 ± 0.006 isotropic EPR line for a tourmaline, otherwise not identified, and concluded that it represents Fe^{3+} replacing B^{3+}.

Other Structure-related Auxiliary Studies

In a paper in which Mn^{2+} and Mn^{3+} ions are dealt with ambiguously, Johnston and Duncan (1973) described an energy-dispersive X-ray diffraction investigation of a so-called tsilaisitic tourmaline (MnO—4.69 weight percent/atomic absorption analysis) from Madagascar, and concluded that the Mn ions (reported as Mn^{3+}) are distributed as follows: 46 ± 3 percent (by weight) in *Y*-sites; 54 ± 3 percent in *Z*-sites.

As already mentioned, Tippe and Hamilton (1971) refined the structural parameters of buergerite utilizing three-dimensional single-crystal neutron diffraction data. Their results, as also noted, agree remarkably well with the X-ray based structure described by Barton (1969). Some of the bond distances calculated by Tippe and Hamilton are of slightly higher precision than those given by Barton.

Yamaguchi (1964*a*, 1964*b*) has reported the results of electron diffraction investigations of tourmaline.

Klemens (1973) included a tourmaline in his investigation of radiation damage in solids and phonon scattering. In an elaboration on the results, he wrote (1984, personal communication) that

> The measurements were made in the range of 4-70 K. The conductivities were low and increased monotonically over that temperature range, indicating that the crystal was very imperfect [low temperature thermal conductivities are extremely defect-sensitive]. From the temperature dependence, I would conclude that the dominant defects are probably (but by no means certainly) a mixture of large-scale defects such as dislocations and thin twinned sheets or stacking faults, in combination with point defects. A perfect crystal would have a conductivity $\alpha 1/T$, where T is the absolute temperature, even around 70 K. One would conclude that the intrinsic conductivity at 70 K should be at least a factor 2 to 3 higher than the measured value, and at room temperature be at least 8×10^{-3} W-cm^{-1}–K^{-1}.

Whether the tourmaline structure does or does not tend to be characterized by an unusual abundance of defects per unit volume is to the present, however, a moot point.

X-RAY DIFFRACTION PATTERNS

The pattern given as Figure 3-10 and the reflections listed in Table 3-6 are for a schorl from Brazil (NMNH #B10627).

JCPDS Cards dealing with tourmaline group minerals are as follows:

Buergerite	25-703*
Dravite	14-76 and 14-76a*
Elbaite	26-964 i
Ferridravite	33-1261
Liddicoatite	30-748
Schorl	22-469 o
Uvite	29-342 and 29-342a i
Chromian dravite	25-1307 i

Some of the most intense reflections for the species are given in Table 3-7. The pattern for which the most indexed reflections (72) are listed is that for dravite (14-76 and 14-76a).

Additional indexed patterns recorded in the literature include the following: "ferric iron tourmaline"—Frondel, Biedl, and Ito (1966); "zoned

Table 3-6
X-ray Diffraction Data for Schorl from NMNH #B10627 for Reflections up to $2\Theta = 50°$ (Copper radiation—$K\alpha_1$; $\lambda = 1.54050\text{Å}$)

*hkl**	*I/I*	2Θ	*d*	*hkl*	*I/I*	2Θ	*d*
101	62	13.85	6.39	232	20	37.77	2.382
021	23	17.76	4.99	511	18	38.27	2.352
300	13	19.18	4.63	113	3	39.37	2.289
211	55	20.98	4.23	520	3	40.61	2.221
220	56	22.185	4.01	502	16	41.09	2.197
012	100	25.62	3.48	431	10	41.57	2.172
131	6	26.295	3.39	303	18	42.56	2.124
410	8	29.49	3.03	422	13	42.66	2.119
122	93	30.17	2.96	223	21	44.12	2.053
321	8	30.71	2.911	152	48	44.27	2.046
312	6	34.17	2.624	161	6	44.65	2.029
051	100	34.675	2.587	440	34	45.33	2.001
042	3	36.01	2.494	342	26	47.23	1.924
241	3	36.51	2.461	351	3	47.61	1.910
003	21	37.64	2.390	413	11	48.52	1.876

*Per JCPDS convention, only the *hkl* values of the Bravais indices are given; that is, *i* is not represented by either a value or an indication of omission.

vanadium tourmaline"—Snetsinger (1966); dravite$_{60}$ schorl$_{40}$—de Camargo and Souza (1970); chromian dravite—Lum (1972); dravite—Holgate (1977); synthetic "end-member" dravite, schorl, elbaite, and tsilaisite—Tomisaka (1968); and synthetic alkali-free tourmaline and synthetic dravite—Rosenberg and Foit (1979).

A suggestion of how X-ray data might be used to establish percentages of elbaite, dravite, uvite, and schorl components of tourmalines was given by Horn and Schulz (1968). The method is based on the intensity ratios I_{051}/I_{131} or I_{051}/I_{321} reflections. The method is, however, less than definitive for many natural tourmalines, even those thought to consist only or largely of the end-members.

Slivko (1971) correlated Al_2O_3 content with *d*-spacing of the $(21\bar{3}2)$ reflection for both elbaite and dravite.

Afonina and Vladykin (1980) published a paper "Determination of the composition of tourmaline based on data from powder X-ray diffraction patterns" (CA—93:75975a) not yet seen by the compiler. On the basis of the content of other papers by Afonina and colleagues (e.g., Afonina et al., 1979), it seems likely that they are using X-ray data to calculate *c* and *a* dimensions and then plotting those values according to ratios on graphs such as those shown on Figure 3-1.

Table 3-7
Seven High-Intensity Reflections for Species of the Tourmaline Group

hkl[*]	*Buergerite*	*Dravite*	*Elbaite*	*Ferridravite*[a]	*Liddicoatite*	*Schorl*[b]	*Uvite*[c]	*Chromdravite*[d]
101	6.33 (45)[e]	6.38 (30)	6.32 (25)	6.63 (9)	6.33 (6)	6.35 (18)	6.4 (50)	6.57 (50)
211	4.20 (40)	4.22 (65)	4.20 (60)	4.32 (7)	4.197 (50)	4.22 (35)	4.23 (70)	4.31 (40)
220	3.96 (52)	3.99 (85)	3.96 (80)	4.05 (9)	3.962 (55)	4.00 (55)	3.99 (70)	4.05 (50)
012	3.47 (48)	3.48 (60)	3.45 (70)	3.61 (8)	3.445 (50)	standard-obscured	3.49 (60)	3.58 (75)
122	2.952 (64)	2.961 (85)	2.931 (90)	3.05 (9)	2.933 (100)	2.953 (45)	2.965 (100)	3.04 (75)
051	2.563 (100)	2.576 (100)[f]	2.560 (100)	2.63 (10)	2.559 (85)	2.581 (70)	2.577 (90)	2.62 (100)
152	2.032 (43)	2.040 (45)	2.029 (50)	2.09 (7)	2.025 (40)	2.039 (45)	2.043 (80)	2.049 (15)

[*] Per JCPDS convention, only the *hkl* values of the Bravais indices are given; that is, *i* is not represented by either a value or an indication of omisson.
[a] From Walenta and Dunn (1979); indices in paper are given in a different form.
[b] 2.039 reflection is recorded as *hkl* = 611.
[c] Chemical analysis indicates $Uvite_{60}Dravite_{30}Schorl_{10}$ (Bouška, Povondra, and Lisý, 1973).
[d] Data from Rumantseva (1983); the 152 reflection is relatively weak; that for, for example, 223 ($d = 2.079$) is of higher intensity (I/I_1-50).
[e] Numbers in parentheses are recorded I/I_1 figures.
[f] Reflection is recorded as 042.

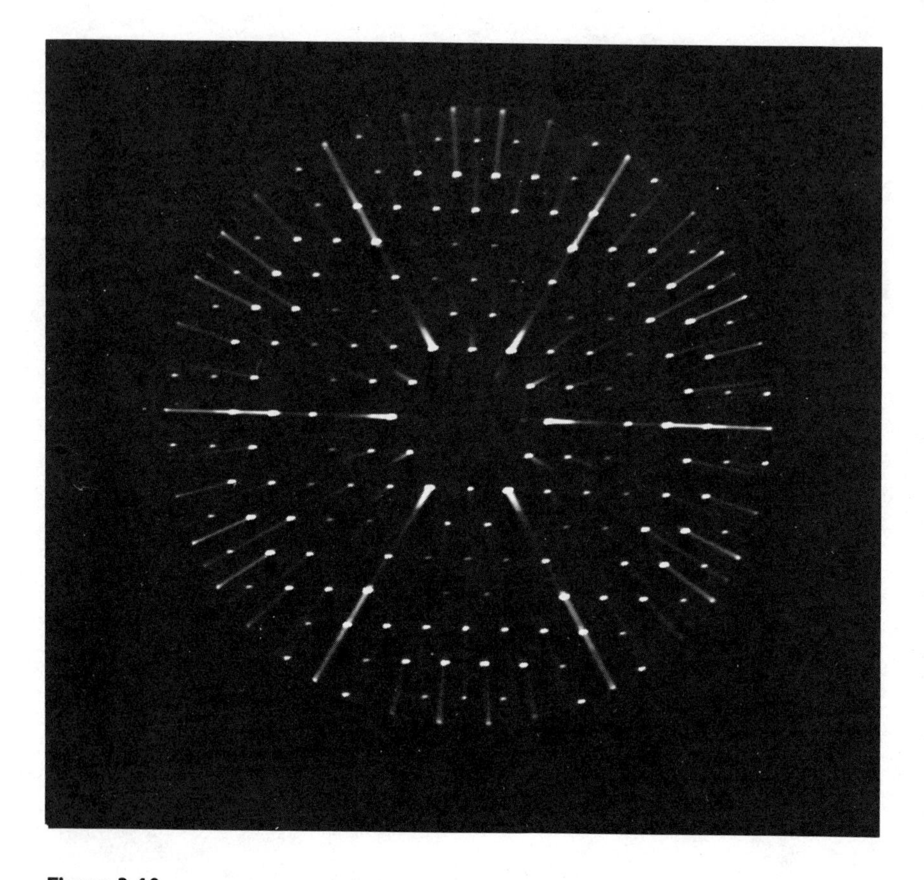

Figure 3-10
Precession photograph of schorl. The grain was oriented with the *c* axis perpendicular to the film; $\mu = 20$; Mo$K\alpha$ with Zr-filter; 0-level. (Photograph courtesy of D. R. Peacor.)

The use of X-ray data for tourmalines in genetic studies has been described by, for example, Makagon, Afonina, and Bogdanova (1977).

NOTES

1. Barton (1969) has defined a ditrigonality value as a measure of the distortion of the ring from hexagonal symmetry. Figures for some of the structurally analyzed specimens are: buergerite—0.038, dravite—0.032, uvite—0.029, elbaite—0.009, and schorl—0.005.
2. Throughout this presentation, apex (pl. apices) is used to indicate a point on a coordination polyhedron that is common to three or more sides with no connotation of relative position, for example, of top versus bottom.

Chapter 4

Chemistry and Alteration

"On the whole, the chemistry of it [tourmaline] is more like a medieval doctor's prescription, than the making of a respectable mineral."
(Ruskin, 1891)

FORMULAE, NAMES, AND RELATIONS

In the structural classification widely used for silicate minerals, members of the tourmaline group are cyclosilicates. As indicated in Table 4-1, the chemical complexity of tourmalines is reflected in the variety of different formulae that have been proposed. Today, however, the consensus is that the group, as a whole, may be expressed by the formula

$$X\ Y_3\ Z_6\ B_3\ Si_6\ O_{27}\ (O,OH,F)_4$$

X is predominantly Na or Ca, may also include K, or may be in a noteworthy part vacant;
Y is predominantly Fe^{2+} and/or Mg^{2+}, (Al $\pm$ Li), or Fe^{3+}, and commonly includes Mn;
Z is predominantly Al^{3+}, Fe^{3+}, or Cr^{3+}, and may also include Mg^{2+} and V^{3+} in noteworthy percentages.

Additional substitutions in the *X, Y,* and *Z* sites, other than those involving trace quantities of many elements, have been suggested by, for example, Novak and Zak (1970), Serdyuchenko (1978), and Povondra (1981): *X*—Mn^{2+}, Mg^{2+}, H_3O^+; *Y*—Ca, Cr^{3+}, Ti^{4+}; *Z*—Fe^{2+}, Mn^{3+}, Ti^{3+}, Ti^{4+}.

Proposed substitution for B in the BO_3 triangles is unproved.

Apparently Ti^{4+}, Al^{3+}, and Fe^{3+} can substitute for Si, but typically do so

Table 4-1
A Few of the Now Unacceptable Formulae That Were Once Proposed for Tourmaline

Formula	*Reference*
$\overset{\text{I}}{R_6}SiO_5 = \overset{\text{II}}{R_2}SiO_5 = \overset{\text{III}}{R_2}SiO_5$ ($\overset{\text{I}}{R}$-Na, Li, K; $\overset{\text{II}}{R}$-Mg, Fe, Ca; $\overset{\text{III}}{R}$-Al, B, Cr, Fe)	Rammelsberg, 1890
$R_9BO_2 . 2SiO_4$ (and there are three tourmaline molecules, giving a Li, an Fe, and a Mg species)	Riggs, 1888
$H_5Al_8B_3Si_6O_{31}$ and $H_{14}Al_5B_3Si_6O_{31}$ (and intermediates such as $H_{11}Al_6B_3Si_6O_{31}$)	Rheineck, 1899
$H_9Al_3(B,OH)_2Si_4O_{19}$ with H replaced by alkalies, Mg, Fe, Ca, etc.	Penfield and Foote, 1899
e.g., $NaHAl_6B_2Si_4O_{21}$	Vernadsky, 1913
e.g., $H_8Na_2Mg_6Al_{12}Si_{12}B_6O_{62}$ (dravite)	Kunitz, 1929
$XY_9B_3Si_6H_4O_{31}$	Machatschki, 1929
$NaR_9B_3Si_6(O,OH)_{30}$	Belov and Belova, 1950

Note: Clarke (1899) suggested a formula, not listed, that would, in the language of Mellor (1925), make tourmalines "triboratotetraluminotetraorthosilicates."

in only small percentages of the tetrahedral sites. Foit and Rosenberg (1979), however, found 17 percent Al^{3+} occupancy in the tetrahedral sites of their vanadium-bearing tourmaline.

As previously mentioned, small percentages of Cl occur in anion positions (e.g., Fuge and Power, 1969, and Kuroda and Sandell, 1953). Also, the number of (OH,F) ions, at least for schorl-dravites, generally equals or is less than four, and larger values in analyses probably reflect inclusion of absorbed H_2O (Povondra, 1981).

The "ideal" formulae of the individual species thus far reported as occuring in nature and accepted as valid species are given in Table 4-2.

"Tsilaisite," ideally considered to be $NaMn_3Al_6B_3Si_6O_{27}(OH,F)_4$, has not been approved by the IMA Commission because no tourmaline with Mn as the major component in *Y* has been found in nature. "Tsilaisites" thus far reported are manganian schorl-elbaites (cf. Makagon, 1972). However, a red manganian dravite (MnO = 4.34 weight percent and MgO = 12.95 weight percent) has recently been reported from Fowler, New York, by Ayuso and Brown (1984).

To date, most specimens termed chrome-tourmaline have proved to be chromian dravites rather than the species chromdravite. Examples are: the chromian dravites from the Urals (Cossa and Arzruni, 1883); Etchinson, Maryland (Gill, 1889); the Krivoy Rog area, U.S.S.R. (Shenderova, 1955); the Konjhar district, Orissa, India (Mukherjee, 1968); Outukumpu and Kaavi, Finland (Peltola et al., 1968); the Northwest Frontier Province of Pakistan (Mian, 1970); and the Line Pit (Lows mine), Pennsylvania/Maryland (Heyl et al., 1977; Foord, Heyl, and Conklin, 1981). Other specimens, such as the vanadian dravite from Silver Knob, Mariposa County, California (Foit and Rosenberg, 1979) and the zincian elbaite reported by Jedwab (1962), should also be so-named, unless the named element is proved to dominate one of the coordination polyhedral sites.

Despite the fact that Rosenberg and Foit (1979) have synthesized tourmalines with compositions somewhat similar to that recorded for the magnodravite of Wang and Hsu (1966), that specimen requires, in the writer's opinion, further study before magnodravite is accepted as a valid species. (Consider that if Wang and Hsu's analysis were recast into a formula with 31 anions: (1) the specimen would be a magno-uvite, not a magnodravite; (2) the B-content is much lower than it should be for a tourmaline; (3) the suggested amount of Mg substitution for Na and Ca in *X*-sites seems unlikely from the standpoint of general coordination requirements—that is, its radius is too small; and (4) even if item (3) is overlooked, after filling the *X*-, as well as the *Y*- and *Z*-, sites with Mg, there still is an excess of 0.89 formula units.)

A few strictly chemical relations recorded in the literature are: (1) substitution of cations with 3+ valence is more common in elbaites than in schorls and dravites (Foit and Rosenberg, 1979); (2) Li substitution is much less common and more restricted in schorls and dravites than Fe and Mg substitution is in elbaites (e.g., Foit and Rosenberg, 1979); (3) Li:F ratios are greater for pink than for blue elbaites (Němec, 1969); and (4) in given pegmatite zones, F in tourmalines is less than that in associated light colored micas. Many additional relations can be seen by studying available analyses.

Table 4-2
"Ideal" Formulae of Individual Species

	X	*Y*	*Z*	
Buergerite	Na	Fe_3^{3+}	Al_6	$B_3Si_6O_{27}(O,OH)_3(OH,F)$
Chromdravite	Na	Mg_3	Cr_5Fe^{3+}	$B_3Si_6O_{27}(O,OH)_3(OH,F)$
Dravite	Na	Mg_3	Al_6	$B_3Si_6O_{27}(O,OH)_3(OH,F)$
Elbaite	Na	$(Al,Li)_3$	Al_6	$B_3Si_6O_{27}(O,OH)_3(OH,F)$
Ferridravite	Na	Mg_3	Fe_6^{3+}	$B_3Si_6O_{27}(O,OH)_3(OH,F)$
Liddicoatite	Ca	$(Li,Al)_3$	Al_6	$B_3Si_6O_{27}(O,OH)_3(OH,F)$
Schorl	Na	Fe_3^{2+}	Al_6	$B_3Si_6O_{27}(O,OH)_3(OH,F)$
Uvite	Ca	Mg_3	Al_5Mg	$B_3Si_6O_{27}(O,OH)_3(OH,F)$

Correlations between chemical composition and properties such as specific gravity, indices of refraction, and birefringence are considered in the sections dealing primarily with those properties.

ANALYSES

Major Elements

The first analysis of tourmaline appears to have been made by Bergmann (1766); he recorded analyses for three specimens as follows:

"Element"	*Tyrol*	*Ceylon*	*Brazil*
Argil	42	39	50
Silex	40	37	34
Calcareous-earth	12	15	11
Iron	6	9	5

Other early analyses are those of Wondraschak and of Vauquelin and Klaproth (Ward, 1931). (For a Brazilian green tourmaline, Vauquelin reported: Silica, 40; Alum, 39; Lime, 3.84; Iron, 12.50; Manganese, 2.00; and Water, 2.66.) The presence of boron in tourmaline was first recorded by Arfvedson (1818) for an elbaite that he misidentified as "crystallized lepidolite." (Discovery of the element lithium was reported from the same specimen.) Fortunately, Berzelius' appendix to Arfvedson's report correctly identified the specimen. According to Hintze (1897), later in the same year, Lampadius and Vogel reported boron in a specimen correctly identified as tourmaline. Nearly a decade passed, however, before Gmelin (1827) published what are now considered to be the first "good" analyses of tourmaline. Probably the most outstanding analyses, considering the time and the procedures then used, are those of R. B. Riggs (1888).

Much of the history of chemical analysis of tourmaline has been reported by Sjögren (1916) and Ward (1931), and a rather different history has been given by Slivko (1962*a*).

Currently, most analyses of tourmaline are made by electron microprobe. These analyses, however, total only 85-90 weight percent; this is true because boron and (OH) cannot be detected (nor can lithium, a major component of elbaites and liddicoatites). Furthermore, valence states cannot be distinguished for, for example, iron or manganese, which are relatively common major constituents. Consequently, determination of B, (OH), Li, and, for example, Fe^{2+} versus Fe^{3+} require use of classical "wet chemical" methods or, perhaps, inductively coupled plasma emission spectroscopy and Mössbauer spectroscopy (see discussion in chap. 3, "Crystal Structure").

In conjunction with the preparation of this report, 788 of approximately

1000 analyses seen were entered into a computer file for study. Partial analyses, obviously poor analyses, and analyses of relatively rare kinds of tourmalines (e.g., a so-called Zn-rich dravite) were not included. Original sources were checked for all except a few of the analyses from relatively obscure reports and repeated by Mateos (1944), Barsanov and Yakovleva (1964, 1965, and 1966), Vladykin et al. (1975) and/or Kuz'min, Dobrovol'skaya, and Solntseva (1979). Publications from which analyses were taken are preceded by asterisks (*) in the references cited.

Each of the 788 analyses was converted into number of ions (subscripts in the formula) on the basis of 31 anions (O, OH, and F). Some of the analyses were deleted so far as inclusion on certain data plots on the basis of considerations of certain aspects of structural analyses (e.g., the fact that B is essentially restricted to BO_3 triangle sites and that those sites contain only B) or of known analytical "pitfalls" (e.g., the fact that analysis for B is "tricky"—it appears that some B is often reported with Si, but not vice versa—and the fact that the presence of F not previously removed can cause loss of B). Results are plotted on Figures 4-2 and 4-3 in the section dealing with solid solution.

Hypothetical analyses for ideal end-members of each recognized tourmaline species plus real analyses for representative natural specimens are given on Table 4-3. For those interested in determining which species they have from chemical analyses without converting the analyses to formulae, theoretical mid-point "ideal" analyses are given for elbaite-schorl, schorl-dravite, elbaite-liddicoatite, etc., in Appendix B.

As can be seen in Table 4-3, analyses of natural tourmalines do not rigidly fit even the general formula. A few attempts to resolve differences are noteworthy.

Barton (1969) corresponded with Ingamells, the analyst, about the apparently excess B in the type buergerite, and consequently suggested that the excess over that required for BO_3 triangles may replace Si in some of the predominantly SiO_4 tetrahedra. In fact, he then gave a Si:B ratio of 5.73:0.27 for those tetrahedral sites.

As previously noted, investigations of Fe^{2+} versus Fe^{3+} contents by Mössbauer spectroscopy indicate that the values determined by conventional means may be in error (e.g., Burns, 1972). This apparent anomaly, however, has not been resolved.

Sampling errors may be involved when zoned tourmalines are analyzed. For example, Leckebusch (1978) has found significant differences even in major constituents in probe analyses across color zones (Fig. 4-1).

"Errors" probably attributable to inclusions are known. For example, Dunn (1977*b*) noted that the high Cr-content in the analysis of the *chromturmalin* of Cossa and Arzruni (1883) probably reflects the

Table 4-3
Chemical Analyses

	1	1a	2	2a	3	3a	4	4a	5	5a	6	6a	7	7a	8	8a
	(Dravite)		*(Uvite)*		*(Elbaite)*		*(Liddicoatite)*		*(Schorl)*		*(Buergerite)*		*(Ferridravite)*		*(Chromdravite)*	
SiO_2	37.60	36.52	37.05	35.96	38.49	37.57	38.20	37.70	34.22	34.88	34.00	33.86	31.85	31.58	32.39	30.75
TiO_2		0.17		0.62												0.13
B_2O_3	10.89	10.32	10.73	11.49	11.15	10.65	11.06	10.89	9.92	8.82	9.85	10.86	9.23	8.98	9.38	9.00
Al_2O_3	31.91	33.41	26.19	26.80	40.82	42.18	37.81	37.90	29.04	35.00	28.85	30.79		2.30		2.92
Cr_2O_3															34.15	31.60
Fe_2O_3											22.59	17.62	42.32	38.37	7.17	7.65
FeO		0.30		0.41		0.19		0.83	20.46	16.06		1.27		7.25		
MnO		0.57				0.24		0.27		0.11		0.13				0.19
MgO	12.61	11.25	16.57	15.20				0.11		1.01		0.13	10.68	5.57	10.86	9.05
CaO		0.42	5.76	5.50		1.20	5.94	4.21		0.28		0.69				0.16
Na_2O	3.23	2.34		0.13	3.31	2.05		0.88	2.94	2.44	2.92	2.46	2.74	2.18	2.78	2.66
K_2O		0.57								0.09		0.07		0.98		
Li_2O					2.39	1.92	3.17	2.48								
F				1.49		0.39		1.72		0.49	1.79	1.86				
H_2O	3.76	3.76	3.70	2.70	3.84	3.38	3.82	2.69	3.42	1.50		0.40	3.18	3.50	3.27	4.43
	100.00	99.63	100.00	100.30	100.00	99.77	100.00	100.06	100.00	100.68	100.00	100.69	100.00	100.71	100.00	100.00

1. Hypothetical analyses for ideal dravite.
1a. Analysis of dravite from type locality, Dobrowa, Carinthia (Austria); analysts—Kunitz and Wülfing (Kunitz, 1929).
2. Hypothetical analysis for ideal uvite (Dunn et al., 1977*a*).
2a. Analysis for neotype uvite from Sri Lanka; analysts—Nelen and Norberg (Dunn et al., 1977*a*).
3. Hypothetical analysis for an ideal elbaite.
3a. Analysis for elbaite from Elba; analyst—Schaller (Schaller, 1913).
4. Hypothetical analysis for an ideal liddicoatite.
4a. Analysis for type specimen liddicoatite from Antsirabe, Madagascar; analyst—Nelen (Dunn, Appleman, and Nelen, 1977); NMNH #135815.
5. Hypothetical analysis for an ideal schorl.
5a. Analysis for schorl from Montofana, Alzo, Italy (Gallitelli, 1937).
6. Hypothetical analysis for an ideal buergerite.
6a. Analysis for type specimen buergerite from Mexquitic, Mexico; analyst—Ingamells (Donnay et al., 1967).
7. Hypothetical analysis for ideal ferridravite (Walenta and Dunn, 1979).
7a. Analysis for type ferridravite from Villa Tuneri, Bolivia; analyst—Dunn [electron microprobe analysis; BO_3 is calculated to fit formula; H_2O is weight loss on ignition (Walenta and Dunn, 1979)].
8. Hypothetical analysis for ideal chromdravite.
8a. Analysis for type chromdravite (Rumantseva, 1983); total includes $V_2O_3 = 1.46$, original analysis was "corrected" to figures given on basis of the assumption that phengite inclusions contained all of the K determined ($K_2O = 6.5$ weight percent).

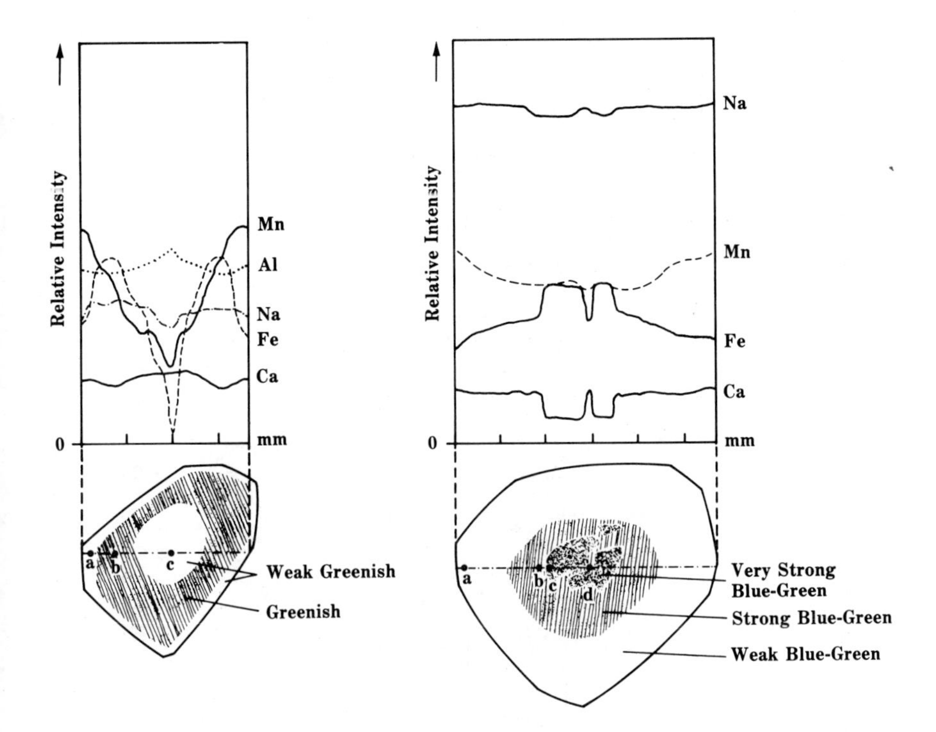

Figure 4-1
Probe analyses across zoned tourmalines indicating how differences in major constituents roughly correlate with differences in color (after Leckebusch, 1978).

presence of magnesiochromite, and Slivko (1965) directed attention to the fact that although some tourmalines contain many fluid inclusions, neither the major nor trace element contents of these inclusions is ever taken into account.

Trace Elements

The results of trace element analyses emphasize Bragg's (1937) statement that "tourmaline [is] one of nature's catch-all or garbage-can minerals" (Tables 4-4A-D). The data upon which these tabulations are based are from the following papers, listed chronologically by author(s): Wild (1931), Warner (1935), Lebedev (1937), Khlopin and Abidov (1941), Gallitelli (1941), Carobbi and Pieruccini (1947), López de Azcona (1947), Rose (1947), Němec (1951), Butler (1953), Ginzburg (1954), Kramer and Allen (1954), Shenderova (1955), Branche and Ropert (1956), Goroshnikov (1956), Ontoev (1956), Damon and

Table 4-4

A. Summary of Recorded Constituents. Major (Stippled) and Trace (Check Marks and Other Notations) Elements Recorded as Occurring in Tourmalines

IA	IIA	IIIB	IVB	VB	VIB	VIIB	VIII	VIII	VIII	IB	IIB	IIIA	IVA	VA	VIA	VIIA	VIIIA
1 H 1.008 (stippled)																	2 He 4.003 ✓
3 Li 6.939 (stippled)	4 Be 9.012 ✓											5 B 10.811 (stippled)	6 C 12.011	7 N 14.007	8 O 15.999 (stippled)	9 F 18.998 (stippled)	10 Ne 20.18 ✓
11 Na 22.99 (stippled)	12 Mg 24.31 (stippled)											13 Al 26.982 (stippled)	14 Si 28.086 (stippled)	15 P 30.974 ✓	16 S 32.07	17 Cl 35.453 ✓	18 Ar 39.948 ✓
19 K 39.102 (stippled)	20 Ca 40.08 (stippled)	21 Sc 44.956	22 Ti 47.90 (stippled)	23 V 50.942 (stippled)	24 Cr 51.996 (stippled)	25 Mn 54.938 (stippled)	26 Fe 55.847 (stippled)	27 Co 58.933 ✓	28 Ni 58.71 ✓	29 Cu 63.546 ✓	30 Zn 65.38 ✓	31 Ga 69.735 ✓	32 Ge 72.59 ✓	33 As 74.922 ✓	34 Se 78.96	35 Br 79.904	36 Kr 83.80
37 Rb 85.47 ✓	38 Sr 87.62 ✓	39 Y 88.906 ✓	40 Zr 91.22 ✓	41 Nb 92.906 ✓	42 Mo 95.94 ✓	43 Tc 98.906	44 Ru 101.07 (W)	45 Rh 102.905 (W)	46 Pd 106.4 (W)	47 Ag 107.868 ✓	48 Cd 112.41 ✓	49 In 114.82 ✓	50 Sn 118.69 ✓	51 Sb 121.75 ✓	52 Te 127.60 (W)	53 I 126.904	54 Xe 131.30
55 Cs 132.905 ✓	56 Ba 137.33 ✓	57 La 138.91 ✓	72 Hf 178.49 (W)	73 Ta 180.948 ✓	74 W 183.85 ✓	75 Re 186.2 (W)	76 Os 190.2	77 Ir 192.2 (W)	78 Pt 195.09 (W)	79 Au 196.967	80 Hg 200.59	81 Tl 204.37	82 Pb 207.19	83 Bi 209.980	84 Po (209)	85 At (210)	86 Rn (222)
87 Fr (223)	88 Ra 226	89 Ac (227)															

58 Ce 140.12 ✓	59 Pr 140.908 N	60 Nd 144.24 N	61 Pm (145)	62 Sm 150.35 N	63 Eu 151.96	64 Gd 157.25 N	65 Tb 158.925 N	66 Dy 162.50 N	67 Ho 164.930	68 Er 167.26 N	69 Tm 168.934	70 Yb 173.04 ✓	71 Lu 174.97
90 Th 232.038 (✓W)	91 Pa 231.036	92 U 238.03 (✓W)	93 Np 237.05	94 Pu (244)	95 Am (243)	96 Cm (247)	97 Bk (247)	98 Cf (251)	99 Es (254)	100 Fm (257)	101 Md (258)	102 No (259)	103 Lr (260)

Note: Elements checked (✓) have been recorded by more than two researchers; those marked with an N were recorded by Neiva (1974); those marked with an encircled W ((W)) were recorded by Warner (1935); those marked with an encircled check and w ((✓W)) were recorded by Neiva and Warner.

Table 4-4

B. Elbaite-liddicoatite. Major (Stippled) and Trace (Check Marks) Elements Recorded

	IIA	IIIB	IVB	VB	VIB	VIIB	GROUP VIII	GROUP VIII	GROUP VIII	IB	IIB	IIIA	IVA	VA	VIA	VIIA	VIIIA
1 H 1.008																	2 He 4.003
3 Li 6.939	4 Be ✓ 9.012											**5 B 10.811**	6 C 12.011	7 N 14.007	**8 O 15.999**	**9 F 18.998**	10 Ne 20.18
11 Na 22.99	**12 Mg 24.31**											**13 Al 26.982**	**14 Si 28.086**	15 P ✓ 30.974	16 S 32.07	17 Cl 35.453	18 Ar 39.948
19 K 39.102	**20 Ca 40.08**	21 Sc ✓ 44 956	22 Ti ✓ 47.90	23 V ✓ 50.942	24 Cr ✓ 51.996	**25 Mn 54.938**	**26 Fe 55.847**	27 Co ✓ 58.933	28 Ni ✓ 58.71	29 Cu ✓ 63 546	30 Zn ✓ 65.38	31 Ga ✓ 69.735	32 Ge 72.59	33 As W 74.922	34 Se 78.96	35 Br 79.904	36 Kr 83.80
37 Rb ✓ 85.47	38 Sr ✓ 87.62	39 Y W 88.906	40 Zr 91.22	41 Nb W 92.906	42 Mo ✓ 95.94	43 Tc 98.906	44 Ru W 101.07	45 Rh W 102.905	46 Pd W 106.4	47 Ag ✓ 107.868	48 Cd W 112.41	49 In W 114.82	50 Sn ✓ 118.69	51 Sb ✓ 121.75	52 Te W 127.50	53 I 126.904	54 Xe 131.30
55 Cs ✓ 132.905	56 Ba ✓ 137.33	57 La ✓ 138.91	72 Hf W 178.49	73 Ta ✓ 180.948	74 W 183.85	75 Re W 186.2	76 Os 190.2	77 Ir W 192.2	78 Pt W 195.09	79 Au W 196.967	80 Hg ✓ 200.59	81 Tl 204.37	82 Pb ✓ 207.19	83 Bi 209.980	84 Po ✓ (209)	85 At (210)	86 Rn (222)
87 Fr (223)	88 Ra 226	89 Ac (227)															

58 Ce ✓ 140.12	59 Pr 140.908	60 Nd 144.24	61 Pm (145)	62 Sm 150.35	63 Eu 151.96	64 Gd 157.25	65 Tb 158.925	66 Dy 162.50	67 Ho 164.930	68 Er 167.26	69 Tm 168.934	70 Yb 173.04	71 Lu 174.97
90 Th W 232.038	91 Pa 231.036	92 U W 238.03	93 Np 237.05	94 Pu (244)	95 Am (243)	96 Cm (247)	97 Bk (247)	98 Cf (251)	99 Es (254)	100 Fm (257)	101 Md (258)	102 No (259)	103 Lr (260)

Table 4-4

C. Dravite-uvite. Major (Stippled) and Trace (Check Marks) Elements Recorded

	IIA	IIIB	IVB	VB	VIB	VIIB	VIII	VIII	VIII	IB	IIB	IIIA	IVA	VA	VIA	VIIA	VIIIA
1 H 1.008																	2 He 4.003
3 Li 6.939	4 Be ✓ 9.012											5 B 10.811	6 C 12.011	7 N 14.007	8 O 15.999	9 F 18.998	10 Ne 20.18
11 Na 22.99	12 Mg 24.31											13 Al 26.982	14 Si 28.086	15 P ✓ 30.974	16 S ✓ 32.07	17 Cl 35.453	18 Ar 39.948
19 K 39.102	20 Ca 40.08	21 Sc 44.956	22 Ti ✓ 47.90	23 V 50.942	24 Cr 51.996	25 Mn ✓ 54.938	26 Fe 55.847	27 Co ✓ 58.933	28 Ni ✓ 58.71	29 Cu ✓ 63.546	30 Zn ✓ 65.38	31 Ga ✓ 69.735	32 Ge 72.59	33 As ✓ 74.922	34 Se 78.96	35 Br 79.904	36 Kr 83.80
37 Rb 85.47	38 Sr ✓ 87.62	39 Y ✓ 88.906	40 Zr ✓ 91.22	41 Nb 92.906	42 Mo 95.94	43 Tc 98.906	44 Ru 101.07	45 Rh 102.905	46 Pd 106.4	47 Ag ✓ 107.868	48 Cd 112.41	49 In 114.82	50 Sn ✓ 118.69	51 Sb 121.75	52 Te 127.60	53 I 126.904	54 Xe 131.30
55 Cs 132.905	56 Ba ✓ 137.33	57 La ✓ 138.91	72 Hf 178.49	73 Ta 180.948	74 W ✓ 183.85	75 Re 186.2	76 Os 190.2	77 Ir 192.2	78 Pt 195.09	79 Au ? 196.967	80 Hg 200.59	81 Tl 204.37	82 Pb ✓ 207.19	83 Bi ✓ 209.980	84 Po (209)	85 At (210)	86 Rn (222)
87 Fr (223)	88 Ra 226	89 Ac (227)															

58 Ce 140.12	59 Pr 140.908	60 Nd 144.24	61 Pm (145)	62 Sm 150.35	63 Eu 151.96	64 Gd 157.25	65 Tb 158.925	66 Dy 162.50	67 Ho 164.930	68 Er 167.26	69 Tm 168.934	70 Yb ✓ 173.04	71 Lu 174.97
90 Th 232.038	91 Pa 231.036	92 U 238.03	93 Np 237.05	94 Pu (244)	95 Am (243)	96 Cm (247)	97 Bk (247)	98 Cf (251)	99 Es (254)	100 Fm (257)	101 Md (258)	102 No (259)	103 Lr (260)

Table 4-4

D. Schorl. Major (Stippled) and Trace (Check Marks) Elements Recorded

	IIA	IIIB	IVB	VB	VIB	VIIB	GROUP VIII			IB	IIB	IIIA	IVA	VA	VIA	VIIA	VIIIA
1 **H** 1.008																	2 ✓ He 4.003
3 ✓ Li 6.939	4 ✓ Be 9.012											5 **B** 10.811	6 C 12.011	7 N 14.007	8 **O** 15.999	9 **F** 18.998	10 Ne 20.18
11 **Na** 22.99	12 **Mg** 24.31											13 **Al** 26.982	14 **Si** 28.086	15 P 30.974	16 S 32.07	17 ✓ Cl 35.453	18 Ar 39.948
19 **K** 39.102	20 **Ca** 40.08	21 ✓ Sc 44.956	22 **Ti** 47.90	23 ✓ V 50.942	24 ✓ Cr 51.996	25 **Mn** 54.938	26 **Fe** 55.847	27 ✓ Co 58.933	28 ✓ Ni 58.71	29 ✓ Cu 63.546	30 ✓ Zn 65.38	31 ✓ Ga 69.735	32 N Ge 72.59	33 ✓ As 74.922	34 Se 78.96	35 Br 79.904	36 Kr 83.80
37 Rb 85.47	38 ✓ Sr 87.62	39 ✓ Y 88.906	40 N Zr 91.22	41 ✓ Nb 92.906	42 ✓ Mo 95.94	43 Tc 98.906	44 Ru 101.07	45 Rh 102.905	46 Pd 106.4	47 Ag 107.868	48 Cd 112.41	49 ✓ In 114.82	50 Sn 118.69	51 Sb 121.75	52 Te 127.60	53 I 126.904	54 Xe 131.30
55 Cs 132.905	56 ✓ Ba 137.33	57 ✓ La 138.91	72 Hf 178.49	73 ✓ Ta 180.948	74 N W 183.85	75 Re 186.2	76 Os 190.2	77 Ir 192.2	78 Pt 195.09	79 N Au 196.967	80 Hg 200.59	81 ✓ Tl 204.37	82 Pb 207.19	83 N Bi 209.980	84 Po (209)	85 At (210)	86 Rn (222)
87 Fr (223)	88 Ra 226	89 Ac (227)															

58 ✓ Ce 140.12	59 N Pr 140.908	60 N Nd 144.24	61 Pm (145)	62 N Sm 150.35	63 Eu 151.96	64 N Gd 157.25	65 N Tb 158.925	66 N Dy 162.50	67 Ho 164.930	68 N Er 167.26	69 Tm 168.934	70 ✓ Yb 173.04	71 Lu 174.97
90 N Th 232.038	91 Pa 231.036	92 N U 238.03	93 Np 237.05	94 Pu (244)	95 Am (243)	96 Cm (247)	97 Bk (247)	98 Cf (251)	99 Es (254)	100 Fm (257)	101 Md (258)	102 No (259)	103 Lr (260)

Note: Elements checked (✓) have been recorded by more than two researchers; those marked with an N were recorded by Neiva (1974).

Kulp (1958), Nishikawa (1958), Slivko (1959*a*), Ivanov and Rozbianskaya (1961), Pehrman (1962), Stepnev (1962), Matias and Karmanova (1963), Giampaolo (1963), Mitchell (1964), Shcherbakov and Perezhogin (1964), Čech, Litomiský, and Novotný (1965), El-Hannawi and Hofmann (1966); Frondel, Biedl, and Ito (1966), Power (1966*a* and 1968), Bank and Berdesinski (1967), Donnay et al. (1967), Dubinina and Kornilovich (1968), Furbish (1968), Mukherjee (1968), Fuge and Power (1969), Shiryaeva and Shmakin (1969), Babu (1970), Getmanskaya et al. (1970), Barsanov and Plyusnina (1971), Chebotarev and Chebotareva (1971), Lum (1972), Němec (1973), Okulov (1973*a*), Rub (1973), Triche, Leckebush, and Recker (1973), Neiva (1974), Beesley (1975), Vladykin et al. (1975), Kulikov et al. (1976), Bridge, Daniels, and Pyrce (1977), Dunn, Appleman, and Nelen (1977), Zadnik (1982), and Conklin and Slack (1983).

Several generalizations have been suggested and/or are evident; for example:

1. Even if the trace elements reported only by Warner (1935) and/or Neiva (1974) are not taken into account, it is quite apparent that, as a group, elbaites contain more trace elements than other tourmalines. This, of course, is to be expected on the basis of the occurrences of essentially all elbaites in complex pegmatites. Especially noteworthy are the common presence of Rb and Sr, the apparently exclusive presence of Hg and Sb, and the absence of W and Tl. (Several other relations between species and trace element contents can be seen by examining the tables.)

2. A few previously suggested generalizations appear to be based on too few data—for example, the statement that Be and Pb occur only in (Li,Al)-tourmalines (Čech, Litomiský, and Novotný, 1965).

3. Some of the "trace" elements occur at levels well above the levels usually so-designated in some of the analyzed tourmalines. Examples recorded, along with weight percentages supporting this observation, include the following:

Cl (0.024-0.102)—range for 31 specimens from southwestern England (Fuge and Power, 1969) and (0.05-0.41) for 25 schorls from northern Portugal (Neiva, 1974);

Zn (0.09-0.86 weight percent ZnO)—elbaite in cleavelandite of Central Asia (Okulov, 1973*a*);

ZnO (>0.5)—blue caps on elbaite from Tourmaline Queen mine, San Diego County, California (recorded by K. L. Williams; E. E. Foord, 1984, personal communication);

ZnO (0.11), Ga_2O_3 (0.11), BaO (1.71), and PbO (0.98)—colorless dravite from the supergene zone of polymetallic deposits in eastern Transbaikal (Dubinina and Kornilovich, 1968);

Rb_2O (to 0.050) and Cs_2O (to 0.064)—pink elbaite in rare-element peg-

matites, U.S.S.R. and Mongolia (Kulikov et al., 1976 and Vladykin et al., 1975);
Sr (0.19) and Ba (0.01)—uvite from Pierrepont, New York (Donnay et al., 1967);
Sn ($\leq$ 0.15)—schorls in tin ores of eastern U.S.S.R. (Ruh, 1973);
Pb (1.1)—schorl-dravite from the Maly Khingan Range, U.S.S.R. (Lebedev, 1937).

4. Trace elements may or may not be responsible for or correlative with coloration of tourmalines (see the discussion in chap. 6); in any case, El-Hannawi and Hofmann (1966) report that trace amounts of Pb, Ba, Be, Cu, Cr, and Ni are more common in red than in yellow, green, or blue elbaites whereas Ge and Zn are more common in green and blue elbaites.

5. The trace-element composition of many tourmalines appears to reflect the composition of the enclosing rocks; for example, the high-chlorine tourmalines listed in paragraph 3 are from high-chlorine granites and the Sn-bearing specimens are schorls from cassiterite-bearing rocks.

6. The He (Khlopin and Abidov, 1941) and Ar (Damon and Kulp, 1958) and at least some of the Pb (López de Azcona, 1947) in some tourmalines may be radiogenic. However, because the quantities of He and Ar appear to be far in excess of that which can be attributed to radioactive decay, Damon and Kulp concluded that these elements occur because of the availability of structural sites suitable for large nonessential atoms, and they agreed with Khlopin and Abidov that the amounts "must represent a sample of the magmatic gases in the immediate environment of the forming crystal and as such can provide useful information on magmatic conditions" (Damon and Kulp, 1958, p.433).

7. Some radioactive gneisses of Sicily and Calabria (southwestern Italy) with 0.4-4.3 $\times$ 10^{-6} g uranium and 15.1-33 $\times$ 10^{-6} g thorium per gram of rock have their radioactive contents centered in tourmaline and/or biotite (Milone-Tamburino and Stella, 1952-1954).

8. Some trace elements may represent contents of fluid inclusions (Slivko, 1965).

9. The trace elements of some tourmalines occur as major constituents of other tourmalines. For example, Li is a trace element in some dravites but a major component in liddicoatites, elbaites, and some elbaite-schorls.

Isotope Studies

Tokudo and Ao (1956) reported a −3.5 percent deviation from the "standard" for the isotopic composition of oxygen in a tourmaline from Hyogo, Japan.

Allais and Curien (1959) gave $^{11}B/^{10}B$ ratios of 3.89 and 3.85 (using different methods), with the latter considered to be the better ratio, for a tourmaline from an undisclosed locality.

Shima (1963) recorded the following $^{11}B/^{10}B$ ratios for six tourmaline specimens from four localities in Japan: 3.997 ± 0.010, 4.024 ± 0.011, 4.099 ± 0.015, 4.036 ± 0.013, 4.038 ± 0.019, and 4.055 ± 0.012.

Matousova (1967), in a review of Li isotopes in several minerals from Czechoslovakia, found $^{7}Li{:}^{6}Li = 12.5 \pm 0.1$ for tourmaline (and for the other Li-bearing minerals she checked).

Esikov and Esikova (1974) list $^{11}B/^{10}B$ ratios for several deposits in Primore as follows:

Tourmaline-cassiterite veins	4.002
Tourmaline-fluorite rocks	4.102
Tourmaline pegmatites	4.058
Tourmaline-quartz greisens	4.119
Tourmaline-quartz veins	4.067
Tourmaline in quartz veins	4.058
Tourmaline alaskite	4.131
Tourmaline hornfels	4.103
Tormaline (schorl) pegmatite	4.000
Tourmaline in granitic pegmatite	4.009
Tourmaline (pink) pegmatite	4.037
Tourmaline-fluorite ore	4.044.

They concluded that the similarities indicate a general absence of fractionation among these occurrences.

Zadnik (1982), using cryogenic techniques, determined the following values for neon isotopes in tourmaline (dravite?!) from Yinnietharra, Western Australia.

$$^{20}Ne \quad -1 \times 10^{-10}\ cm^3\ STPg^{-1}$$
$$^{20}Ne/^{22}Ne - 2.12 \pm 0.23$$
$$^{20}Ne/^{22}Ne - 0.57 \pm 0.07$$

He found "the largest departures from atmospheric values [for any non-radioactive material] . . . in neon trapped in . . . pegmatite beryl and tourmaline" and concluded that "These anomalies are apparently produced by the Wetherill (1954) [*Phys. Rev.* 96, 679] reactions: $^{18}O(\alpha,n)\ ^{21}Ne$, $^{19}F(\alpha, n)\ ^{22}Na(\beta+)^{22}Ne$" (p. 302).

Taylor and Slack (1984), in their investigation of textural, chemical, and isotopic relations in tourmalines from Appalachian-Caledonian massive sulfide deposits, report that

Tourmalines from tourmalinites and massive sulfide deposits are typically characterized by heavier δ ^{18}O values (9.5—15.5 ‰) and less variable, heavier δ D values (generally −45 to −60 ‰) than tourmalines from unmetamorphosed igneous pegmatites. The latter typically have δ $^{18}O < 9.5$ per mil and δ D ≤ -60 ‰. Oxygen isotope fractionations between quartz and tourmaline ($\Delta = 1.27$-4.01) do not correlate directly with inferred grade of subsequent metamorphism, although in areas of high metamorphic grade the Δ values approach those of igneous quartz-tourmaline pairs. The δ ^{18}O values of quartz and tourmaline are believed to reflect variation in the oxygen isotope compositions of the host lithologies. The similar, narrow range in hydrogen isotope composition for tourmaline in all studied deposits is compatible with a modified seawater origin for the tourmaline- and sulfide-forming fluids (p.1703).

SOLID SOLUTION

Solid solution—the occupancy of equivalent structural sites by different ions—is, of course, to be expected in members of any mineral group with cation sites in which substitutions are possible. Its existence in the tourmaline group was recognized early in this century (e.g., Schaller, 1914) and was fairly well defined by Kunitz in 1929. In tourmaline, solid solution may involve partial to complete substitution in one or all of the *X*-, *Y*-, and *Z*-sites, and possibly also in the OH ($= O_1$) site. In some cases, it appears to be based largely on substitution in only one site whereas in others it appears to have required coupled substitutions. Some of the complexities are not yet well understood.

At present, however, it is widely agreed that elbaite and schorl, schorl and dravite, elbaite and liddicoatite, and dravite and uvite constitute solid-solution series (see Figs. 4-2 and 4-3). The elbaite-schorl and schorl-dravite series are dependent upon *Y*-site substitutions: (Al^{3+}, Li^{+}) for Fe^{2+} and Fe^{2+} for Mg^{2+}, respectively. The elbaite-liddicoatite series is based on Na^{+}:Ca^{2+} differences in the *X*-site, which are generally accompanied by charge-compensating differences between Al^{3+} and Li^{+} in the *Y*-sites (i.e., Al:Li tends to be greater in elbaite than in liddicoatite). The dravite-uvite series is also based on Na^{+}:Ca^{2+} differences in the *X*-site, in this series apparently charge-compensated by partial substitution of Mg^{2+} for Al^{3+} in *Z*-sites.

On the other hand, nearly all investigators (e.g., Vladykin et al., 1975) have concluded that there is essentially no solid solution between elbaite and dravite or, by extension, between liddicoatite and uvite. Nonetheless: many elbaites contain Mg; many dravites contain Al to the extent that it must occupy *Y*-sites; and a few dravites contain small percentages of Li. So, there must be a partial elbaite-dravite series and, by extension, probably a partial liddicoatite-uvite series (see Figs. 4-2 and 4-3).

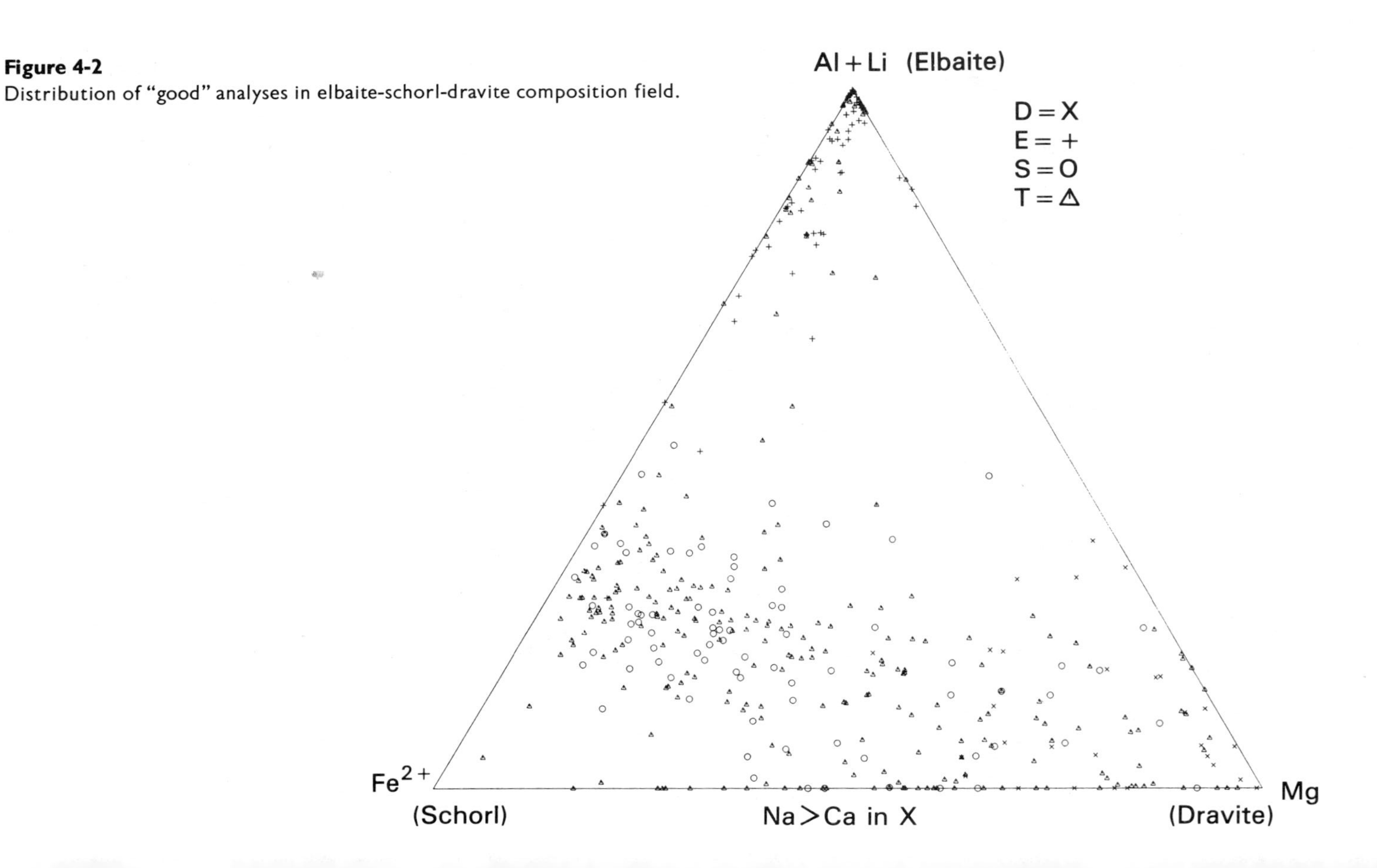

Figure 4-2
Distribution of "good" analyses in elbaite-schorl-dravite composition field.

Analyses of a few Fe^{2+}-rich specimens indicate the probable existence of the Ca-analog of schorl. There are also analyses that indicate the existence of three additional, at least partial, solid-solutions series: those between the Ca-analog of schorl and (1) schorl, (2) liddicoatite, and (3) uvite (see Fig. 4-3*B*).

Ferridravite and dravite appear to constitute another solid-solution series (Walenta and Dunn, 1979); the ferroan dravite reported by Frondel, Biedl, and Ito (1966) would probably be an intermediate member. This series, if it exists, depends on substitution of Fe^{3+} for Al^{3+} in *Z*-sites. Chromdravite may have similar relations with dravite and/or ferridravite.

Buergerite, thus far known only from one locality, has not been shown to be the end-member of any natural solid-solution series. Several analyses, however, indicate that a schorl-buergerite series may exist. Frondel, Biedl, and Ito (1966) have suggested that a tourmaline from the Urals, as described by Cossa and Arzruni (1883), may be a Cr-analog of buergerite.

Foit and Rosenberg (1975), by introducing the term "aluminobuergerite," did not imply a buergerite-"aluminobuergerite" series.

Hermon, Simkin, and Donnay (1973) state that "complete solid solution exists between any two endmembers, except for the pair dravite-elbaite, the two iron-free forms" (p. 125). Both their "complete . . ." and "between any two . . ." are probably overstatements. Nevertheless, considering all of the dependent and interdependent substitutions that might take place, there are indeed several possible end-members and series. One in particular appears to be "there for the finding" as a natural species: Li-poor tourmaline with Al predominant in both the *Y*- and *Z*-sites. It also seems likely that several other end-members ($\pm$ series) may occur in one or a few places.

The general lack of tourmalines that are intermediate between elbaite and dravite and the consequent conclusion that there is at most only limited elbaite-dravite solid-solution series has led to a number of suggestions. Two examples are:

1. Donnay (in Hermon, Simkin, and Donnay, 1973) suggested that the common O_3-O_6 edges of adjacent *Y* and *Z* octahedra can assume appropriate lengths for sharing between members of the elbaite-schorl and dravite-schorl series but not for what would be members of the elbaite-dravite series. The basis of this conclusion was that Fe, with its different valences and radii, can replace both (Al $\pm$ Li) and Mg thus permitting the required flexibilities required for the series involving schorl, whereas (Al, Li) and Mg cannot interexchange to provide the required O_3-O_6 edge-length adjustments for the elbaite-dravite and liddicoatite-uvite series.

2. Slivko (1959*a*) and Foit and Rosenberg (1979), however, saw no need to explain the lack of intermediate members between elbaite and dravite on the basis of the absence of Fe. They suggested, instead, that the apparent lack of solid solution reflects fractionation of Mg and Li during petrogenesis; that is,

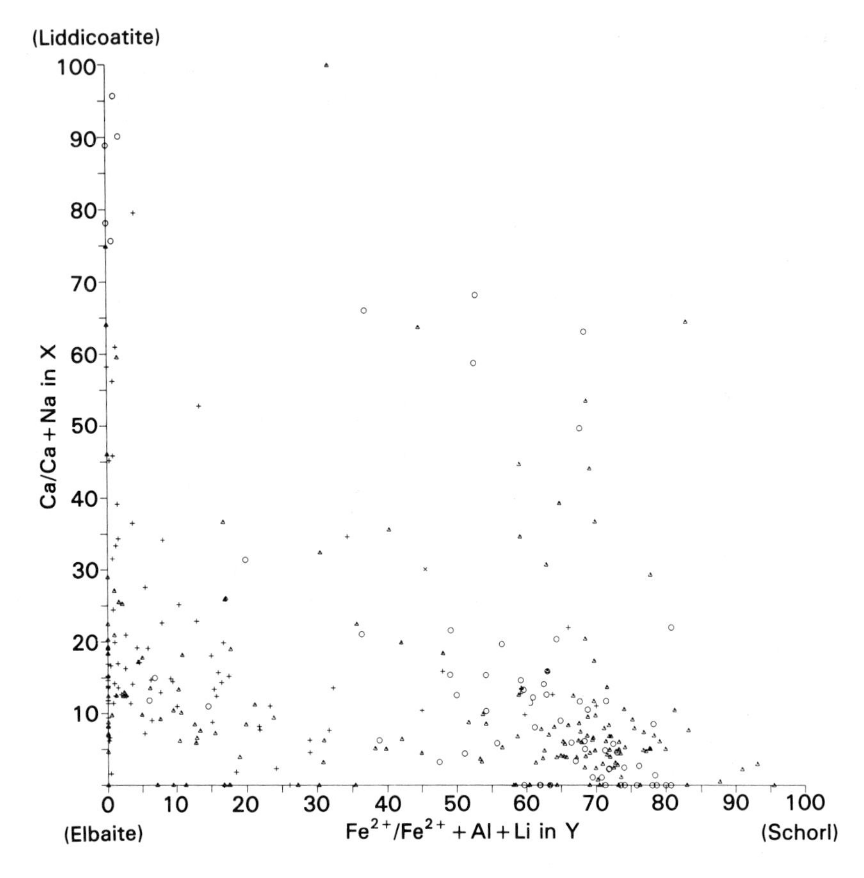

Figure 4-3 (Symbols same as for Fig. 4-2)
Distribution of analyses in liddicoatite-elbaite-schorl-Ca-schorl field

they concluded that the conditions needed for the presence of Mg versus Li are so different from the geochemical standpoint that the two elements are hardly ever simultaneously available for incorporation into tourmaline. Depending on one's point of view, the apparent paragenesis of the amphibole holmquistite ($Li_2(Mg, Fe)_3Al_2Si_8O_{22}(OH)_2$) can be used either to support or vitiate this conclusion. In any case, the possibility of mixing Mg and Li could, and should, be checked by experimental studies.

Alternative solid-solution relationships have been suggested. Five are:

1. Slivko (e.g., 1959*a,* 1962*a*) adds an elbaite-"tsilaisite" ($Na(MnAl_2)Al_6(BO_3)_3Si_6O_{18}(O,OH)_4$ in this system) series and notes the general lack of specimens indicating solid-solution between dravite and "tsilaisite," as well as between dravite and elbaite (Fig. 4-4).

2. Povarennykh (1972) suggests a scheme that differs primarily in terminology. His scheme is shown in Figure 4-5.

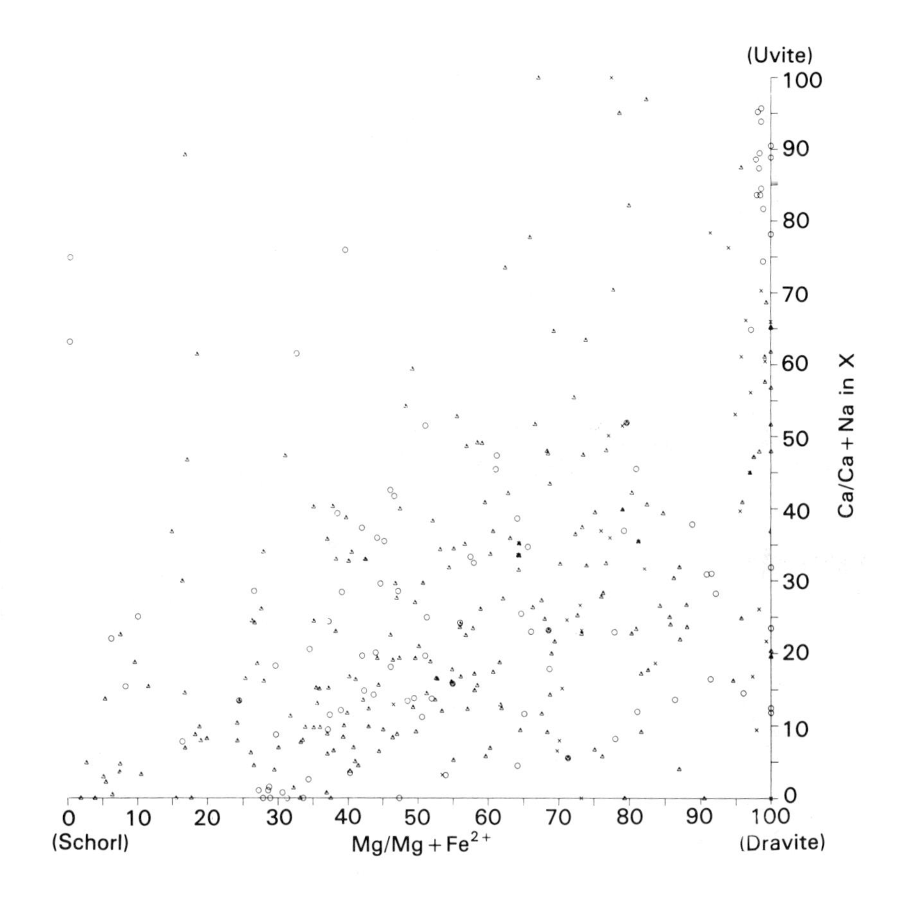

(left), and Ca-schorl-schorl-dravite-uvite field *(right)*.

3. Timokhina (1976) suggests solid-solution involving

$$Fe^{2+}, OH^{-} \leftrightarrow Fe^{3+}, O_2$$

and

$$Ca^{2+}, Al^{3+} \leftrightarrow Si^{4+}, Na^{+}.$$

Foit (1984, personal communication states that the first substitution, a dehydroxylation reaction, does take place but is of only limited extent.

4. Foit and Rosenberg (1977) suggest that members of the tourmaline group should be viewed as belonging to solid-solution systems like the one shown in Figure 4-6. As shown, the hypothetical end-members have compositions based on one or a combination of three kinds of coupled substitutions:

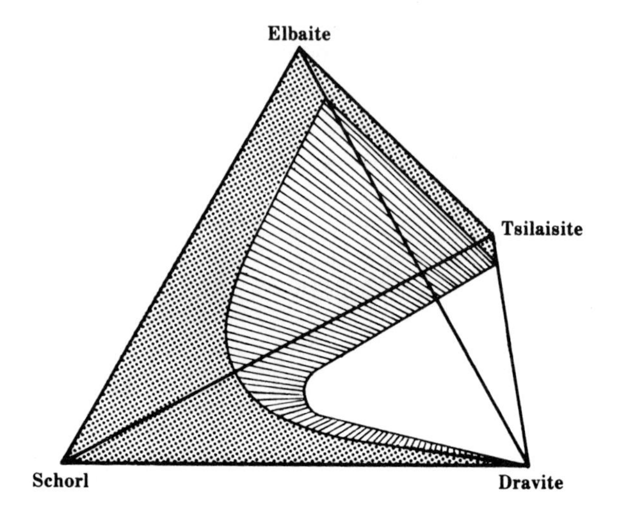

Figure 4-4
Solid-solution series setup suggested by Slivko (e.g., 1962*a*).

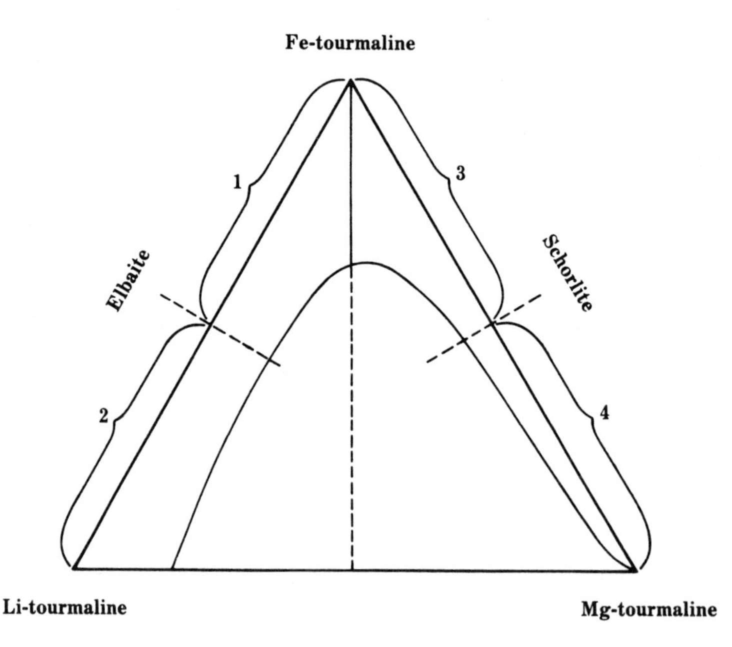

Figure 4-5
Solid-solution series and nomenclature scheme suggested by Povarennykh (1972).

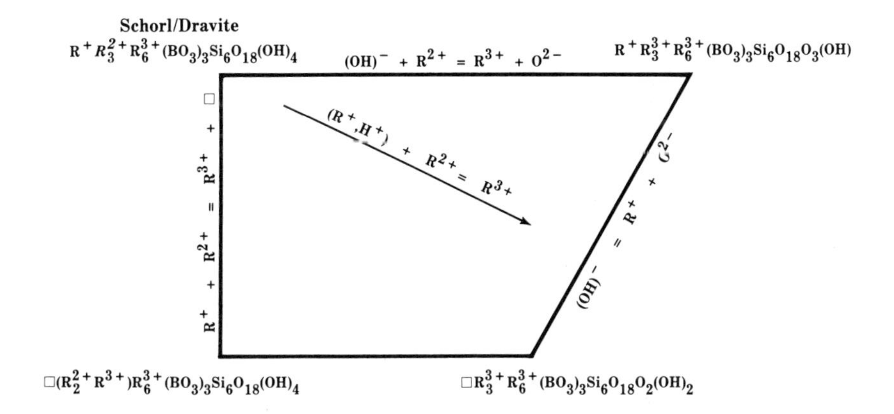

Figure 4-6
Solid-solution scheme suggested by Foit and Rosenberg (1977).

(1) dehydroxylation substitution—$(OH)^{-} + R^{2+} + O^{2-}$; (2) alkali-defect substitution—$R^{+} + R^{2+} = R^{3+} + \square$ and (3) octahedral-defect substitution—$Li^{+} + O^{2-} = (OH)^{-} + \square$. The first two kinds of substitution may be shown to obtain for tourmalines of the schorl-dravite series and all three types of substitution appear to take place in tourmalines of the elbaite-schorl series. In addition, a combination of substitutions (1) and (2) gives intermediate compositions between end-member schorl-dravite (= $R^{+}R_{3}^{2+}R_{6}^{3+}(BO_3)_3Si_6O_{18}$-$(OH)_4$) and a hypothetical end-member $R_{1-x}^{+}R_{3}^{2+}R_{6}^{3+}(BO_3)_3Si_6O_{18}O_{3-x}(OH)_{1-x}$ (cf. their aluminobuergerite—Foit and Rosenberg, 1975). In addition, they show that substitution from both schorl/dravite and elbaite toward this hypothetical end-member is rather extensive (Fig. 4-7). Their chief explanation for the apparent existence of this system is that it "results in improved local charge balance."

5. Povondra (1981) concludes that tourmalines of dravite-schorl compositions can be described best by relating them to the composition triangle with schorl, dravite, and Al-buergerite at its apices, and thus, to the formula

$$R_{1-x}^{+}(R_{2x}^{3+}R_{3-2x}^{2+})R_{6}^{3+}B_3Si_6O_{27}(OH)_{4-y}O_y.$$

Substitution is then explained by

R^{3+}, vacancy <=> R^{2+}, R^{+}, and H^{+} leading to loss of alkalies and formation of vacancies and of members poor in H^{+}. The mechanism of penetration of R^{2+} into the X positions can be explained by a similar deprotonation substitution. Combination of all the substitution reactions leads to the final members in further miscibility series, alkali-deficient R^{3+} tourmaline, proton-deficient R^{3+} tourmaline and alkali and

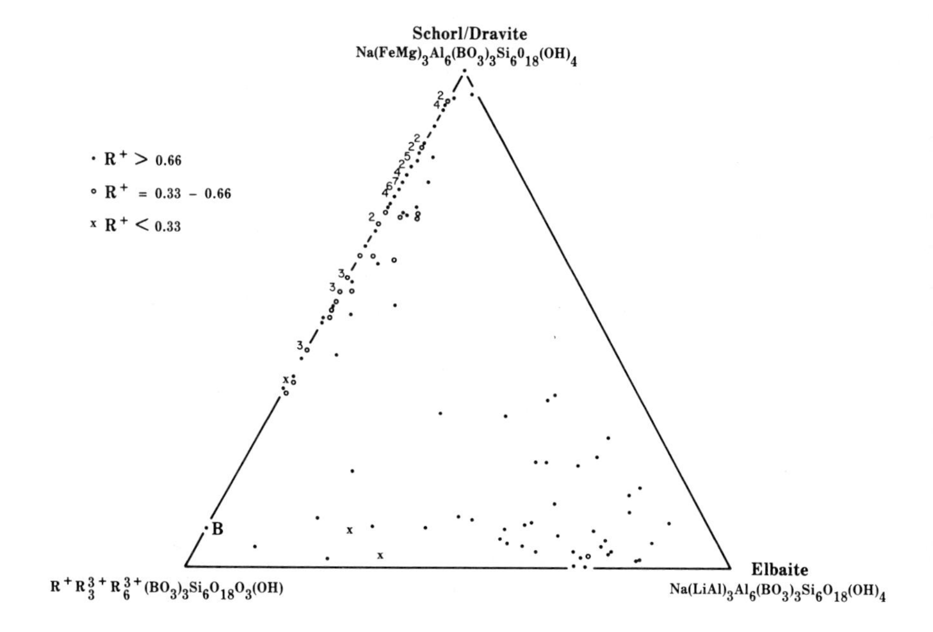

Figure 4-7
The rather extensive substitution from schorl/dravite and elbaite toward the hypothetical end-member R . . . plotted by Foit and Rosenberg (1977).

proton-deficient Mg^{2+} tourmaline. Combination of both substitutions leads to the final alkali- and proton-deficient R_2^{3+} tourmaline [and] . . . If the negligible miscibility of these schorls with elbaite and tsilaisite is ignored, the natural phases can be considered as solid solutions of 4 phases, schorl, dravite, alkali- and proton deficient R^{3+} tourmaline and alkali- and proton deficient Mg^{2+} tourmaline (p. 223).

The writer's considerations of chemical data indicate that the last two suggestions, in particular, warrant further consideration.

ALTERATION

Solubility

At room temperature and pressure, tourmalines are insoluble in individual acids except HF, in which they are slightly soluble. However, Brammall and Leech (1937) reported that dusts of "some varieties" of tourmaline are partially dissociated in cold H_2O and render the water alkaline. Furthermore, according to Winchers, Schlect, and Gordon (1945), when heated to 300°C under pressure (which is unspecified), tourmaline (not otherwise defined) dissolves in both HCl and HI.

It is of at least passing interest that as a result of laboratory experiments performed to test the solubilities of diverse B-bearing substances in organic compounds secreted from sugar beet roots, Titiz (1977) found that: the B of tourmaline (schorl, in his experiments) may be released—that is, removed—by chelate formation with diols and polyols, especially glucose, pyrocatechol, and fructose. The process appears to depend upon the diols and polyols being more effective complexing agents then the tourmaline structure. According to T. J. Delia (1983, personal communication), the release of the B is "very likely related to delocalization of electrons in the organic complex."

For both wet chemical analyses and etching, tourmalines are usually fused or melted with a flux. Fluxes used include: a mixture of potassium bisulfate and fluorite (e.g., Dana, 1892); potassium hydroxide (Kulaszewski, 1921); sodium hydroxide (Basset, 1956); alkali carbonates (e.g., Deer et al., 1962); a mixture of NH_4Cl and $CaCO_3$ (Kalenchuk, 1964); and a mixture of ammonium fluoride, potassium hydrofluoride, and orthophosphoric acid (Weiss, 1969).

Bezborodov, Zaporozhtseva, and Moiseeva (1940) stated that of all the minerals common in glass sands, tourmaline (not otherwise identified) and magnetite are most easily assimilated during glass production.

In checking the reactivity of several minerals with NaOH within a differential thermal analysis (DTA) apparatus, Kim (1972) found that tourmaline (again, not identified to species) plus NaOH mixtures undergo endothermic reactions at approximately 70°, 260°, and 300°C, and noted that of the minerals checked the tourmaline was the only one that gave endothermic rather than exothermic or no reactions near the melting point of NaOH (300°-350°C).

Etching

Differentiation between natural etching and "sculptured growth" is frequently difficult (Slivko, 1966*a*). Nonetheless, it seems safe to say that one or more faces of some tourmaline crystals exhibit what appear to be natural etch pits. Among features recorded as due to etching are the pits on the faces of the brownish green tourmaline crystals in the nepheline-bearing pegmatites of Seiland, Norway (Hoel and Schetelig, 1916), and the "vicinal hillocks" on the basal surface of a Brazilian tourmaline described and figured by Seager (1952). (See also Fig. 2-12 and the subsection "Surface Features" in chap. 2.)

The history of etch tests on tourmaline, the general procedures used in etching experiments, and conclusions based on his own and others' experiments are given by Honess (1927). Early experiments include those of Erofeyev (1871), Baumhauer (1876), Solly (1884), Ramsay (1886), Traube

(1896), Walker (1898), von Worobieff (1900), and Kulaszewski (1921). The solvents that appear to give the best results are HF, red-hot fusions of NaOH, and a 2:1 mixture of $KHSO_4$:CaF_2. The faces most susceptible to definitive etching are those of simple indices such as (0001), $(10\bar{1}1)$, $(11\bar{2}0)$, and $(10\bar{1}0)$.

Although a few of the early workers correlated the etch patterns with a lower than appropriate symmetry (e.g., Ramsay, 1886), most workers correctly interpreted the figures as indicating a ditrigonal polar symmetry (see Fig. 4-8 and 4-9). In addition, Wooster (1976) showed how the shapes and orientations of etch-forms are influenced by crystal structure and how they can be used to determine the pole-wise orientation of a hemimorphic structure, for example, the "absolute" orientation of tourmaline as first determined on the basis of X-ray studies by Barton (e.g., Barton and Donnay, 1968).

Also noteworthy so far as etching experiments, Kulaszewski (1921) found: triangular elevations on (0001) and hexagonal depressions on $(000\bar{1})$ of etched crystals (Fig. 4-9); a trigonal arrangement of light figures reflected from the surface of etched spheres with differences between those from the antilogous and analogous ends (Fig. 4-10); and etching to take place much more rapidly at the antilogous than at the analogous end (see Fig. 2-4). In addition, Sergeev (1963) described alternative procedures using HF and illustrated examples of such etching of some zoned tourmaline specimens.

The whole subject of solubility and etching brings up the question as to how one can remove, for example, enclosing lepidolite from tourmaline crystals without harming the crystals. Wilson (1977) outlines the following

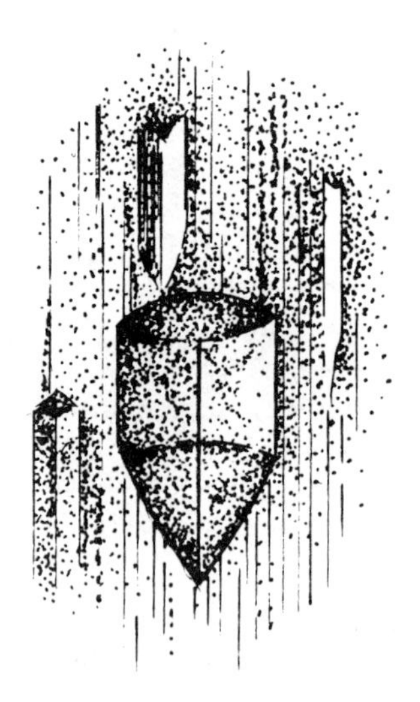

Figure 4-8
Etch pit on trigonal prism face produced by NaOH fusion for ½ hour (after Honess, 1927).

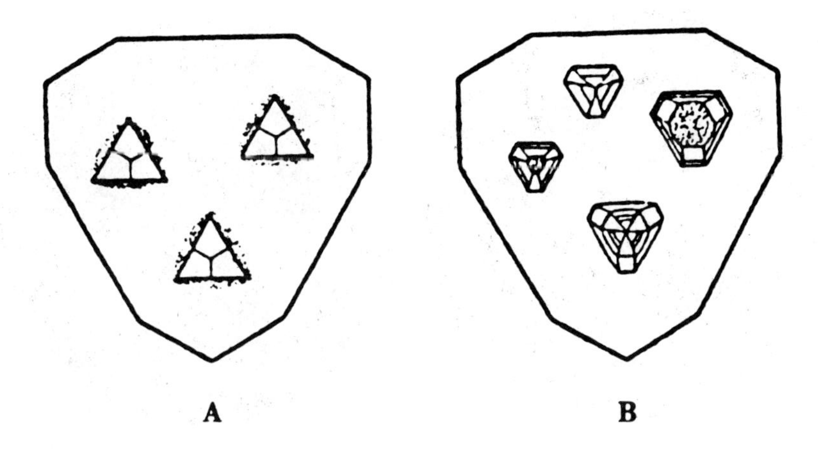

Figure 4-9
Etch figures on *A*, (0001) and *B*, (000$\bar{1}$) of red tourmaline from pegmatite at Wölkenburg (Saxony), produced by KOH fusion (from Kulaszewski, 1920).

procedure: cover the exposed tourmaline with wax; treat the specimen with hydrofluoric acid; and, after unwanted mineral material has been removed, dissolve the wax coating with carbon tetrachloride. He cautions

> However, here we have two of the most dangerous chemicals you might ever be inclined to use. Hydrofluoric acid may seem deceptively weak in most reactions but it is perhaps the most vicious acid when in contact with human skin; the fumes can be equally dangerous. Carbon tetrachloride is not much better. Although short contact with the skin can be tolerated, the fumes are exceedingly toxic; people who have consumed an alcoholic drink after working with carbon tetrachloride have been known to die in seconds; the two chemicals combine in the bloodstream to form an even more lethal and faster acting poison! . . . This means that all work with these chemicals is done under a fume hood, wearing protective eye-shields, and wearing appropriate laboratory gloves. [Also] Don't forget to *avoid* glass beakers when using hydrofluoric acid . . . they dissolve in it. Teflon, Nalgene, or other non-glass beakers work fine (p. 389).

Weathering and Other Natural Alteration

Tourmalines, especially schorls, are highly resistant to nearly all conditions of weathering. Consequently, "fresh" tourmaline occurs in many highly weathered rocks and soils the world over. In fact, tourmaline generally persists during warm, moist weathering conditions that produce, for example, laterites (Sirotin, 1966) and "kaolin weathering crusts" (Kazanskii, 1961). Furthermore, in the laboratory, in an investigation in which diverse B-bearing

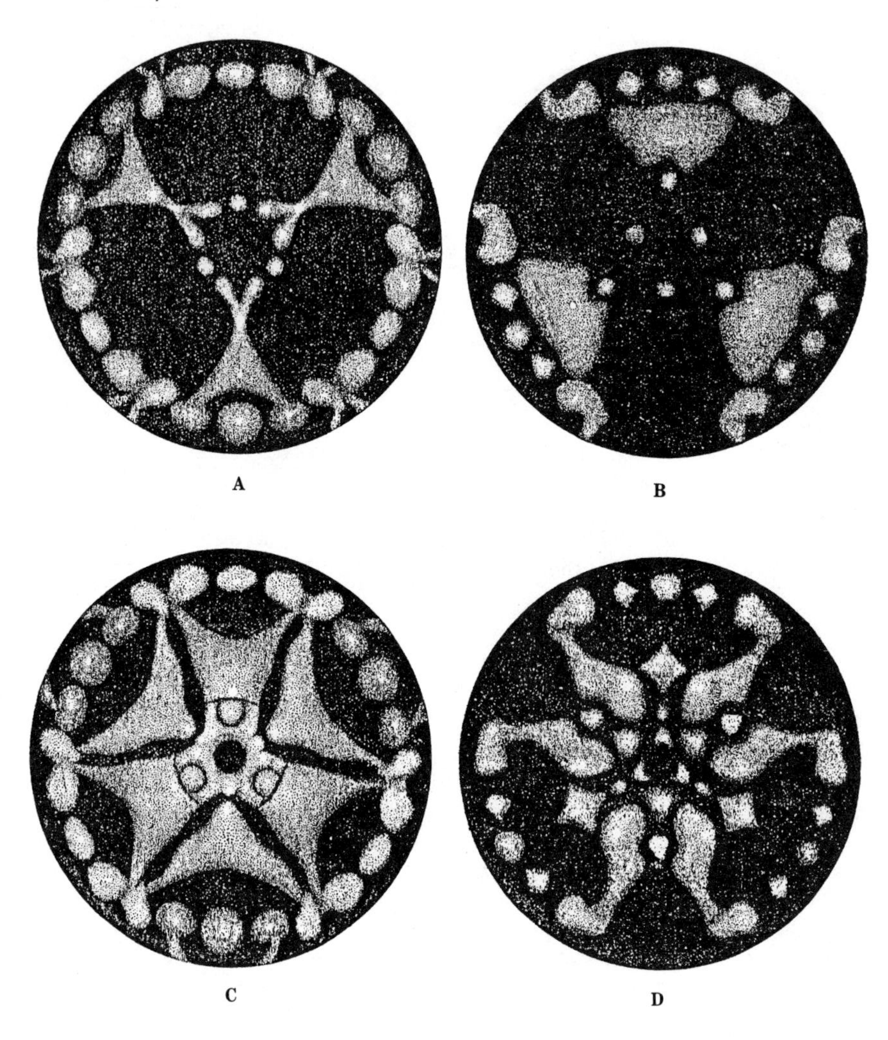

Figure 4-10
Light figures (i.e., patterns seen by observing light through thin, etched plate of crystal): *A* and *B*, antilogous side; *C* and *D*, analogous side. *A* and *C*, HF solution; *B* and *D*, KOH fusion. All plates of Minas Gerais, Brazil, tourmaline (from Kulaszewski, 1920.)

minerals and fertilizer components were subjected to artificially produced weathering conditions, Graham (1957) found that tourmaline is stable (i.e., it released no boron) under all of the conditions checked.

Nonetheless, altered tourmaline does occur and some of it may have been altered in response to surface or near-surface weathering processes. An apparent example is the tourmaline with reduced color saturation, refractive indices, and birefringence that occurs in the bauxitized and kaolinized schistose country rocks surrounding the Yakovlev iron ore deposits in the Ukraine (Makarov and Kondrat'eva, 1965). The tourmalines that have lost their lusters and become opaque, apparently as the result of partial alteration to montmorillonite, that occur sporadically in many pegmatite masses (e.g., in some of those of the Pala district, San Diego County, California) may be another example. It also appears that, at least in this district, pink elbaite altered more readily than spatially associated green elbaite. Most natural alteration of tourmaline—possibly including that just mentioned—appears, however, to have been caused by higher temperature processes.

As might be expected from etching experiments, alteration of tourmaline may be quite different on different crystal faces (see, for example, Bischoff, 1855). The differences on one versus the other end of crystals are also recorded in nineteenth century reports (e.g., Blum, 1843; Mallet, 1866; and Gumbel in Roth, 1879). The greater, and in some cases the only, alteration appears to have been on antilogous ends, although this cannot be ascertained definitely from some of the reports. Alteration also commonly differs from zone to zone in zoned tourmalines (e.g., Teschemacher, 1842; Bischoff, 1855; Leidy, 1871; Penfield, 1893; and Bastin, 1911). For example, some prismatic crystals have highly altered, even hollowed out, interior zones whereas their outsides have remained apparently unaltered. In fact, some such hollowed-out crystal interiors have served as the loci of later crystallization of, for example, lepidolite, uraninite (Teschemacher, 1842), or some secondary mineral such as cookeite (Penfield, 1893).

Three tourmaline-alteration product pairs are of rather widespread occurrence: (1) elbaite-lepidolite (e.g., Riggs, 1888; Fraser, 1930; and Quensel and Gabrielson, 1930); (2) elbaite-damourite (e.g., Riggs, 1888); (3) elbaite-cookeite (e.g., Penfield, 1893; and Quensel, 1937). Also, combinations occur, for example, elbaite-lepidolite *and* cookeite (Tarnovskii, 1960).

In addition, in chronological order by author(s): Roth (1879) presented a rather comprehensive review of the early literature (e.g., Bischoff, 1855; Rammelsberg, 1850; and Rose, 1836) about both the partial and complete alteration and the transformation of tourmalines. The products recorded are cookeite, cordierite, fluorite, limonite, "lithomarge" (kaolin ± halloysite), micas [commonly "pinite" but typically lepidolite for pink and green tourmalines (elbaites)], oligoclase, orthoclase, rhodizite, ripidolite (chlorite), and talc. A few of these recorded associations, however, may represent inclusion relationships or "intergrowths" with, rather than replacements of, tourmaline.

Döll (1886) recorded partial replacement of a pegmatitic tourmaline by pyrite and "arsenical pyrite."

Quensel and Gabrielson (1939) reported zoned tourmalines with altered pink cores composed of myrmekitic or symplectic intergrowths of quartz and albite but intact green shells, and Quensel (1937) recorded similar replacement involving cookeite and spodumene.

Hitchen (1935) described tourmaline crystals from Fitchburg, Massachusetts, that are replaced by quartz along a central core, parallel to *c*. Similarly described tourmaline-quartz pairs from southeastern Tuva, U.S.S.R., (Shaposhnikov, 1959) and northern Nigeria (McCurry, 1971*a*) have been interpreted as replacement pseudomorphs. McCurry described the relationship as follows: the quartz cores, some of which are discontinuous, consist of a mosaic of irregular, strained grains most of which have their *c* axes at about 30° to the *c* axis of the tourmaline; a few of the cores exhibit striated faces that appear to be negative tourmaline faces; the tourmaline, both green and brown, appears to have had its green zones selectively replaced by the quartz. So far as origin, it is of interest that she rejected simultaneous crystallization with the rather cryptic remark "the evidence presented here suggests that this is unlikely."

Hess (1943) reported tourmaline crystals from Goshen, Massachusetts, that are altered to "silky cymatolite," a mixture of albite and muscovite (cf. phenomenon described by Michel-Lévy and Kurylenko, 1952).

Jahns and Wright (1951) described the alteration of pink and colorless tourmalines of the pegmatites of San Diego County, California, as "a progressive decrease in hardness and toughness, loss of luster and transparency, and an increasingly clayey appearance" and elsewhere in the same report they noted "pale to deep pink clay minerals . . . were formed by the alteration of tourmaline."

Goni and Guillemin (1964), in a paper about the pegmatites of Alto Ligonha, Mozambique, reported rubellite replaced by feldspar (especially mirocline), violet colored muscovite, and leuchtenbergite (a variety of clinochlore), with the clinochlore constituting a network of microfissures within some of the tourmaline crystals. They also noted that some of the partially replaced tourmalines contain crystals of manganotantalite.

Dubinina and Kornilovich (1968) suggested that colorless tourmaline and plumbojarosite-beudantite formed as the result of the decomposition of dark-colored tourmaline in the supergene zones of deposits in eastern Transbaikal.

Fletcher (1977) reported that at depths within the Ilkwang breccia pipe near Pusan, Korea, the tourmalines of "sunburst" have "lost their resinous luster and have a 'charcoal-like' appearance." He did not elaborate upon the change.

In addition: Frondel (1935) suggested that "The well-known tourmaline shells—perimorphs—from Roe's Spar Bed, Crown Point, New York, enclosing a core of felspar, and from Newcomb, New York, enclosing a core of calcite are probably skeletal growths and are not the results of interior replacement, or eutectic crystallization" (p. 356).

Jaroš (1936), in a discussion of pseudomorphs of muscovite after tourmaline, noted "common alteration of tourmaline to micaceous materials may be related to the 'kaolin-type' nucleus of tourmaline, the central core of the structure being equivalent to part of a Mg- or Fe-kaolinite sheet" (p. 164).

Prescott and Nassau (1978) have recorded relationships involving elbaite from Corrego do Urncum, Minas Gerais, Brazil, which they interpreted to indicate solution (alteration) of original tourmaline crystals followed by the deposition of another generation of tourmaline overgrowth.

Another feature more or less pertains to this section—Michel-Lévy and Kurylenko (1952) reported what they termed a physical modification whereby fibers are developed perpendicular to the *c* axis of original grains when tourmaline is heated to temperatures above approximately 1000°C. They observed such features in laboratory heated specimens and in one natural specimen, the latter a grain in a gneiss xenolith from a trachyte mass at Mont Dore, France.

DISSOCIATION

In the laboratory, Robbins and Yoder (1962) found tourmaline to dissociate to cordierite (etc.) above 865°C and at pressures up to 2000 bars, and to kornerupine and sapphirine (etc.) above 895°C and 5000 bars (see Fig. 4-11). It is noteworthy that tourmaline-cordierite and tourmaline-kornerupine assemblages have indeed been reported (e.g., Ussing, 1889), and tourmaline (dravite)—along with sapphirine, plagioclase ($An_{\sim 10}$), kyanite, corundum, and garnet—has been reported in a "kornerupine rock" (Schreyer, Abraham, and Behr, 1975).

Fuh (1965) carried on some hydrothermal experiments that he interpreted to establish the dissociation

$$\text{Tourmaline} \rightarrow \text{spinel (hercynite)} + \text{vapor} \pm \text{cordierite}.$$

Němec (1981) recorded "indicolite (which he says is a schorl) reaction rims around amblygonite" in a pegmatite near Velké Meziříčí (Bohemia).

So far as tourmaline replacement of other minerals, tourmalinization other than that involving pseudomorphism *per se* is treated in Chapter 11, "Occurrences and Geneses."

PSEUDOMORPHISM

In his compendium about pseudomorphs, Landgrebe (1841) listed "Glimmer [mica] in Turmalin umgewandelt." In his classic book on the subject,

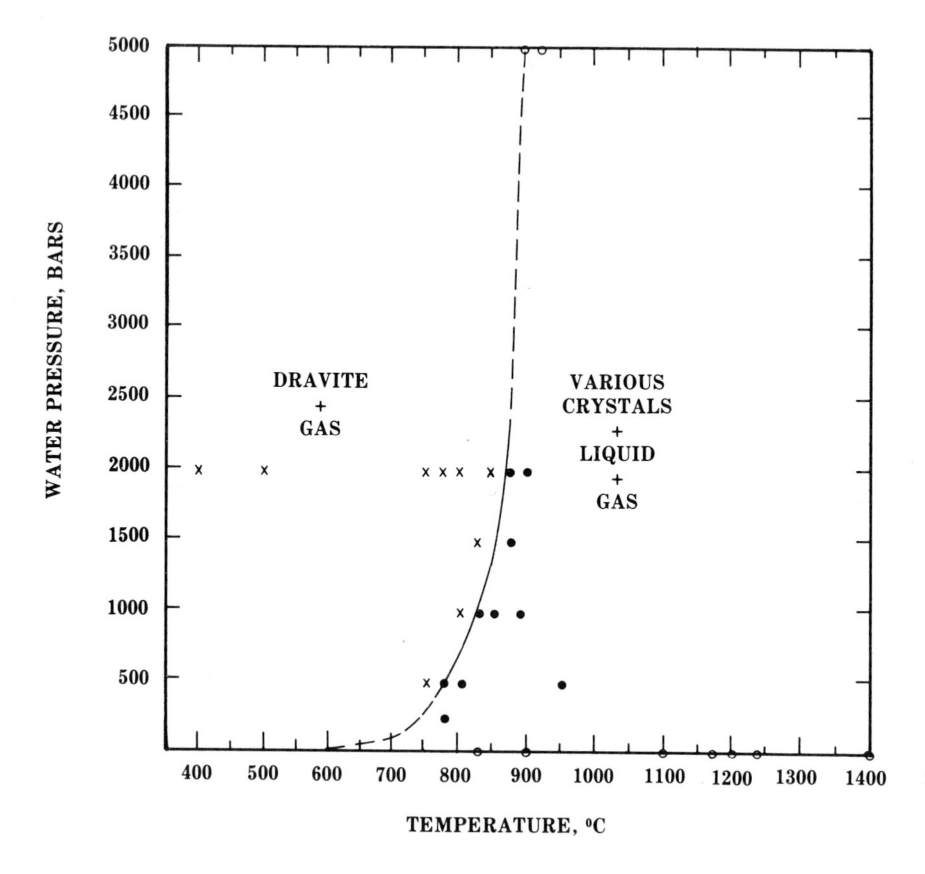

Figure 4-11
"Preliminary P-T diagram of the system dravite-water" (after Robbins and Yoder, 1962).

Blum (1843) listed "Glimmer nach Turmalin" from granite and from a chlorite schist from three German localities and "Speckstein [talc] nach Turmalin" in granite from Rozena.

Delesse (1859) reported limonite and mica after tourmaline and, in an opposite sense, tourmaline after orthoclase. He also noted that tourmaline zoned around *c* was once considered to constitute a pseudomorph; as an example, he mentioned a tourmaline from Chesterfield, Massachusetts, which had a red core within a green shell; he gave no reference to such records, however, and the writer has not found any.

The previously mentioned replacement ("pseudomorphos") of pyrite and arsenopyrite after schorl, as described by Döll (1886), is not pseudomorphism under any current definition of the term.

Cordierite pseudomorphs after tourmaline have been reported by Jaroš (1936) and also by Volborth (1955).

Some of the previously mentioned alteration of tourmaline to cookeite has resulted in the formation of cookeite pseudomorphs after elbaite, and some of these pseudomorphs contain relict tourmaline (Quensel, 1937).

Clay minerals, otherwise unidentified, form pseudomorphs after some of the tourmalines of San Diego County, California (Jahns and Wright, 1951).

Tourmaline replaced by tetrahedrite is recorded by Genkin (1954).

Elbaite that is partially to wholly replaced by lepidolite and/or muscovite plus or minus a clay mineral as pseudomorphs is relatively common. Such have been reported, for example, from pegmatites of the Samokov district of Bulgaria (Breskovska and Eskenzai, 1959-1960), from eastern Siberian pegmatites (Tarnovskii, 1960), from northeastern Brazil (Bhaskara-Rao and de Assis, 1968), and from San Diego County, California (Jahns, 1953). Many of the pseudomorphs from San Diego County consist of composite groups of mica grains that appear to have filled either a tourmaline shell or a complete mold.

Encasement "pseudomorphs" of muscovite after tourmaline (e.g., Harvard Museum #97669) have, with exceptions, basal planes of the muscovite parallel or nearly parallel to the faces of the underlying, typically partially replaced, tourmaline; that is, the mica plates appear as if they were plastered on the tourmaline crystals. Similar "pseudomorphs" are recorded from pegmatites of the Chandra Valley in the Himalayas (Mallet, 1866).

Whereas nearly all of the pseudomorphs after tourmaline that are reported in the literature and all of those seen by the writer involve the species elbaite, sericite and muscovite pseudomorphs after tourmalines of the schorl-dravite series have also been reported: Kosoĭ (1939) described, from Karelian pegmatites, large black crystals that have been replaced by sericite within which there are small tourmaline crystals of a later generation; Malireddi and Gordienko (1968) described and illustrated sheathlike muscovite that appears to have propagated inward from the surface of an original tourmaline crystal, and also muscovite that appears to have started along central canals and branched out within original tourmaline crystals to form incomplete replacement pseudomorphs after the tourmaline.

Tourmalinization has resulted in the formation of pseudomorphs of tourmaline after several diverse minerals. Examples are the schorl-quartz pseudomorphs after orthoclase phenocrysts in the Dartmoor granite recorded by Brammall and Harwood (1925); the radial aggregates of "pneumatolytic" tourmaline after relatively large crystals of perthite (Fig. 4-12) and also after "primary" tourmaline from a loose boulder found near Luxullian, in Cornwall, England, described and illustrated by Wells (1946); and the tourmaline pseudomorphs after K-feldspar phenocrysts in the sericitized quartz latite of the "porphyry tin" deposits of the San Pablo stock of Japo, Bolivia, described

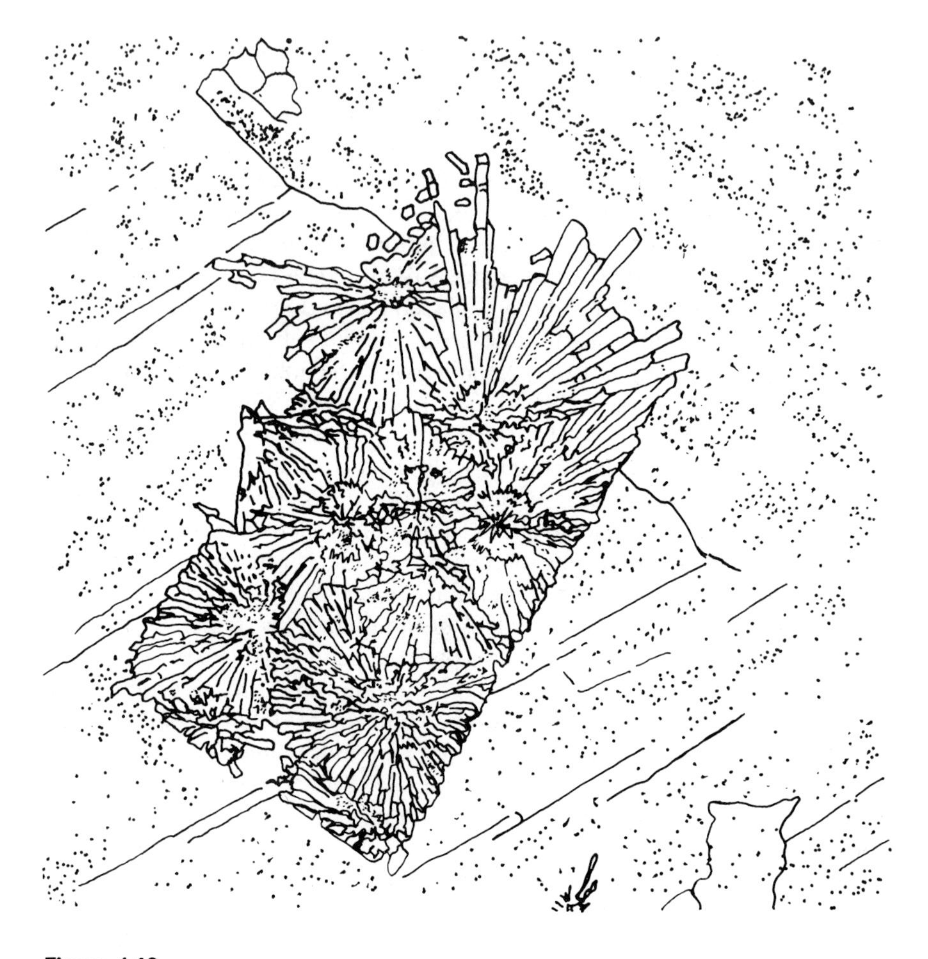

Figure 4-12
"Pneumatolytic" tourmaline pseudomorph after perthite (after a sketch of a thin-section by Wells, 1946).

by Sillitoe, Halls, and Grant (1975). There also are the rather poor but recognizable tourmaline pseudomorphs after brown corundum (no locality given) in the Harvard University collection (#90768). In addition, replacement tourmaline rather frequently has faithfully reproduced diverse mineralogic and petrographic features—for example, the "micro-structure of mica flakes [in rocks] with striking fidelity" (Brammall and Harwood, 1925).

Chapter 5

Inclusions and "Intergrowths"

Inclusions of tourmaline in other minerals, inclusions of other minerals in tourmaline, fluid and multiphase inclusions in tourmaline, and "intergrowths" involving tourmaline are described under this heading. The term "intergrowths" is enclosed in quotation marks to indicate its broad use to include spatially associated minerals that originated by, for example, replacement as well as those proved to have formed as a result of simultaneous growth. Summary tabulations are given in Tables 5-1, 5-2, and 5-4.

One of the more interesting relationships between tourmaline and its inclusions is shown in Figure 5-1. The most commonly reported relationship involving tourmaline as an inclusion is pictured in Figure 5-2. Additional data and interpretations relating to inclusions are given in Chapters 8, 10, and 11: "Synthesis," "Tourmaline as a Gemstone . . ." and "Occurrences and Geneses," respectively.

Some pseudomorphs after tourmaline include tourmaline relicts. These relations, mentioned in Chapter 4, are not dealt with in this chapter.

TOURMALINE INCLUSIONS IN OTHER MINERALS

The following text is keyed to entries in Table 5-1.

1. Dravite occurs as radiating tuffs and other "little groups"along twinning planes in albite and as "long needles" in "isolated crystals" of albite at Dinashead, Cornwall (Agrell, 1941).

2. Tourmaline as inclusions in beryl (var. emerald) have been recorded from the Habachtal mine, Austria (Gübelin, 1956), the Urals (Sinkankas, 1981), and Zambia (Koivula, 1982). A black, doubly terminated, lath-like tourmaline crystal included in an aquamarine crystal, with the *c* axes of the

Table 5-1
Tourmaline Inclusions in Other Minerals

Host Mineral	*Tourmaline Reported*	*Reference*
1. albite	dravite	Agrell, 1941
2. beryl (emerald)	tourmaline	Gübelin, 1956; Sinkankas, 1981; Koivula, 1982
(aquamarine)	tourmaline	Atkinson and Kothavala, 1983
3. biotite	tourmaline	Brammall and Harwood, 1925
4. calcite	tourmaline	Koch, 1957
5. fluorite	tourmaline	Messina, 1940
6. hydroxyl-herderite	elbaite	Cassedanne and Lowell, 1982
7. hydromuscovite	tourmaline	Koch, 1957
8. mica (e.g., muscovite)	tourmaline	Leidy, 1871
9. orthoclase	tourmaline	Brammell and Harwood, 1925
10. plagioclase	dravite	Bridge, Daniels and Pyrce, 1977
11. pyrite	tourmaline	Smith, 1952
12. quartz	tourmaline	e.g., Mawe, 1818; Kunz, 1902; MacCarthy, 1928; Agrell, 1941; Parker, 1962; Dietrich, de Quervain, and Nissen, 1966; Stadler, 1967; Gross, 1972; Gübelin, 1976; Fletcher, 1977
13. siderite	tourmaline	Koch, 1957
14. topaz	elbaite	Cassadanne and Lowell, 1982
15. "ores" (including pyrrhotite and sphalerite)	tourmaline	Koch, 1957

two minerals perpendicular to each other, was reported from the Zanskar Range of Kashmir (Atkinson and Kothavala, 1983).

3, 9. "Primary" tourmaline occurs as inclusions in biotite in the Dartmoor (England) granite (Brammall and Harwood, 1925).

4, 7, 12, 13, 15. Colorless fiberlike crystals and greenish needlelike crystals of tourmaline within calcite, hydromuscovite, quartz, siderite, and "ores" from Nagybörzsöny (Czechoslovakia?) are reported and illustrated by Koch (1957).

5, 12. Feltlike masses of blue tourmaline crystals occur as inclusions in fluorite and quartz of the Baveno granite (Messina, 1940).

6, 14. Hydroxyl-herderite crystals pierced by as well as perched atop

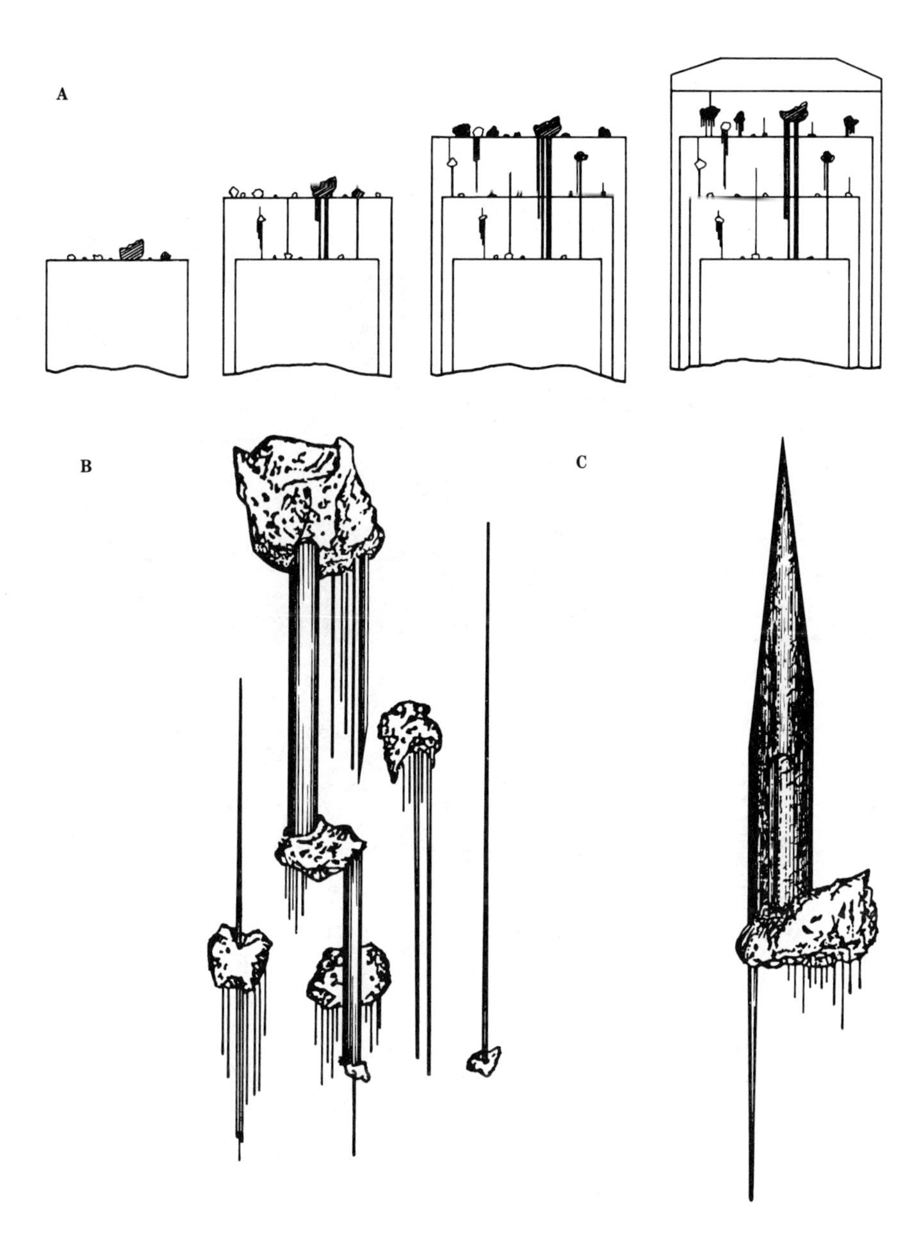

Figure 5-1
Schematic drawing of solid inclusions in tourmaline crystals showing interactions between solids (e.g., quartz and feldspar fragments) and their now enclosing tourmaline host crystals. Vertical lines and conical cross-section in *C* represent cavities that are typically hollow beneath and gas-filled above their associated solids. Displacements range up to 3 mm. *A* and *B* from the Urals; *C* from Siberia. (After Slivko, 1966*a*.)

Figure 5-2
Polished plate of tourmalinated quartz from Brazil; from the collection of John Sinkankas.

elbaite crystals are present in the Virgem da Lapa (Brazil) pegmatites (Cassedanne and Lowell, 1982). Also, lepidolite coatings occur on many of the elbaite crystals and topaz encloses one or both of these minerals in the same masses.

8. Muscovite from Mt. Mica, Maine, contains radiating groups of "compressed crystals" and "friable" green tourmaline (Leidy, 1871). (See also the succeeding section "Intergrowths".)

8, 10. Some of the well-known dravite crystals of Yinnietharra, Western Australia, occur as inclusions in both plagioclase and phlogopite (Bridge, Daniels, and Pyrce, 1977).

9. "Primary" tourmaline occurs as inclusions in orthoclase phenocrysts of the porphyritic Dartmoor (England) granite (Brammall and Harwood, 1925).

11. Blade-shaped tourmaline crystals approximately ⅛-¾ inch (~3-20 mm) long occur in pyrite at the Flalökken mine on Vassfjellet, southern Tröndelag, Norway (H. H. Smith, 1952).

12. "Tourmalinated quartz" occurs widely (see, for example, references cited on the table). That from Jefferson County, Montana, is particularly interesting in that some of the colorless and smoky quartz crystals have amethystine terminations and the tourmaline inclusions occur only in the colorless and smoky portions of the crystals (Kunz, 1902). Also, the inclusions, which range from "delicate needles" to nearly ¼ inch (5.5 mm) across, show no preferred orientation although some are arranged to produce "phantoms."

Some tourmalinated quartz was originally described as "rutilated quartz" (e.g., MacCarthy, 1928). Specimen #108826 in the Harvard collection consists of clusters of needlelike tourmaline crystals closely resembling those within quartz crystals from the same Brazilian locality.

In some cases, tiny tourmaline needles, just as some rutile needles, have been found to render their quartz host blue (e.g., Parker, 1962, Dietrich, deQuervain, and Nissen, 1966; and Stadler, 1967).

As a result of his statistical study of 109 inclusion-bearing quartz crystals from Metathorn (Switzerland), Gross (1973) concluded that tourmaline inclusions show a preference for quartz crystals that exhibit some right-left twin law; that is, tourmaline inclusions occur in such crystals rather than in untwinned or Dauphine-twinned quartz crystals.

The tourmalinated quartz from Governador Valadares (Minas Gerais), Brazil, contains helvite tetrahedra atop some of the schorl(?) filaments within the quartz (Dunn, 1975*b*).

Another feature that is noteworthy here, in that it involves tourmaline partially included within quartz, is shown by some specimens from Brazil that are in the Smithsonian Institution (e.g., NMNH #R-17238). Individual tourmaline crystals or parallel groups of tourmaline crystals have individual quartz crystals growing off their terminated ends with the *c* axes of the tourmaline substrate and the overgrown quartz at approximately 18 degrees to each other.

MINERAL INCLUSIONS IN TOURMALINE

A few rather interesting or noteworthy relations follow (they are keyed to entries on Table 5-2).

3, 5, 6, 7, 14, 21, 24, 25, 28. The inclusions recorded by Krynine (1946) occur within authigenic overgrowths. He also recorded quartz, feldspar(?), muscovite, and fluid inclusions in such overgrowths.

4. Apatite inclusions occur within tourmalines, and vice versa, in a pegmatite mass of Miami, Zimbabwe (formerly Southern Rhodesia) (Frankel, 1950).

4, 7, 16, 18, 21, 28. Apatite, fluorite, plagioclase, phlogopite, rutile and zircon occur as inclusions in the dravite of the Yinnietharra, Western Australia, deposits (Bridge Daniels, and Pyrce, 1977).

Table 5-2
Mineral Inclusions in Tourmaline

	Mineral	*Tourmaline Reported*	*Reference*
1.	actinolite	"green tourmaline"	Anderson, 1980
2.	albite	tourmaline	Slivko, 1966*a*
3.	anatase	tourmaline	Krynine, 1946
4.	apatite	dravite	Frankel, 1950; Bridge, Daniels, and Pyrce, 1977; Schmetzer et al., 1977
5.	brookite	tourmaline	Krynine, 1946
6.	cassiterite	tourmaline	Krynine, 1946
7.	fluorite	dravite	Bridge, Daniels, and Pyrce, 1977; Dolomanova, Loseva, and Tsepin, 1974
8.	garnet	schorl; dravite	LaCroix, 1910; Novak and Zak, 1970
9.	hematite	tourmaline	Valeton, 1955
10.	hydroxyl-herderite	elbaite	Cassedanne and Lowell, 1982
11.	K-feldspar	tourmaline	Slivko, 1966*a*; Dolomanova, Loseva, and Tsepin, 1974
	"orthoclase"	tourmaline	Williams, 1876
12 & 13.	kyanite within sillimanite	dravite/uvite/schorl	Clifford, 1958
14.	magnetite	tourmaline	Krynine, 1946
15.	manganotantalite	"rubellite"	Goni and Guillemin, 1964
16.	mica		
	lepidolite	elbaite	Dunn, 1974
	muscovite	tourmaline	e.g., Burgelya, 1961; Jacobson and Tilander, 1982
	phlogopite	dravite	Bridge, Daniels, and Pyrce, 1977
17.	pharmacosiderite	tourmaline	Čech et al., 1975
18.	plagioclase	dravite	Bridge, Daniels, and Pyrce, 1977
	oligoclase	schorl	Sitdikov, 1968
19.	pyrite	tourmaline	Gübelin, 1979

Table 5-2 *(continued)*
Mineral Inclusions in Tourmaline

Mineral	Tourmaline Reported	Reference
20. quartz	tourmaline	e.g., Burgelya, 1961; Lister, 1979; Shaposhnikov, 1959; Slivko, 1966*a*
21. rutile	dravite	Bridge, Daniels, and Pyrce, 1977
	tourmaline	Krynine, 1946; Burgelya, 1961
22. sericite	zincian elbaite	Jedwab, 1962
23. "sulphides"	Cr-dravite	Lum, 1972
24. titanite	tourmaline	DuParc and Sigg, 1914; Krynine, 1946
25. topaz	tourmaline	Krynine, 1946
26. uraninite	tourmaline	Stella and Tamburino, 1952
27. uranmicrolite-microlite	elbaite	Foord, 1976
28. zircon	dravite	Bridge, Daniels, and Pyrce, 1977
	tourmaline	Corin, 1940/1941; Schmetzer et al., 1977
29. "carbonaceous particles"	tourmaline	Krynine, 1946
30. coal	tourmaline	Perozio, 1959
31. "organic compounds"	tourmaline	Kranz, 1967
32. "unknown"	tourmaline	Beesley, 1975; Schmetzer et al., 1977

8. An almandine trapezohedron, approximately ¾ inch (2 cm) in diameter, "implanted" in a black tourmaline crystal, approximately 3½ × 2 inches (~9 × 5 cm), from a pegmatite of Tongafeno, Madagascar, is recorded by LaCroix (1910). Garnet grains "up to 0.4 mm grow into" dravite in quartz veins near Chvaletice, eastern Bohemia (Novak and Zak, 1970).

9. The hematite inclusions are within tourmaline in vugs in New Red Sandstone in Piesberg, northwestern Germany; both minerals are interpreted to have formed as a result of lateral secretion (Valeton, 1955).

11, 15, 16. Some so-called inclusions are better described as fillings within shells of tourmaline. Examples are: the feldspar, "orthoclase," inclusions of Williams (1876); the manganotantalite crystals within partially replaced pink elbaite from the pegmatites at Alto Ligonha, Mozambique (Goni and Guillemin, 1964); and the fine-grained muscovite cores within dark green tourmaline crystals from a pegmatite mass in Clear Creek County, Colorado (Jacobson and Tilander, 1982).

16. Small lepidolite crystals within elbaite from a pegmatite of the Nuristan Province of Afghanistan, are "randomly oriented" (Dunn, 1974).

16, 20, 21. Burgelya (1961) reported small inclusions of muscovite and quartz and idiomorphic crystals of rutile in some of the tourmalines of the Bakal (U.S.S.R.) iron deposit.

17. The barium-pharmacosiderite inclusions are microscopic (0.2 mm) cubes, some also having tetrahedral faces, within tourmaline grains in a mylonitized tourmalinite from southern Bohemia, now Čechy (Čech et al., 1975). The authors suggested that the pharmacosiderite formed as a result of hydrothermal decomposition of originally present arsenopyrite.

20. Shaposhnikov (1959) reported a quartz crystal within a tourmaline crystal from a pegmatite of southeastern Tuva (U.S.S.R.). Lister (1979) described tourmaline crystals with cores of quartz grains that are neither texturally nor optically continuous, from the vicinity of Cornwall (England) and concluded primarily on the basis of chemical and textural relations, that they were formed by skeletal growth of the tourmaline shells rather than by, for example, replacement (cf. McCurry, 1971*b*).

21. Apparently many gem quality uvites contain rutile inclusions (Dunn, 1977).

23. "Fine inclusions of sulphides" occur in chromian dravite from Kaavi, Finland (Lum, 1972).

24. DuParc and Sigg (1914) described tourmaline (dravite according to their analysis) that is literally riddled with inclusions of "sphene" (titanite); the grains are from a serpentinite in the Urals.

26. Tabular brown uraninite crystals are included in the tourmaline of a pegmatite at Delianova, Italy (Stella and Tamburino, 1952).

27. The uranmicrolite inclusions in elbaite crystals of San Diego County, California, are surrounded by halos and radial fractures interpreted to express metamictization ("radiation burns") of the tourmaline (Foord, 1976). In pink elbaite, the halos are darker pink; in green and colorless elbaite, they are pink.

28. The zircon inclusions recorded by Corin (1940-1941) occur in tourmaline grains constituting a tourmalinite of Remagen, Germany. They are surrounded by pleochroic halos.

32. Solid inclusions of unknown identity, some of which are fine needles with a patterned arrangement "reminiscent of graphic granite or Egyptian hieroglyphics," occur in some of the Newry (Maine) tourmalines (Beesley, 1975). A spiral inclusion of an unidentified material winds parallel to *c* within a dravite-uvite crystal described and illustrated by Schmetzer et al. (1977).

FLUID AND MULTIPHASE INCLUSIONS IN TOURMALINE

Both fluid inclusions and two-phase inclusions were recorded rather early for tourmaline gemstones (Bauer, 1896). Three-phase inclusions have been reported for several tourmalines (Fig. 5-3), for example, for those from Newry, Maine (Beesley, 1975).

Weiss (1953) described "primary type" cavities (i.e., those with negative crystal shapes) in tourmalines from five pegmatites of the Black Hills of South Dakota.

In a preview article, Slivko (1965) lamented that there then was little information in the literature about fluid and multiphase inclusions in tourmaline and summarized available data about these inclusions in multicolored tourmaline (elbaite?!) from the Central Urals and Borshchovochnyi Kryazh (Soviet Union) as follows:

1. Primary, secondary, and pseudo-secondary inclusions occur.
2. Inclusions are up to a few centimeters long.
3. Liquid inclusions are common in tourmaline: (a) two-phase liquid inclusions are most common; (b) single phase liquid inclusions are rare; (c) multiphase (liquid and solid) inclusions are rare; (d) multiphase (liquid-solid-gas, with as many as six solid phases) are widespread.
4. A common solid phase appears to be an avogadrite-ferruchite.
5. Homogenization of the primary inclusions occurs in the liquid phase at ~250°-300°C.
6. Utilization of the data obtained must be predicated on considerations of whether the inclusions have or have not remained sealed since formation and whether the original filling represents only one or more than one period in the growth of the containing crystal.

The following contents were found in a tourmaline from the Borschovochnyi Range (Dolgov and Shugurova, 1968): (SO_2, H_2S, NH_3, Cl)—18.3 %;

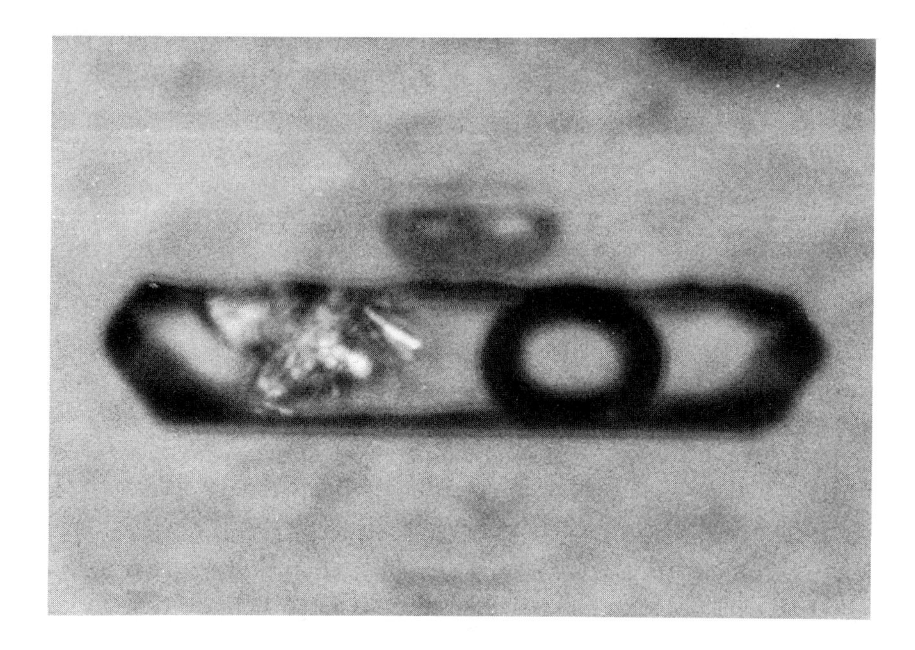

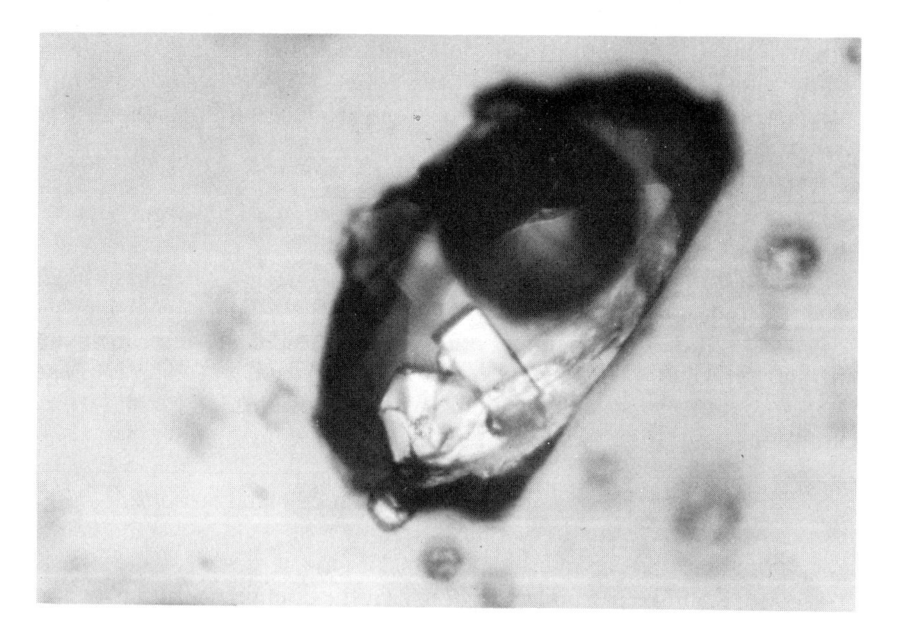

Figure 5-3
Multiphase inclusions in tourmaline from San Diego County, California; length of both inclusions is ~0.1 mm. (Photomicrographs courtesy of E. E. Foord.)

CO_2—63.7%; CO—0.0%; O_2—4.78%; H_2—0.0%; and N_2 + rare gases—13.1%. The "bubble" was 0.457 mm in diameter and contained only gas.

In one of the more recent studies of these inclusions, Slivko (1969*a*) analyzed carbon dioxide-aqueous extracts from pegmatitic schorls and reported the following components: Na, K, Li, Ca, Mg, Fe, Mn, HCO_3, CO_3, Cl, F, and $HSiO_3$, with HCO_3, CO_3, Na, Ca, and $HSiO_3$ the major constituents (cf. Dashdavaa, 1970). In addition, he noted that the absolute and relative quantities of the components differ markedly from zone to zone within some of the individual crystals.

In schorl crystals from Koretz pegmatites, Slivko (1969*b*) found all primary inclusions to homogenize in a liquid phase at temperatures of 260°-320°C and pressures of 1-1.5 kbar. (This, of course, excepts the liquid CO_2 inclusions, which homogenized at 16°-24°C.) In a later report, Slivko (1974) noted that some of the primary cavities exhibit well-defined analogous ($\bar{c}$) versus antilogous (c) terminations and that the included materials consist of such things as Na halides, carbonic acid, and carbonates, and have pH values that range with position within individual crystals, within single pegmatite masses, and from one type of pegmatite to another.

Pomârleanu and Movileanu (1971) found primary, "pseudo-secondary," and secondary fluid inclusions in pegmatitic tourmalines from the Carpathians of Romania. Both primary and secondary inclusions were monophasic (either liquid or gas), biphasic (liquid + gas, liquid + crystal, or gas + glass), and triphasic (liquid + gas + solid).

Dolomanova, Loseva, and Tsepin (1974) described vacuoles in tourmalines from cassiterite deposits of Zun-Undur, Transbaikal (Siberia), that contain K-feldspar and fluorite, and vacuoles in tourmalines from another deposit (Satrlovaya Gora) that contain sulfates (or sulfides) and chlorides of alkali and alkali earth elements.

Gübelin (e.g., 1979) described and illustrated fluid and multiphase inclusions in tourmaline gemstones (see Fig. 5-4) and directed special attention to their shapes and distribution. For example, he stated that typical inclusions are threadlike ("trichites"), either individuals or in groups that comprise interlaced networks, and that they commonly cross color-zone boundaries with no interruption.

Foord (1976) reported that for the tourmalines of the Mesa Grande district of San Diego County, California, fluid inclusions are especially "evident within catseye tourmaline as well as within overgrowth material (repaired terminations)" (p. 166). Considering the fact that the cat's-eye material there is typically on broken terminations, the relations appear to indicate either a ready availability of fluids for inclusion or some favorable kinetic conditions promoting the relative concentration of fluid inclusions during late stage growth.

Taylor, Foord, and Friedrichsen (1979) gave the data in Table 5-3 for fluid

Table 5-3
Fluid Inclusions from a San Diego County (California) Pegmatitic Tourmaline

Type of Inclusion	*Homogenization temperature, °C*[a]	*Freezing point depression, °C*	*Salinity (equiv. wt. % NaCl)*
I. 10-20 vol. % vapor	260-265	−2.9	4.7
II. 5-10 vol. % vapor	285-295	−2.8	4.6

Source: From Taylor, Foord, and Friedrichsen, 1979.
[a]Not pressure-corrected.

inclusions in pink tourmaline from a "pocket" in a pegmatite of the Himalaya dike system of San Diego County, California.

"INTERGROWTHS" INVOLVING TOURMALINE

The term intergrowth should, in the writer's opinion, be restricted to mineral pairs and groups known to have "grown together"—that is, minerals formed simultaneously in some intimate, typically regular, intermixture. However, the constituents of many so-called "intergrowths" probably grew sequentially—and, thus, should be described by inclusion terminology—or were formed as a result of replacement. It is likely that some of the "intergrowths" described in this section do not consist of simultaneously formed constituents; they are included because they were termed or implied to be intergrowths by the referenced investigators.

Just as for the other subsections in this chapter, the following text is keyed to the appropriate table (Table 5-4).

1, 10, 12. Filiform schorl-dravite occurs "intergrown with and in quartz, and andalusite, and lazulite" in pyrophyllite deposits of central North Carolina (Furbish, 1968).

2. In a discussion about diataxy (i.e., oriented intergrowths), Mücke (1975) described and illustrated a tourmaline crystal within a beryl crystal from a pegmatite of Amkazobé, Madagascar (Fig. 5-5). He directed special attention to their mutual tolerances, which are 0.25 percent for prisms and 3.7 (not 10.5, as recorded) for c (i.e., $4c$ for tourmaline versus $3c$ for beryl).

3. Bluish tourmaline, termed schorl, is intimately "intergrown" with fine-grained bertrandite in Be-bearing veins at the South Crofty mine, Cornwall, England (Clark, 1970).

4. "Intergrowths" of brownish green "nearly black" tourmaline and biotite, interpreted by Hoel and Schetelig (1916) to represent simultaneous crystallization of the two minerals, occur in nepheline-bearing pegmatites of Seiland,

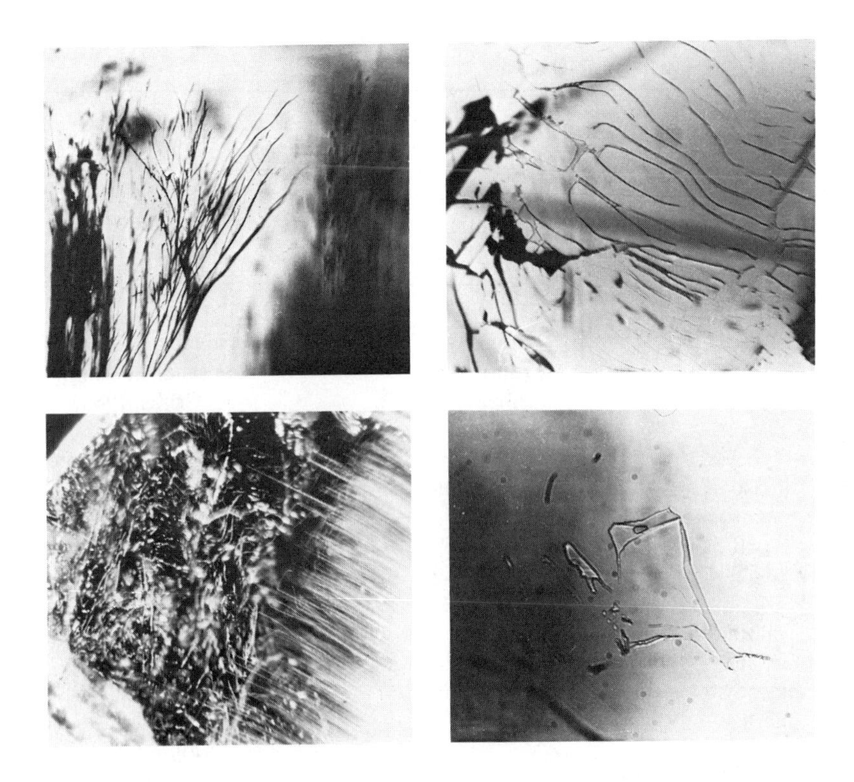

Figure 5-4
Inclusions in tourmaline. *Upper left,* broomlike concentration of trichite inclusions in green tourmaline (10x); *upper right,* characteristic trichite fibers in green tourmaline (7.5x); *lower left,* transition zone in a crystal between green part with dense packing of fibers (lower right) and pink part with irregular arrangement of trichites (7.5x); *lower right,* a more highly magnified, twisted trichite inclusion in green tourmaline that reveals it to be a two-phase inclusion (25x). (Photomicrographs courtesy of E. Gübelin.)

Norway. This is an especially interesting occurrence from the standpoint that several investigators have suggested that the two minerals are incompatable.

5, 9. White asbestiform dravite is intimately intergrown with light greenish hyalophane and a penninite/clinochlore chlorite in veins in a pyrite-manganese deposit near Chvaletice, eastern Bohemia, now Čecky (Novak and Zak, 1970).

6. A tourmaline cat's-eye that consists of an intimate mixture of tourmaline and diopside is reported by Mitchell (1967). (This material would probably be better described as tourmaline with diopside inclusions.)

Table 5-4
"Intergrowths" Involving Tourmaline

Other Mineral	*Tourmaline Reported*	*Reference*
1. andalusite	schorl/dravite	Furbish, 1968
2. beryl	tourmaline	Mücke, 1975
3. bertrandite	schorl	Clark, 1970
4. biotite	tourmaline	Hoel and Schetelig, 1916
5. chlorite	dravite	Novak and Zak, 1970
6. diopside	tourmaline	Mitchell, 1967
7. eosphorite	elbaite	McCrillis, 1983, personal communication
8. "feldspar"	tourmaline	Fersman, 1915
microcline	tourmaline	Krokström, 1946
"orthoclase"	tourmaline	Williams, 1876
plagioclase	dravite	Bridge, Daniels, and Pyrce, 1977
9. hyalophane	dravite	Novak and Zak, 1970
10. lazulite	schorl/dravite	Furbish, 1968
11. muscovite	tourmaline	e.g., Frondel, 1936
12. quartz	schorl	e.g., Newhouse and Holden, 1925
13. tourmaline ("achroite")	brown & green tourmaline	Babu, 1970
14. sulfides	dravite	Slack, 1980

8. Fersman (1915) has reported an "intergrowth" of tourmaline and "feldspar" from Murzinka (in the Urals) in which the *c* axes of the two minerals are perpendicular and a prism face of the tourmaline is parallel to {010} of the feldspar. The microcline-tourmaline "intergrowths" recorded by Krokström (1946) may represent microcline partially replaced by tourmaline.

8. "Poikilitically aligned laths" of plagioclase (An_{29}) are subparallel to the *c* axes of the associated dravite in masses that are made up of an ~50:50 mixture of the two minerals in the Yinnietharra district of Western Australia (Bridge, Daniels and Pyrce, 1977).

11. Tourmaline-mica (typically muscovite) "intergrowths," most of which are thought to have been formed by simultaneous crystallization of their two constituents, are recorded from pegmatites of several localities. Some of the pairs, however, have been interpreted as, and may well be, inclusions of tourmaline within mica. Examples include those from Paris, Maine (Shepard, 1830); Grafton, New Hampshire, and Haddam, Connecticut (Mügge, 1903);

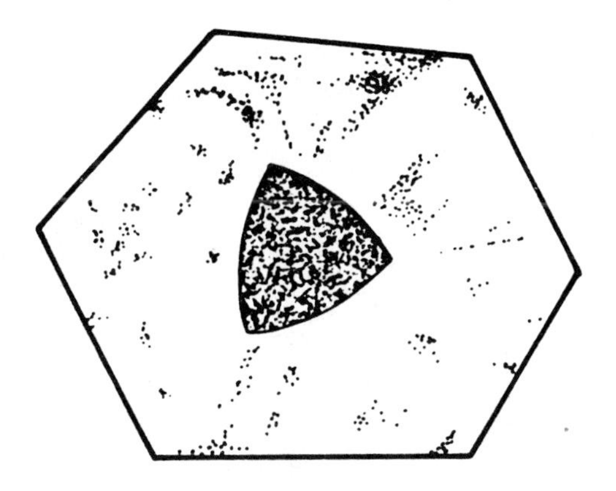

Figure 5-5
Diataxis intergrowth of tourmaline (dark) and beryl. (Sketch based on photograph of Mücke, 1975.)

near Sondalo, northwestern Italy (Linck, 1899); Gilsum, New Hampshire, and New York, New York (Frondel, 1936); and Miami, Zimbabwe (formerly Southern Rhodesia) and Olgiassa, Italy (Frankel, 1950). Several features are apparently common to the well-described occurrences:

(a) The tourmaline crystals, which occur as both isolated individuals and in groups (commonly radiating), are enclosed between basal cleavages of the mica.
(b) A majority of the tourmaline crystals are prismatic and between 0.5 and 15.0 mm long, 0.05 to 5.0 mm wide, and less than 0.5 mm thick.
(c) The most common, but not the only, planes of flattening are $(11\bar{2}0)$, $(10\bar{1}0)$, and (0001).
(d) The crystals have preferred crystallographic orientations, some of which are clearly recognizable—for example: tourmaline crystals flattened on $(11\bar{2}0)$ with *c* parallel to a ray of the mica's percussion figure; similar crystals with *c* parallel to a ray of the mica's pressure figure; tourmaline crystals flattened on (0001) with $(11\bar{2}0)$ faces parallel to rays of mica's pressure figure; and crystals aligned along growth planes of the mica. Also, it is suggested that other regular relations may be detected if more specimens are analyzed statistically (see, for example, Mügge, 1903, and Frondel, 1936).
(e) The surrounding mica shows no evidence of having been disturbed.

Dmitriev (1955) has interpreted apparently similar tourmaline-muscovite pairs to represent muscovite that contains tourmaline crystals formed by later metasomatism (see also Jérémine and Sandréa, 1957). He thinks that

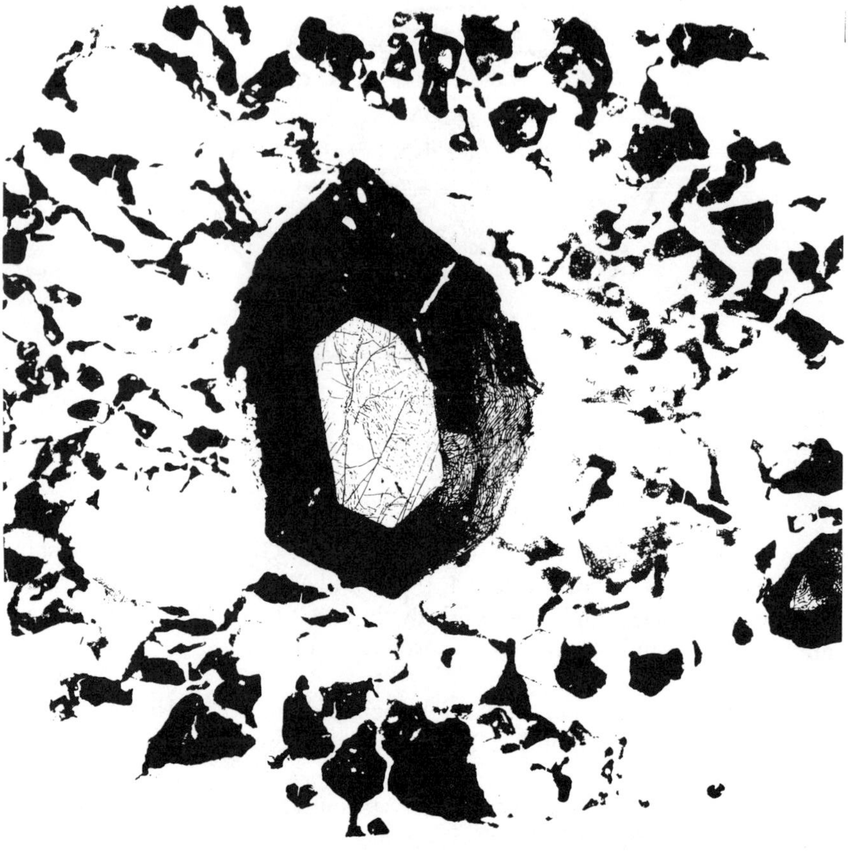

Figure 5-6
A so-called graphic intergrowth of schorl(?) [dark] and quartz [light] that occurs as an irregular-shaped mass surrounded by feldspar, from Mt. Rubellite, Hebron, Maine. Large central tourmaline crystal and a few of the smaller crystals are zoned with bluish cores surrounded by a brownish outer zone; colors are those seen macroscopically by light transmission through the section. "Photograph" is a contact print made directly from thin-section of specimen (3x).

the lathlike shapes and their occurrence between muscovite cleavage planes (with {10$\bar{1}$0} parallel to {001} of the host), the "splitting" of ends of some tourmaline crystals by muscovite, and the presence of mica inclusions in some of the larger tourmaline crystals are better explained under this alternative hypothesis of origin.

Another tourmaline-muscovite "intergrowth" of interest is described from a muscovite-rich pegmatite of Brissago, Tocino, Switzerland (Stuker, 1961); scales of schorl occur with their (0001) either parallel to or perpendicular to

(001) of the enclosing muscovite. The extremely thin scales of tourmaline exhibit good {11$\bar{2}$0} cleavage (see Fig. 2-21), which Stuker interprets to have been formed in response to tectonism.

12. Quartz-tourmaline "intergrowths," some of which have been described as "graphic intergrowths," have been recorded from several areas; see, for example, Bastin (1911), LaCroix (1922), Brammall and Harwood (1925), Newhouse and Holden (1925), Hitchen (1935), Jahns and Wright (1951), Slivko (1955*b*), Breskovska and Eskenazi (1959-1960), Shiryaeva and Shmakin (1969), and Němec (1981). The one illustrated in Figure 5-6 is a mass with a roughly circular cross-section, about 2½ inches (~6 cm) in diameter, which was originally surrounded by feldspar at Mt. Rubellite, Hebron, Maine. It consists of a central schorl crystal, which is zoned, with several smaller grains of schorl, all with their *c* axes nearly parallel, within a mass of white quartz. The writer has also seen cobbles of such "intergrowths" in channel-fill conglomerates in the Cambrian Potsdam Sandstone of Northern New York.

Some "intergrowths" having similar macroscopic appearances have been found loose in the glacial drift in central Michigan. These masses, however, are quite different; they consist of mosaics of small ($<$ 2 mm, greatest dimension), more or less randomly oriented quartz grains surrounding relatively large (up to 15 mm long), commonly fractured and/or composite schorl laths that do not have a parallel crystallographic orientation.

The most complete and best illustrated descriptions of diverse kinds of these schorl-quartz "intergrowths" is that of Brammall and Harwood (1925).

13. Babu (1970) reported an achroite (i.e., a white-gray tourmaline, otherwise not defined) that occurs as prismatic laths "intergrown" with pale brown and green tourmalines associated with pink elbaite in a pegmatite from near Ajmer, India.

14. Slack (1980) reported "Brown or golden yellow dravite . . . intergrowths commonly with pyrrhotite and chalcopyrite but rarely with sphalerite, galena, or pyrite" at the Black Hawk mine near Blue Hill, in the Penobscot Bay area of Maine, U.S.A.

Chapter 6

Color and Optical Properties

> "The causes which give rise to the great variety of colors among the tourmalines . . . form interesting themes of inquiry; and they are yet subjects of controversy among scientific men, and probably always will be."
> Hamlin (1873)

COLOR

Color-to-Species Relationships

There are tourmalines of virtually all colors, including black, white, brown, and colorless, and many individual crystals exhibit two or more hues and/or color saturations (see the color plates). None of the colors, except the bronzy brown of buergerite, is restricted to one species. However, the other species also have one or more colors that are characteristic. On the following list, the characteristic colors, as discerned macroscopically, are in bold face type. (See also Table 6-1.)

Buergerite—**bronzy brown;**
Chromdravite—**dark green** (nearly black);
Dravite—**brown,** red, yellow, green, blue, colorless, white, gray, black (chromian and vanadian dravites are green, bluish green or nearly black);
Elbaite—**pink, green,** colorless, red, orange, yellow, blue, violet, white, black, brown;
Ferridravite—**black;**
Liddicoatite—same as elbaite;
Schorl—**black,** yellow, green, blue(?), brown;
Uvite—same as dravite.

Table 6-1
Color-to-Tourmaline Species Correlation

Color	*Species*[a]
red	elbaite[b] [dravite from Osarara, Kenya (Dunn, Arem, and Saul, 1975; cf. Bank, 1974*a*); uvite from Fowler, New York (Ayuso and Brown, 1984)]
orange	elbaite
yellow	dravite[c], elbaite [schorl(?) from veins in southwestern England (Power, 1968)]
green[d]	elbaite, chromian and vanadian dravites [dravite from Edwards, New York; uvite from Franklin, New Jersey; schorl-dravite from fluorite deposit of Siberia (Getmanskaya et al., 1970)]
blue	elbaite [dravite from Stupna, Czechoslavakia (Čech, Litomisky, and Novotný, 1965) and from East Africa (Dunn, 1978); schorl/dravite from fluorite deposit of Siberia (Getmanskaya et al., 1970)[e]]
violet	elbaite
colorless	elbaite, dravite
white	elbaite, dravite
gray	dravite
black[d]	schorl, ferridravite [some of dravite from Yinniethara, Western Australia (Bridge, Daniels, and Pyrce, 1977); elbaite, which is smoky purple when strongly illuminated, from Minas Gerais, Brazil (Prescott and Nassau, 1978); elbaite, which is blue in splinters, San Diego County, California; uvite from, for example, Pierrepont, New York]
brown	dravite, buergerite (actually bronzy brown) [schorl from several localities (see, for example, Deer, Howie, and Zussman, 1962); greenish brown elbaite from Madagascar]

[a]A few "exceptions" recorded or seen by compiler are given in square brackets.
[b]Elbaite and the apparently relatively rare liddicoatite are indistinguishable on the basis of color so elbaite in this tabulation means elbaite/liddicoatite.
[c]Dravite means dravite/uvite in this tabulation.
[d]Chromdravite is "dark green, nearly black" (Rumantseva, 1983).
[e]Dunn (1979, personal communication states that considering many hundreds of analyzed tourmalines he has found no macroscopically discernible blue tourmaline that can be referred to schorl composition. Several schorls, however, are blue in fine powders and/or thin sections.

Some people who have dealt with many tourmaline specimens think they can tell the source of many specimens—in some cases as to district, in others as to mine within a district—on the basis of their color and/or color zoning pattern. They are probably correct, in most cases. "In most cases" because exceptions can sometimes be found. On the other hand, it is imprudent even to guess the source of many tourmaline gemstones; too many of them have been subjected to one or another treatment to modify their original colors.

This, of course, is especially disturbing for persons involved in research relating to the colors of tourmaline. In fact, investigations of treated stones, not known to be such, may, in part, account for some of the apparently inconsistent results recorded in the literature.

Color Zoning

What may be the first reference to tourmaline in the literature is so judged on the basis of the fact that tourmaline commonly exhibits color zoning: Theophrastus (~315 B.C.) may have been referring to tourmaline when he mentioned a stone from the Island of Cyprus that was emerald green in color at one end and light colored at the other.

In any case, most described and/or illustrated multicolored tourmaline crystals are elbaites. However, in addition: some dravite (see, for example, Agrell, 1941), especially chromian dravite (see, for example, Dunn, 1977*b*), and uvite crystals exhibit different hues and/or color saturations; some dravite-schorls exhibit zoning both concentrically around *c* and perpendicular to *c* (Getmanskaya et al., 1970, and McCurry, 1971*b*); most macroscopically opaque tourmalines, including schorl, can be seen to be color zoned when viewed in thin slices (see, for example, Jahns and Wright, 1951, and Black, 1971); and, of course, liddicoatite, like elbaite, is apparently commonly multicolored. Also, what appear to be composite crystals of, for example, schorl and dravite have relations that result in a feature that may be termed color zoning (Fig. 6-1).

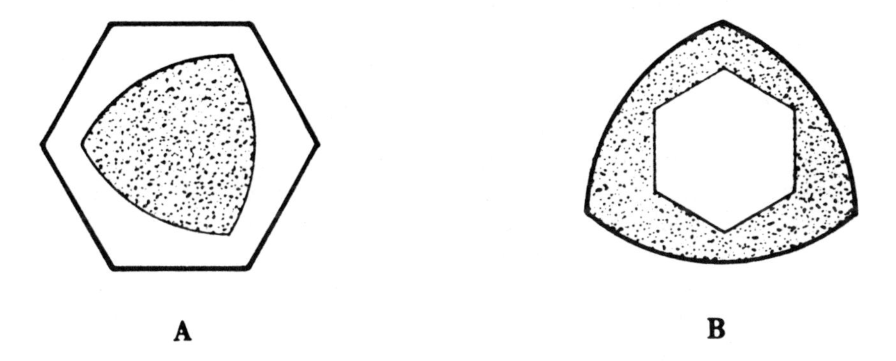

Figure 6-1
Zoning involving both color and habit differences: *A*, Harvard Museum #112558, from Madagascar with predominantly black internal zone and brown to nearly colorless external zone; *B*, specimen in collection of E. E. Foord, from Brazil with pink internal zone and green external zone. Both sketches are approximately natural size.

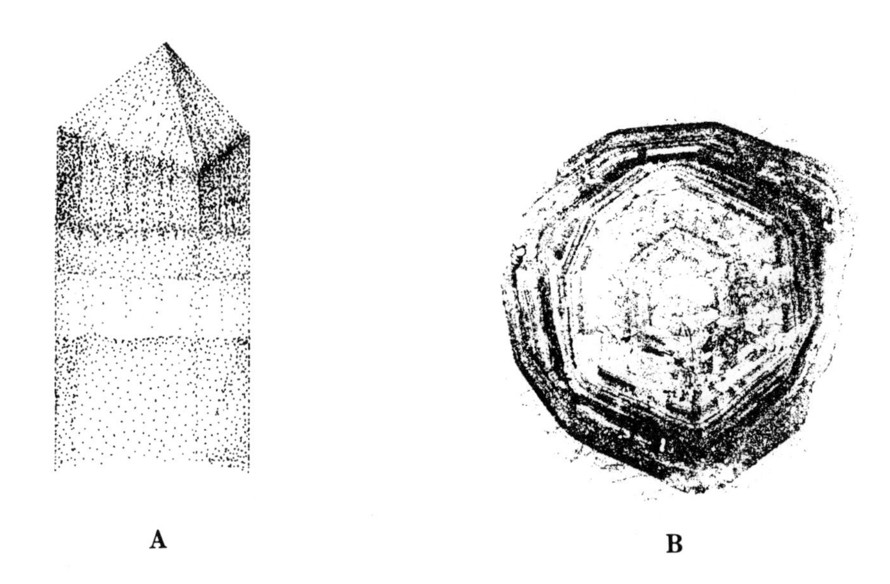

Figure 6-2
Zoning perpendicular to *c* (*A*) and parallel to *c* (*B*). (*A* is a sketch after Hamlin, 1895 and *B* is from D'Achiardi, 1897.)

Some multicolored tourmalines exhibit one or more fairly regular patterns of color zoning (Fig. 6-2*A* and *B* and plates); other multicolored tourmalines have irregular or patchy ("random") arrangements of their diverse hues (Fig. 6-3). The so-to-speak regular patterns of zoning are: perpendicular to *c* (i.e., roughly parallel to pedion faces); parallel to *c* (i.e., roughly parallel to prism faces such as $\{10\bar{1}0\}$ and $\{11\bar{2}0\}$; and roughly parallel to pyramid faces. The last type has been recorded for relatively few specimens (see, for example, Karnojitzky, 1890; D'Achiardi, 1893; Tarnovskii, 1961; Dunn, Appleman, and Nelen, 1977; and Althaus, 1979). Some of the zoning recorded as parallel to prisms is very likely parallel to pyramids (and vice versa); in fact, many zoned crystals exhibit both of these zoning patterns (Plate V). Also, zoning with different color saturation in first-order prism sectors rather than in second-order prism sectors has been reported (Pollard and Wagner, 1973).

Transitions between differently colored zones have been reported to range from abrupt ("razor-sharp") to gradational over a few microns to a few millimeters (Fig. 6-4*A*). Bradley and Bradley (1953), in reporting sharp contacts between zones, state that such contacts are characteristic of crystals whose color difference is "a strong pink to a strong green." Actually, they occur between all sorts of diverse color combinations. An example of a gradational transition is from medium green through greenish yellow to nearly colorless.

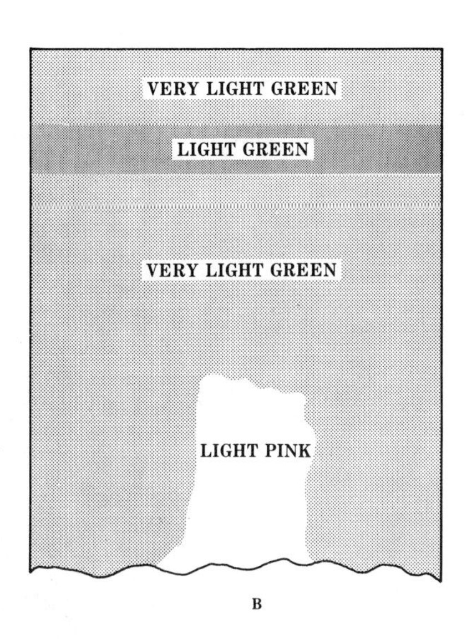

Figure 6-3
Irregular zoning as exhibited by elabites from Newry, Maine: characteristically, the pink to green changes, particularly in specimens from this locality, represent physical breaks (e.g., what appear to be fractures) wherever the zoning is not parallel to a common crystallographic plane; *A*, a 10 × 8 mm section of NMNH #133814; *B*, a 10 × 8 mm section of NMNH #133815; both sections are parallel to *c*. (After Dunn, 1975*a*.)

Multicolored tourmalines with different macromosaic textures corresponding to color differences (Fig. 6-4) have been described by, for example, Young et al. (1969). Wagner et al. (1971) have also described and illustrated these features and, after a brief discussion of their origin, concluded that twinning, exsolution, compositional differences, and deformation can be ruled out as responsible, but "sudden drastic changes in conditions controlling growth are possible causes" (p. 114).

Pollard and Wagner (1973) described the boundaries of the color zones in an elbaite that exhibits light green in first-order prism sectors and dark green in second-order prism sectors (referred to above) as being the loci of microfaults and "lattice misorientations," each involving rotation of greater than 40 seconds of arc around an axis perpendicular to the boundaries.

Foord and Mills (1978) reported abrupt color changes in elbaite from Minas Gerais, Brazil; San Diego County, California; and Oxford County, Maine; to be manifested by biaxiality and an extinction pattern resembling that of the grid twinning of microcline (Fig. 6-4*B*). No one, however, appears to have discovered any correlation between any particular texture, type of boundary, biaxiality, or extinction pattern and any particular color zoning.

When zoned crystals are considered in general, a few color arrangements

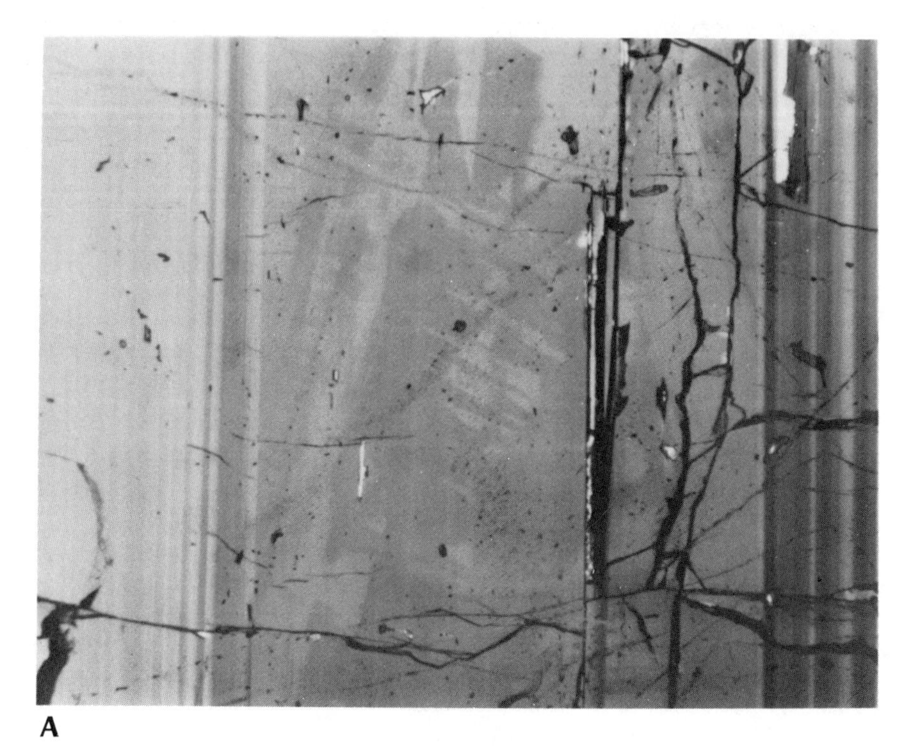

A

Figure 6-4
A, zoning in a dravite (~dravite$_{65}$schorl$_{35}$) from Cecil County, Maryland; width of field ~2.25 μ (photomicrograph courtesy of Stephanie Mattson). *B*, zoning in an elbaite from San Diego County, California, showing different textures corresponding to different color zones; long dimension is ~3 mm (photomicrograph courtesy of E. E. Foord). *C*, "X-ray diffraction of 0003 Bragg reflection from two-color tourmaline plate . . . showing source-image distortion" (from Young et al., 1969).

appear to be common, others have been described only once or a few times, and regularities appear to characterize some—but not all!—individual deposits and/or districts. A relatively common arrangement for elbaite crystals zoned parallel to {0001} is green near and including the analogous end, pink near and including the antilogous end, and nearly colorless through an intervening zone. A common arrangement for elbaite crystals zoned parallel to prism faces is one in which the pink to green or reverse color sequence extends from the central to and through the marginal zone (Plate V). Tourmalines exhibiting the zoning parallel to prism faces with pink centers and green rims are widely referred to as "watermelon tourmalines." And, in a similar vein, prisms that are colorless, or nearly so, and tipped with black have long been termed "mohrenkopf." (Some of the black-tipped

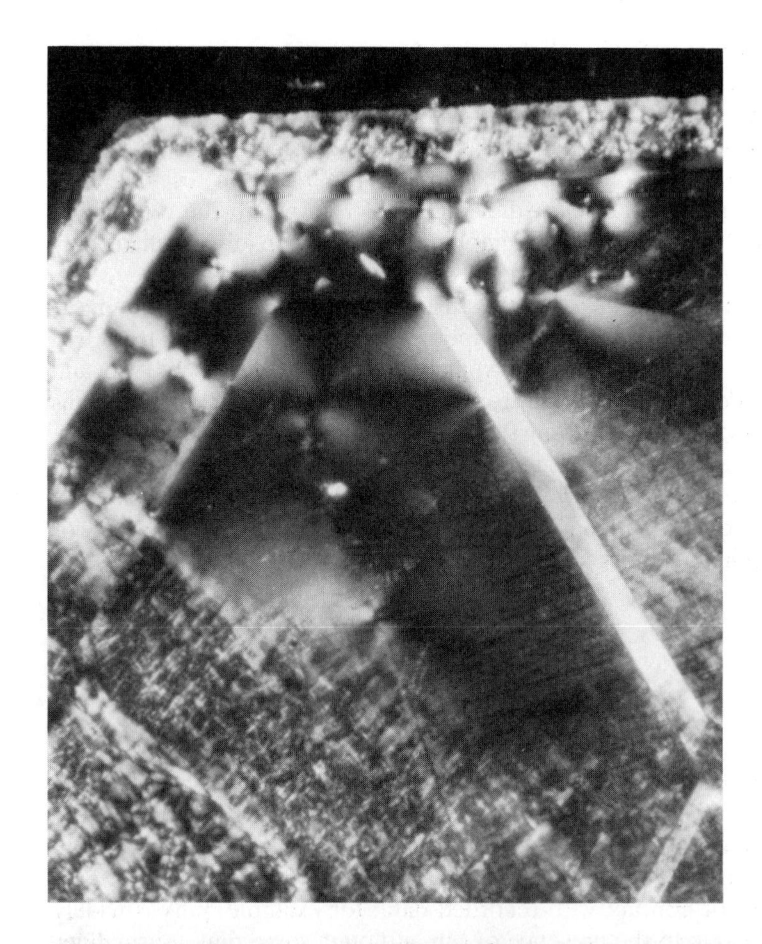

B

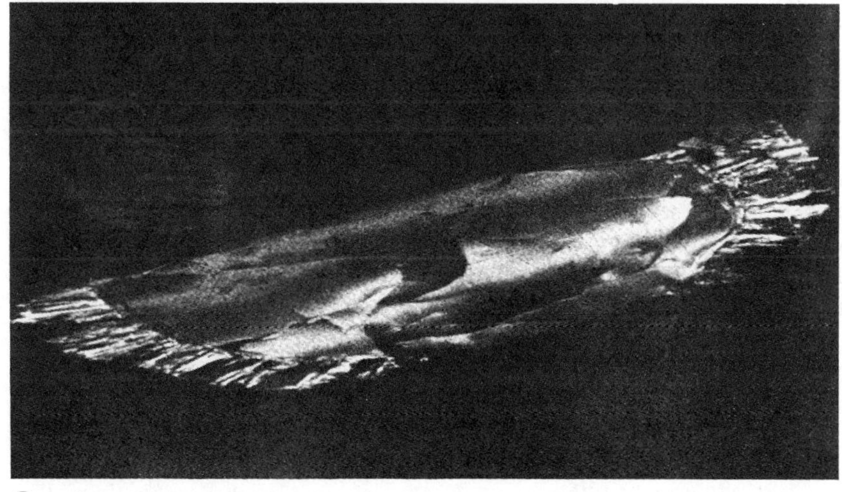

C

crystals consist of individual light-colored crystals the extended growth of which resulted in the formation of numerous parallel individuals of the darker tourmaline.)

Within individual pockets, all crystals, large to small, tend to exhibit zoning with essentially identical sequences. Apparently in response to a sequence of critical changes within the fluid from which the tourmaline was deposited, the color "stratigraphy" commonly involves remarkably complex and, in some cases, extremely closely spaced zones.

It warrants emphasis, however, that in general all sorts of arrangements of color occur; see, for example, descriptions of the zoned crystals from Utö (Sjögren, 1916), from northeastern Afghanistan (e.g., Leckebusch, 1978, and Bariand and Poullen, 1979), and from San Diego County, California (Jahns and Wright, 1951). The following statement of Jahns and Wright is representative: "Many of these crystals are green on one end and pink on the other, . . . many others grade from black to green or pink, or even from black to colorless. Nearly all combinations are known, and some crystals show five or more alterations of colors in their length" (p. 36).

Causes of Colors

Diverse concentrations of major elements, of trace elements, or of combinations of major and/or trace elements, structural imperfections ("color centers"), electron charge transfers, and combinations of all of these phenomena have been suggested as responsible for one or another of the colors of tourmaline. If applied generally, some of these causes are mutually exclusive and/or conflict with recorded data; for example, pink is widely thought to be due to the presence of Mn, although some pink tourmalines have been recorded to contain no—not even trace amounts of—Mn. However, considering the compositional variability of tourmaline, it may be that many, perhaps most, of the hypotheses are correct for the specimens studied, and that at least some of the data:hypotheses discordancies merely reflect the tendency to make unwarranted generalizations from specific cases. Also, a few of the apparent discordancies may even depend on the subjectivity involved in describing the colors as perceived macroscopically; many tourmalines have colors that are intermediate between easily distinguishable hues; for example, there are purplish and henna reds as well as "ruby reds" and yellowish and bluish greens as well as "emerald greens."

Many pertinent observations and several chemical controls were suggested before the 1960s. For summaries, see Stamm (1926), Carobbi and Pieruccini (1947), Deer, Howie, and Zussman (1962) and Table 6-2.

Optical absorption spectra are given in Figure 6-5. Charge transfers (etc.) assigned to the peaks are reviewed for, for example, blue-green tourmalines by Smith (1978*b*). While reviewing the assignments that are repeated in

Table 6-2
Compositional Controls, Excluding Charge Transfer *per se*, Suggested as Causes for Colors

Suggested Control	*Color*	*Reference*
	red (including pink)	
Fe^{3+}		Quensel and Gabrielson, 1939 (cf. Bradley and Bradley, 1953)
Mn, Cs, and Li		Carobbi and Pieruccini, 1947 (cf. Chaudhry and Howie, 1976)
Mn^{2+} and/or Mn^{3+}, but no "appreciable" Fe		Bradley and Bradley, 1953 (cf. Vinokurov and Zaripov, 1959)
Mn^{3+}		Slivko, 1959 (cf. Vinokurov and Zaripov, 1959)
$Mn^{3+} \pm Fe^{2+}$		Grum-Grzhimailo, 1948 (cf. Dunn, 1975*a*)
	orange (see text)	
	yellow	
Fe^{3+}		Grum-Grzhimailo, 1956
	green ("Cr-free")	
Fe^{2+}		Quensel and Gabrielson, 1939 (cf. Dunn et al., 1977*a*)
	green (Cr-bearing)	
Cr		Lum, 1972 (cf. Dunn, 1977*b*)
Cr (and/or V)		Schmetzer and Bank, 1979
	blue	
Cu		Carobbi and Pieruccini, 1947 (cf. Leckebusch, 1978)
Fe^{2+} and Fe^{3+}		Grum-Grzhimailo, 1956 (cf. Leckebusch, 1978)
Fe and Mn		Deer, Howie, and Zussman, 1962
	purple (see text)	
	black	
Fe		Ward, 1931 (cf. Manning, 1969*b*)
	brown	
Fe^{3+}		Slivko, 1957 (cf. Dunn et al., 1977*b*)
Ca (low content)		Dunn et al., 1977*b*

Note: Some of the references are only examples.

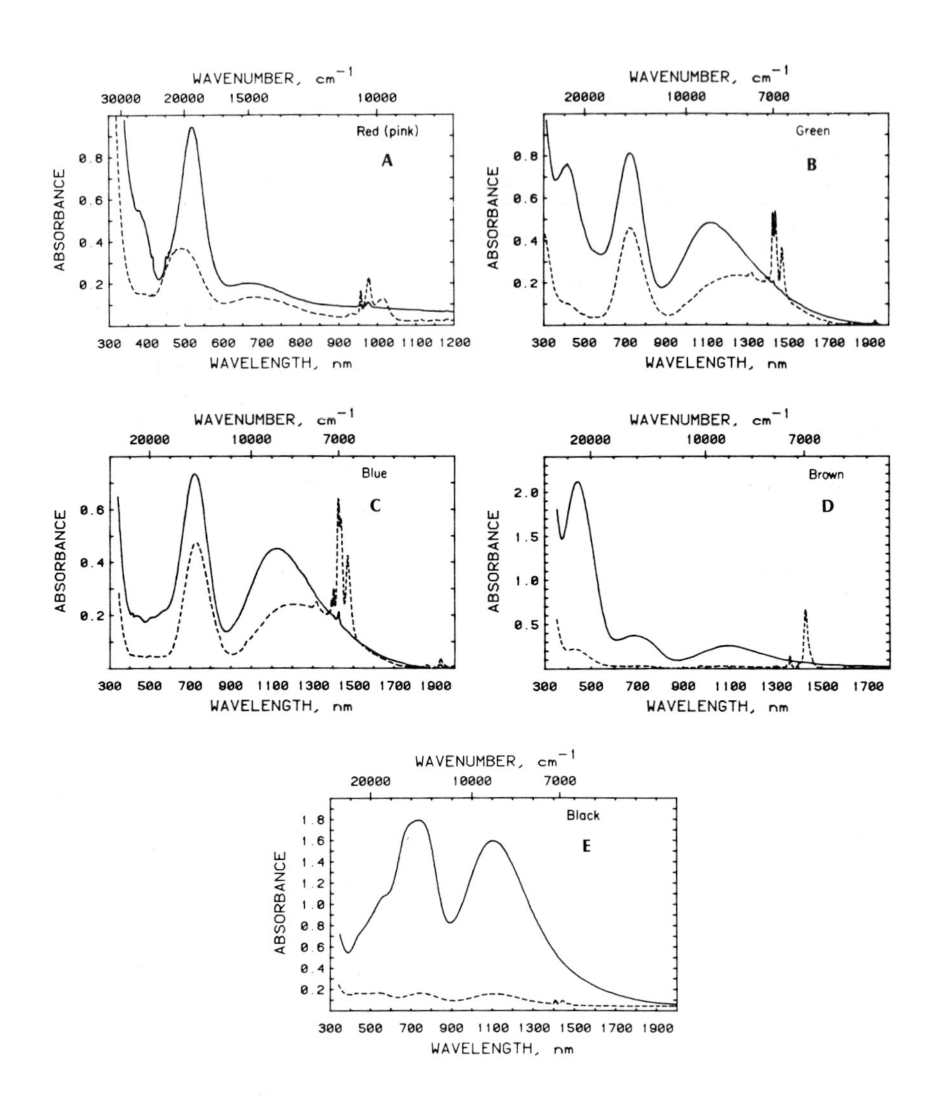

Figure 6-5
Polarized absorption spectra for tourmalines. Solid line—E ⊥ c; dashed line E || c. Notice that scales differ. (Spectra and data courtesy of S. Mattson and G. Rossman.)

Specimen	*Thickness (mm)*	*Location*	*FeO*[a]	*MnO*	*TiO*
A: red elbaite	0.166	San Diego County, Calif.	1.87	0.05	0.01
B: green elbaite	0.733	Nuristan, Afghanistan	4.32	1.67	0.04
C: blue elbaite	0.958	Nuristan, Afghanistan	1.62	1.71	<0.01
D: brown dravite	0.68	Yinnietharra, Australia	0.46	<0.01	1.09
E: black shorl[b]	0.45	Riverside County, Calif.	13.3	0.28	0.05

[a]Total iron as FeO
[b]Black in hand-specimen but blue in very thin sections.

the following text, it should be kept in mind that even some of the investigators who measured and interpreted the data have prefaced their pertinent statements by disclaimers, such as: "The complex structure of tourmaline inevitably introduces a considerable speculative element into spectral interpretation" (Faye, Manning, and Gosselin, 1974). There have been "several mutually contradictory assignments of the main absorption bands . . . [by] different authors (and sometimes the same authors at different times . . .)" (Smith and Strens, 1976). In fact, the current state of knowledge about the diverse colors of tourmaline is perhaps still described best by the statement "the optical absorption spectra continue to defy a satisfactory unified interpretation" (Faye, Manning, and Gosselin, 1974). Nonetheless, it does appear that most investigators are now of the opinion "that single-ion explanations . . . are not able to explain many of the aspects of tourmaline color. [and] The role of interactions between cations is extremely important in Fe-containing tourmalines."(G. R. Rossman, 1984, personal communication).

All of these things should, of course, be kept in mind while considering the absorption spectra and the following summaries.

Red—So far as the suggested direct chemical controls, many mineralogists still attribute the pink color of tourmalines to Mn, be it Mn^{3+}, Mn^{2+}, a combination, or to charge transfer involving Mn (e.g., Leckebusch, 1978), and they are probably right so far as the pink color of most tourmalines. However, it is noteworthy that: as a result of an investigation of the magnetic susceptibility of pink tourmalines from the U.S.S.R., Vinokurov and Zaripov (1959) concluded that Mn^{3+} is not a component of pink tourmaline; Donnay (1969) reported a pink elbaite crystal that appears uniform in color to the naked eye to have a Mn-content that differs "through a range down to 43% of the maximum over a distance of 0.45 mm," but she still attributed the pink color to Mn; Chaudhry and Howie (1976) found $Fe^{2+} >> Fe^{3+}$ in both the pink and the green tourmalines from Devonshire (England) and neither Cs nor Mn present in the pink tourmalines; Dunn (1975*a*), on the basis of his study of Newry (Maine) pink elbaites noted that color saturation does not correlate with Mn-content and that these tourmalines contain no appreciable Fe; and Nassau (1975), on the basis of chemical analyses of 85 specimens of pink tourmalines, showed that there is considerable variety: "Irradiation and heating indicates the possibility of at least seven causes of pink . . . [and] This complexity may explain the diversity of previous assignments for the origin of the pink color." And, he also concluded, *inter alia,* that Cr, V, and Ti concentrations are too low to be of direct significance so far as the color of at least some tourmalines in each of the groups, that Fe is too low to be of direct significance except possibly in one group, and that Mn is not present in significant quantities in one of the groups.

As can be seen, some of these data appear to negate even the charge transfer suggestions that require the presence of Mn. It is, however, important to remember that although Mn^{3+} is present in much smaller amounts than Mn^{2+} in most tourmalines, and it is also true that a trace of Mn^{3+} is a much stronger chromophore than a relatively large amount of Mn^{2+}. This is true because Mn^{3+} has spin-allowed absorption bands whereas Mn^{2+} has spin-forbidden absorption bands.

Of the suggested controls, another remains: on the basis of electron paramagnetic resonance spectroscopic studies, Bershov et al. (1969) concluded that electron-hole centers (i.e., oxygen vacancies with trapped electrons) account for both the color and fluorescence of pink elbaite. This conclusion does not, however, appear to be consistent with at least some of the recorded fluorescence data (see Table 6-4).

Orange—Orange is seldom seen except as one of the dichroic colors of some pink tourmalines. At least some peach-colored gemstones are apparently originally yellow stones that have undergone an increase in red, or pink stones that have undergone an increase in yellow as a result of irradiation treatments (Nassau, 1984).

Yellow—A golden yellow gem-quality tourmaline ($dravite_{78}$ $uvite_{22}$) from Kenya gave an absorption spectrum that was published by Hanni, Frank, and Bosshart (1981). They concluded that the color is due to a 0.3-0.8 weight percent Ti-content. George Rossman (1984, personal communication) thinks that $Fe^{2+} \rightarrow Ti^{4+}$ charge transfer causes the yellow, brownish yellow, yellowish brown, and brown colors of tourmalines, and that those low in Fe are yellow whereas those higher in Fe are brown. It also seems that some of the purportedly "natural yellow" gemstones may be irradiated pink tourmaline (Nassau, 1975).

Green—There appear to be at least two distinct kinds of green tourmalines, those that contain appreciable Cr and/or V and those that are essentially Cr- and V-free.

For both types, Vargas and Tupinamba (1962) offered the rather incongruous conclusion that the role of "impurities such as iron, chromium, and other ions . . . must be limited to the supply of trapping centers for entities (such as holes and electrons)" that serve as color centers (p. 69).

Both Lum (1972) and Dunn (1977) reported the intensity of the green color of chromian dravites to vary directly with the Cr-content. It seems well established that even small amounts of Cr can produce green.

In a note about gem-quality green tourmalines from East Africa, Schmetzer and Bank (1979) reported that all specimens they checked have higher V- than Cr-contents and thus are properly vanadian dravites and uvites; apparently the Cr-content is extremely low because these specimens do not give the sharp absorption peak, usually attributed to Cr, at 14,600 cm^{-1}.

For essentially Cr-free green tourmalines: The $\sim$14,000 cm^{-1} peak is probably best assigned to Fe^{2+} d - d transition (Faye, Manning, and Gosselin, 1974, and Smith, 1978*b*), whereas the relatively weak absorption band at 22,000 to 24,000 cm^{-1}, which also contributes to the green color of some elbaites, is probably a response to $Fe^{2+} \rightarrow Ti^{4+}$ charge transfer (Faye, Manning, and Gosselin, 1974). Stephanie Mattson (1984, personal communication) considers the 22,000 band plus the Fe^{2+} *d*-*d* bands to be absolutely necessary to produce the green color in the absence of Cr.

Dunn (1975*a*) reported the green elbaites from Newry, Maine, to have Fe-contents that vary directly with green color-saturation. Dunn et al. (1977) reported the green uvite from Franklin, New Jersey, to have an Fe-content (<0.28 weight percent) that is nearly the same as that of the colorless to white uvite from DeKalb, New York. The Cr-content was $.001 \pm .005$ weight percent for the Newry specimens and was not recorded for the uvites.

Also, as noted previously, Chaudhry and Howie (1976) stated that $Fe^{2+} >> Fe^{3+}$ in both the green and pink tourmalines (elbaites) included in their investigation. And, Čech, Litomiský, and Novotný (1965) reported that the spectrographic analyses of green and colorless parts of a crystal they analyzed were essentially identical.

Attention is also directed to the review and analysis of Smith (1978*b*).

Blue—Faye, Manning, and Gosselin (1974) concluded that the color of blue tourmalines is "strongly influenced by the $Fe^{2+} \rightarrow Fe^{3+}$ process at $\sim 18{,}500\ cm^{-1}$" (p. 379). Dunn (1975*a*) recorded the blue elbaites of Newry, Maine, to have Fe- and Mn-contents higher than those of the pink and green elbaites from the same locality. Leckebusch (1978) concluded that although "several indicolites [of those studied] do not show a connexion between colour depth and iron distribution," (p. 63), blue is always due to Fe, either the presence of Fe^{2+} *per se* or $Fe^{2+} \rightarrow Fe^{3+}$ charge transfer.

Purple—The deep purple tourmaline from Susice, Czechoslovakia, gave spectroscopic data that led Čech, Litomiský, and Novotný (1965) to suggest that Mn is responsible for the color. However, they also directed attention to the fact that the "deep purple" elbaite from Ctidruzice contains more Fe and Mg than other elbaites they studied. Perhaps the color is due to a combination of red ($Mn^{3+?!}$) and blue (Fe . . .).

Colorless—As previously mentioned, Čech, Litomiský, and Novotný (1965) reported essentially identical spectroscopic analyses for colorless and green parts of an individual crystal. Also, although the colorless uvite from DeKalb, New York, has an Fe-content of only 0.24 weight percent, this amount is nearly the same as that for green uvite from Franklin, New Jersey (Dunn et al., 1977).

In the writer's opinion, a thorough investigation of the chemistry (etc.) of colorless and nearly colorless tourmalines is needed before several of the suggested causes of the diverse colors can be considered valid.

Black—Intense blue and intense dark green, as well as black, tourmalines may appear macroscopically black. Also, it is well known that superimposition of several colors yields black. Nonetheless, in the past, it has generally been tacitly assumed that black is attributable to Fe-content. For example, as a result of his study of schorls, Ward (1931) stated that intensity of black correlates with indices of refraction, which were then assumed to vary directly with Fe-content.

With the advent of assigning different absorption peaks to transitions (charge transfers, etc.), the cause of the black has become muddled. In 1969, Wilkins, Farrell, and Naiman assigned the 9,200 and 13,800 cm^{-1} bands to Fe^{2+}-Fe^{3+} transitions and the bands at ~20,000 cm^{-1} to the presence of Fe^{3+} in both *Y*- and *Z*-sites, and Manning (1969*a*) concluded that the black color is due to "transitions within Fe^{2+} [$^3T_2 \rightarrow {}^5E(D)$] and Fe^{3+}, and to transitions within another ion, possibly Mn^{3+}." Later, however, Manning (1969*b*) concluded that the black color is due "mainly to strong Mn^{3+} *d* - *d* bands superimposed on the low energy wing of intense charge-transfer absorption centered in the ultraviolet," and in his abstract he stated:

> Optical absorption studies demonstrate that the extinction coefficients of bands marking the transition $^5E_g \rightarrow {}^5T_{2g}$ in octahedral-Mn^{3+} are at least an order of magnitude greater in black tourmalines than in pink. Ultraviolet charge-transfer absorption sweeping into the visible-region is more prominent in black tourmalines. It is proposed that the stronger absorption in black tormalines is due to distortion of Mn^{3+} octahedra brought about by substitution of adjacent Si^{4+} ions by Fe^{3+} and Al^{3+}. Two relatively sharp bands at ~20,000 cm^{-1} (500 nm) in spectra of black tourmalines have been assigned to transitions to the $^4T_1$ and $^4T_2$ levels in tetrahedral-Fe^{3+} (p.57).

And, still later, Manning was one of the coauthors (Faye, Manning, and Gosselin, 1974) of a paper in which it was concluded that the color of black tourmalines is "strongly influenced by the $Fe^{2+} \rightarrow Fe^{3+}$ process at ~18,500 cm^{-1}, and . . . also by strong uv-centered $O^{2-} \rightarrow Fe^{3+}$ charge-transfer." In addition, they noted that "In general, the colour of a tourmaline, as determined by the intensity of the Fe^{2+} Fe^{3+} band at ~18,500 cm^{-1}, is a good indicator of the degree of oxidation of iron according to the sequence: green-brown elbaites ≃ brown dravites $<$ blue elbaites $<<$ black schorls" (p. 379)

Brown—Deer, Howie, and Zussman (1962) stated that in the schorl-dravite series intensity of color decreases from brown to light brown to colorless with a decrease in Fe-content. Dunn et al. (1977*b*) noted, similarly, that the brown saturation in uvites varies directly with Fe-content. (They also reported that the brown dravites from Yinnietharra, Western Australia, have lower Ca-contents than associated black dravites.) Black (1971) recorded Ti-enrichment in the deep olive brown core of zoned schorls from New Zealand. Manning (1969*a*) concluded that the controls by $^5T_2 \rightarrow {}^5E$, etc. that were suggested by Faye, Manning, and Nickel (1968) are insignificant so far as their effect on color. As mentioned under the subheading "Yellow," George Rossman (1984, personal communication) is of the opinion that $Fe^{2+} \rightarrow Ti^{4+}$ interaction and Fe-content are important so far as the production of both the yellow and brown colors in tourmalines.

Miscellany—In their description of liddicoatite, Dunn, Appleman, and Nelen (1977) called attention to the fact that the Na:Ca ratio is essentially the same across differently colored zones. This fact, along with the fact that liddicoatite cannot be distinguished from elbaite on the basis of color, corroborates the preceding discussions that implicitly indicate that Na:Ca ratios do not, at least directly, control color in either of these species (cf. Schreyer, Abraham, and Behr, 1975).

Smith (1978*a*) has reported optical absorption spectra for some green (Fig. 6-6), blue, black, and brown tourmalines at diverse temperatures ranging from 6 K to 800°C, upon heating in air and in hydrogen.

For what it is worth—probably very little so far as application to natural tourmalines—according to Taylor and Terrell (1967), the following ingredients led to the listed colors in the tourmalines they synthesized.

iron metal	green
iron metal + Fe_2O_3	blue
cobalt hydroxide	pink
cobalt metal	mauve
nickel metal	green
manganese metal + MnO	gray
chromium carbonate	light green
aluminum metal + V_2O_5	gray

As a result of their synthesis experiments in which they produced tourmalines with K and Ca in the *X*-sites and V, Cr, Co, Ni, Cu, and Zn in *Y*-sites, Darragh, Gaskin, and Taylor (1967) concluded that the colors are due to structural defects and the valency state of Mn.

Relations Between Color and Other Properties

Two additional generalizations relating to color warrant repeating:

1. Color of black tourmalines, as seen in thin-section, varies with optical parameters such as indices of refraction (Ward, 1931).
2. Color saturation correlates with indices of refraction for both pink and green elbaites from Newry, Maine: that is, the more intense the color, the higher the indices of refraction (Dunn, 1975*a*).

Dichroism

Dichroism in tourmaline was recorded in the early literature by, for example, Haüy, (1822) and Mohs (1824); it was described rather well by Haidinger (1845) and D'Achiardi (1897); and it was correlated with kind of tourmaline by Wülfing (1900). Dichroism in the infrared and ultraviolet regions was studied and described before 1900 by Merritt (1895) and Agafonoff (1896), respectively.

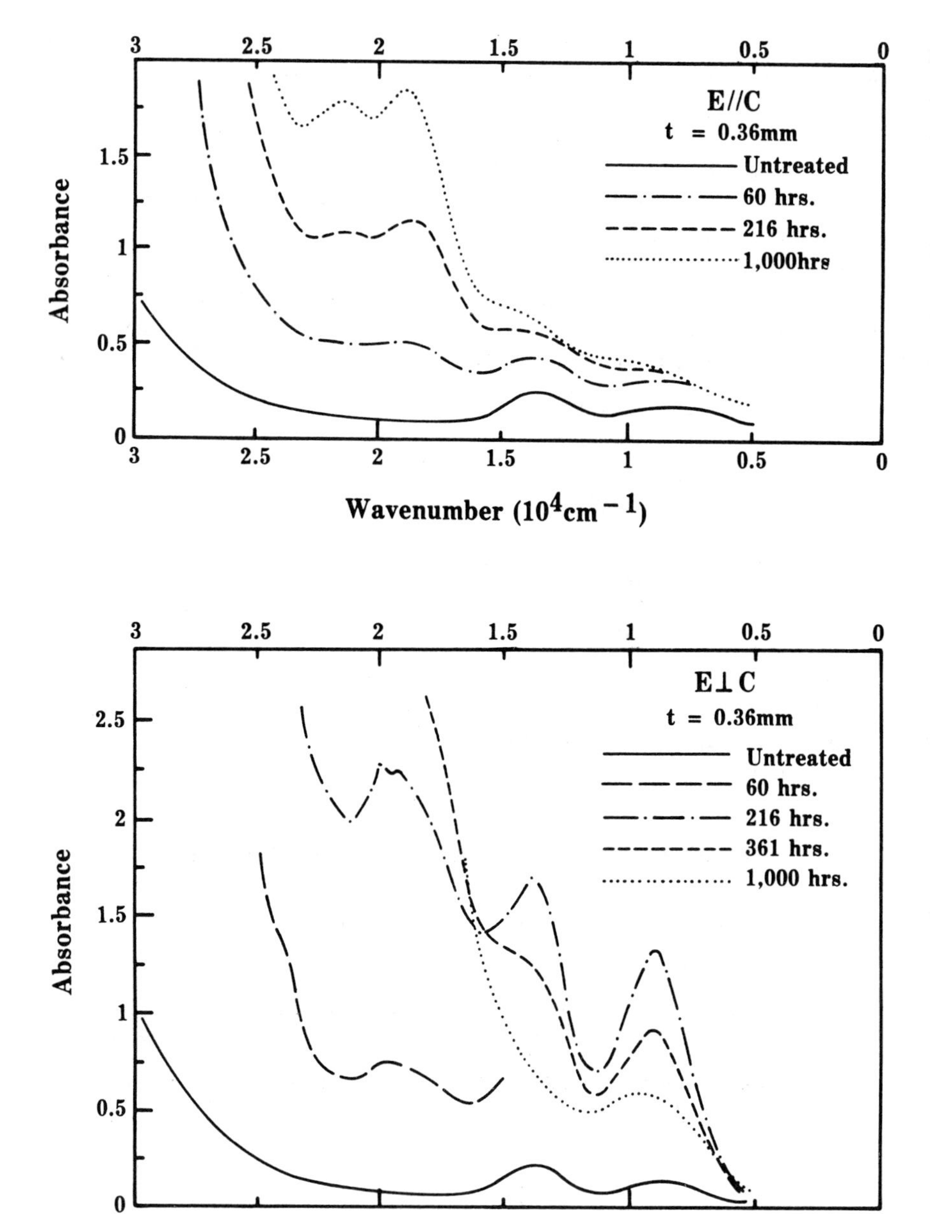

Figure 6-6
Absorption spectra for untreated and heat-treated green tourmaline (after Smith, 1977).

Dichroic colors in the visible range are given in the section dealing with optical properties. As mentioned there, dichroism is discernible macroscopically in some tourmalines, a fact that must be considered when such stones are cut as gems (see chap. 10). Most colorless and even some light pink and light green tourmalines, however, appear both macroscopically and microscopically to be nondichroic.

Except for peaks in the ~1400 nm region, which are generally assigned to OH vibrations, the absorption bands are greater for E $\perp$ *c* than for E $\parallel$ *c* for nearly all spectra.

The maximum dichroism of pink and blue tourmalines decreases and shifts to longer wavelengths with increasing temperature; for pink and blue tourmalines, the changes appear to be irreversible above 425°C and 275°C, respectively (Vedeneeva and Grum-Grzhimailo, 1934).

Slawson and Thibault (1939) used tourmaline, not otherwise defined, from Swakopmund, Namibia, (formerly Southwest Africa) to show how quantitative measurements of dichrosim could be made with a Glan spectrophotometer. Results for that tourmaline (apparently an elbaite) are given in Table 6-3.

Dichroism obviously derives from the same phenomena as color; see, for example, Manning (1969) and Black (1971). Faye, Manning, and Nickel (1968) recorded polarized and unpolarized optical absorption spectra in the 7,000-30,000 cm^{-1} (1430-330 nm) range for blue and green tourmalines, and concluded that the visible pleochroism of the specimens "is due to a strongly pleochroic band [between 13,700 and 13,900 cm^{-1}] that is . . . due to the electronic interaction of adjacent Fe^{2+} and Fe^{3+} ions so located in the crystal lattice that certain of their *d*-orbital lobes overlap" (p. 1174).

Wilkins, Farrell, and Naiman (1969) have explained tourmaline dichroism

> by the combination of two mechanisms. The d^n states mixing with the charge transfer states provide the basic mechanism for making the optical transitions allowed, while the trigonal component of the local point symmetry in the crystal then provides the mechanism for the dichroism . . . [that is,] the selection rules of the trigonal field shows that the E $\parallel$ *c* terms are weak, while the E $\perp$ *c* terms are strong (p.55).

Townsend (1970), as a result of the most extensive study of the dichroism of tourmalines, concluded that the dichroism "arises partially from charge-transfer transitions within next-nearest neighbor cations through shared octahedral edges in the 001 layer" (p. 2481), and can be promoted by any combination of transition-metal ions.

Most subsequently recorded conclusions (e.g., Faye, Manning, and Gosselin, 1974) are in line with Townsend's suggestions. In addition, however, Faye et al. (1974) concluded that "uv-centered $O^{2-} \rightarrow Fe^{2+}$ and $O^{2-} \rightarrow Fe^{3+}$ charge-transfer processes [also] influence the dichroism in the visible and

Table 6-3
Dichroism as Measured with a Glan Spectrophotometer (from Slawson and Thibault, 1939)

Wavelength $\mu\mu$	*Percent Transmission* ω *ray*	*A.D.*	*Wavelength* $\mu\mu$	*Percent Transmission* ω *ray*	*A.D.*
427.5	2.71	.04	538	10.24	.10
445	3.60	.05	545	10.77	.09
465	5.20	.05	567	12.01	.10
480	6.77	.07	585	11.59	.10
489	7.85	.06	600	9.66	.08
495	8.76	.10	615	7.88	.07
502	9.76	.09	630	6.45	.09
510	10.02	.06	650	5.24	.06
520	9.60	.08	670	4.45	.07
530	9.64	.08	685	3.99	.06

Notes: Percent transmission—ω ray is the itensity of ω-ray as compared to the intensity of ϵ-ray taken as unity. A.D. is average deviation.

near-infrared regions," and these "processes are intensifed by the replacement of OH^- by O^{2-}, for example when Ti^{4+} is substituted into octahedral sites . . ." (p.375).

Streak

The streak for most tourmalines is colorless/white or a pastel of the body color of the specimen; for example, it is light green for dark green elbaite. The dark gray or brown streak of ferridravite and the yellow-brown streak of buergerite are exceptions.

Effect of Temperature and Irradiation on Color

The effects of heat and irradiation are treated more or less separately in this section (see also pertinent remarks in chap. 10 on gemstones). Here again there are apparently conflicting data.

Temperature—In 1830, in reporting his "blowpipe tests" on red, green, and blue tourmalines from New England, Shepard noted that pink tourmaline loses color at red heat, green tourmaline "pales" at red heat, and blue tourmaline retains its color at "low redness" but at higher temperatures turns yellowish gray. This information was verified by, for example, Karnojitsky (1890) and elaborated upon so far as its application to gem treatment by

Bauer (1896), who noted, among other things, that whereas heating to about 700°C renders some green stones unsuitable for use as gemstones it lightens others (e.g., the dark green stones from Namibia, formerly Southwest Africa) to a desired emerald-green color (cf. Kálmán, 1941). Later investigations indicate that slow heating will produce some of the desired results at less elevated temperatures, especially the changing of brownish red stones to pink; see, for example, Sinkankas (1976) and Nassau (1984).

Nonetheless, as should be expected, different tourmalines with similar original colors may react quite differently. For example, in their attempts to lessen the color saturation of the red dravite from Kenya, Dunn, Arem, and Saul (1975) saw no change as a result of heating for eight hours at each of three temperatures (200°, 400°, and 600°C).

Also, heating in different environments may yield different results. Wilkins, Farrell, and Naiman (1969) found that blue and green tourmalines exhibit no visible color change "When an oxidized thick section . . . was reduced in hydrogen at 800°C for 8 hr.," but that the same tourmalines may be given a reddish brown hue, albeit strictly a surface effect, when oxidized. (Actually, physical removal of the surface coating of "oxy-tourmaline" restores the original appearance.) Also, Smith (1978*a*) reported that when heated in air some green tourmalines turn reddish brown, whereas when heated in hydrogen they first become colorless and then, upon additional heating, blue. He proposed that the first effect, resulting from the heating in hydrogen, is caused by reduction of Fe^{3+} (in both *Y*- and *Z*-sites) and that the further heating causes oxidation ($Fe^{2+} \rightarrow Fe^{3+}$). The blue products give absorption bands near 19,100 and 21,600 cm^{-1}, which he attributed to Fe^{3+}-Fe^{3+} pairs.

Somewhat surprisingly, Warner (1935) found that powders change more slowly and less completely than crystals.

Newberry and Lupton (1918) heated three tourmalines, along with several other minerals, to see the effect on their luminescence as well as on their color. They found that both a dark green tourmaline from Cornwall and a pink tourmaline from Canada turned white and opaque, that subsequent exposure to radiation produced no additional changes, and that cathode rays caused a "very fine orange glow" on the heated pink tourmaline but no noticeable luminescence on its unheated precursor. Bershov et al. (1969) reported that both the color and the fluorescence of pink elbaite are lost upon heating "above 400°C," but upon subsequent irradiation by X-rays or γ-rays both are restored.

Vargas and Tupinamba (1962) reported that the increased color saturation (green in the specimens tested) that is irradiation-induced in the laboratory can be reduced by heating just as natural color saturation can.

As previously mentioned, Vendeneeva and Grum-Grzhimailo (1934) found that with increasing temperature, absorption bands shift irreversibly to longer wavelengths for both pink (above 425°C) and blue (above 275°C) specimens. Later, they (1948) and Slivko (1955*b*) attributed the weakening of color saturation upon heating rose-colored tourmaline to a reduction of the number of color-centers.

Grum-Grzhimailo and Klimusheva (1960) investigated the effects of reduced temperature (103 K) on broad bands in the absorption spectra and found that those in the near-infrared become slightly higher while those in the visible (blue-green range) are displaced slightly toward shorter wavelengths.

Irradiation—Contradictory results notwithstanding, the general rule for most tourmalines appcars to be that irradiation causes effects opposite those caused by heating, and, in some cases, irradiation will counteract previously imposed effects of heating.

The following coverage is, for the most part, in chronological order according to date of cited report.

Early observations were recorded by Miethe (1906), Doelter (1909), and Simon (1908). It was found, for example, that exposure to 10 mg of RaBr for 32 days turned brownish red tourmaline to a deep red color (Simon, 1908) whereas other crystals were not, or only slightly, changed (Doelter, 1909).

Newberry and Lupton (1918) included three tourmalines in their study of the effects of α, β, and γ radiation (from radium) on minerals. They reported that (1) a dark green tourmaline from Cornwall (England) exhibited no visible effects when exposed to radium either before or after heating; (2) a pink tourmaline from Canada became slightly darkened when exposed to radium before being heated; (3) cathode rays caused a "very fine orange glow" (luminescence) from heated pink tourmaline; and (4) a colorless tourmaline from Elba became pink after one day's exposure to 50 mg of radium, and the intensity of color increased during the following four days [of exposure?] (but not thereafter) until it resembled "rubellite."

Lind and Bardwell (1923) exposed a pink and a green tourmaline to radium salts and saw no macroscopically discernible changes. Stamm (1926) exposed diverse tourmalines to 125 mg of RaBr for 25 days and studied the effects on their absorption, infrared, and ultraviolet spectra; he reported that by using 125 mg RaBr for 25 days: Brazilian pink, Madagascar brown, and Brazilian blue-green tourmalines were essentially unchanged whereas Brazilian brown-green tourmaline became darker and California colorless tourmaline became pink (cf. Rinne, 1927). Newcomet (1941) stated that exposure of green tourmaline from Brazil to radium made it become "slightly darker." Pough and Rogers (1947) reported that after exposing some pink and green tourmalines to X-rays, both the pink and dark green specimens took on a dark rose purple color in 10 and 8½ hours, respectively, whereas a light green stone became yellowed within 1 hour and 20 minutes.

Also in an investigation of the possible use of tourmaline in thermal neutron dosimeters, Vargas and Tupinamba (1962) found that the green tourmaline they studied underwent a marked increase in color saturation as the result of neutron irradiation, that the intensity increased with time or irradiation ("a linear relationship exists between the neutron dose and the logarithm of the change of the absorption coefficient, between 3.0×10^{13} and 3.6×10^{14} nvt for a crystal plate [cut $\perp$ *c*] 1.00 mm thick" p. 67), and that subsequent heating diminished the color saturation. They also interpreted their data to support the conclusion that the 700 nm and 1200 nm absorption bands are due to color centers "resulting from the $^{10}B(n,\alpha)^{7}Li$ reaction".

A particularly remarkable change was recorded by Crowningshield (1974), who noted that exposure to gamma radiation of some faceted pale green tourmalines from Maine caused one of them to become tri-colored with one end pink, the other blue, and the middle portion green. George Rossman (1984, personal communication) notes that he has witnessed the same change as a consequence of irradiating green tourmalines from Afghanistan.

Nassau (1975) irradiated more than 500 specimens in a cobalt-60 γ-ray cell for 14 to 72 hours at room temperature. The changes recorded can be categorized as follows (with the numbers in parentheses giving the number of pertinent observations):

I. Development or intensification of red
   a) colorless or pale colors → pink or red (165)
   b) pale green → gray (18)
   c) yellow → peach (3)
   d) blue → purple (7)
II. Development of yellow
   a) pink → yellow or orange (55)
   b) pale green → yellow-green (15)
III. No significant change (diverse colors) (279)

In addition, Nassau noted that "Considerable variation occurred even within individual specimens. For example, one green pencil from Topsham, Maine, turned into a green-pink bicolor, while another . . . turned gray . . . . a green-pink bicolor from San Diego, California, turned to a uniform pink, and so on" (p. 711). Furthermore, when he checked the stability of the colors formed as a consequence of irradiation, by heating them to 260°C and 400°C for up to 24 hours and by exposing them to the sun for five weeks, Nassau found a wide range of results. For example, some of the deep pink and red stones (Ia on the preceding list) retained their intensified color whereas others returned to their original colors after only a few hours at 260°C or after the five-week exposure to sunlight. The results for one particularly interesting group of pink elbaites are shown on Figure 6-7.

Dunn, Arem, and Saul (1975) reported that no changes occurred when the red dravite from Kenya was irradiated with CuKα X-rays and ultraviolet

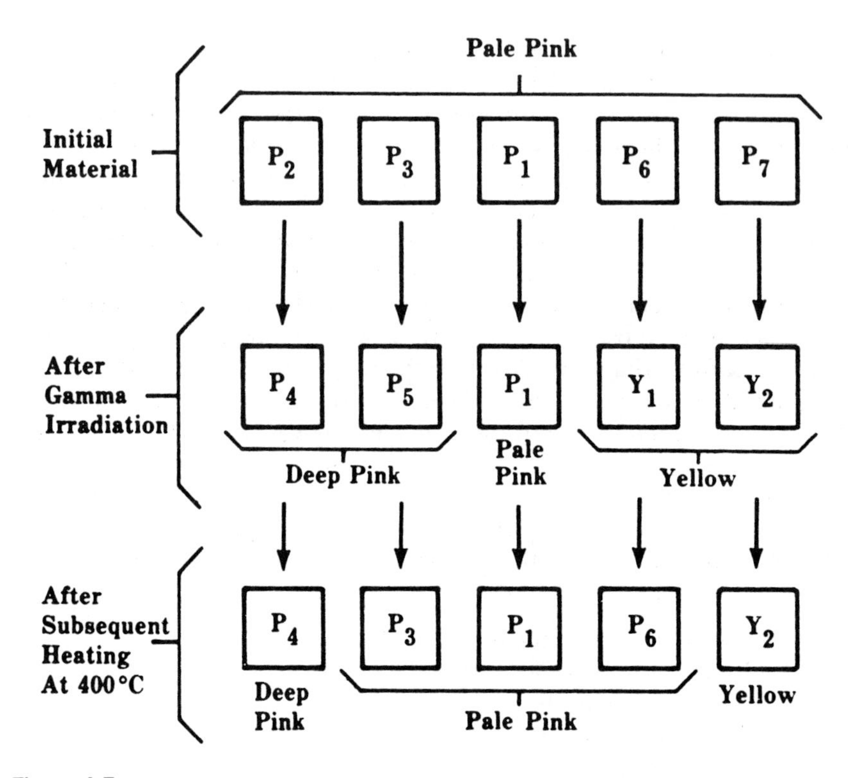

Figure 6-7
"Color changes observed in 85 pale pink elbaite tourmalines on irradiation with 10 megarads gamma rays and subsequent heating for 24 hours at 4000°C" (after Nassau, 1975).

radiation. Perhaps this specimen had already undergone natural irradiation; cf. those of category Ia on the list that retained their intensified red color. Or, considering the fact that this is an uncommon tourmaline, the red color may depend upon the Fe^{3+} rather than any Mn^{3+} content.

Voskresenskaya et al. (1979) investigated the Mössbauer spectra of irradiated and heated Zn-bearing elbaites from central Asian pegmatites and concluded that: irradiation by small doses of γ-rays causes part of the Fe^{3+} to be reduced to Fe^{2+}; larger doses cause increases in Fe^{3+}; and raising of temperature causes $Fe^{2+} \rightarrow Fe^{3+}$ charge transfers via both natural and irradiation-induced defects in the vicinity of the Fe ions.

It also is worthy of note that Rossman (1982) stated

The colors of many attractive minerals in gem pegmatites appear to be due, in part, to a history of exposure to ionizing radiation. Tourmaline is pink from Mn^{2+}, and much deeper red from Mn^{3+}. Radiation in the laboratory can transform Mn^{2+} to Mn^{3+} suggesting the possibility that the same process happens in nature (p. 189).

Luster

Tourmaline is frequently characterized as vitreous to resinous on prism faces and dull on pyramid and pedion terminations. This is an oversimplification in that luster is strongly influenced by the physical state of the surface—smooth, rough, growth irregularities, etch pits, etc.—and by alteration. In fact, many tourmaline crystals have terminal faces that are just as vitreous as their prism faces and some crystals have different terminal faces with different lusters (e.g., $32\bar{5}1$—dull and $10\bar{1}1$—vitreous).

Beesley (1975) noted that tourmaline crystals from Newry, Maine, have an "exceptional brilliance, possibly due to the presence of cesium." In addition: most buergerite has a high luster on all faces; ferridravite has been characterized as "resinous, more or less splendent" (Walenta and Dunn, 1979); and, as previously mentioned, schorls and elbaites with equilateral-triangular cross-sections range from vitreous to mirrorlike.

Chatoyancy

Some fibrous tourmaline exhibits chatoyancy: that of good quality and color can be cut *en cabochon* to make fine cat's-eyes. The chatoyancy depends upon the presence of hollow channels or fiberlike inclusions of other minerals and/or fluids that are typically aligned parallel or subparallel to *c*.

It is widely thought that the hollow tube type predominates. However, Eppler (1958) reported chatoyancy caused by inclusions of unidentified fiberlike crystals; Mitchell (1967) recorded a cat's-eye that is an intimate mixture of tourmaline and diopside; and Graziani, Gübelin, and Lucchesi (1982) found some of the channels ("growth tubes") to contain cookeite, prosopite, pyrrhotite, or tourmaline, itself. Furthermore, Bhaskara-Rao and de Assis (1968) stated that the chatoyancy of some elbaites from northeastern Brazil can be attributed to channel leaching.

In addition, some tourmaline crystals appear to exhibit chatoyancy because of striae on, for example, their prism faces. If the striae are left as the base, this quasi-chatoyancy, including a cat's-eye effect, may be seen on cabochons cut from such crystals.

Luminescence

Some tourmaline fluoresces under short-wave ultraviolet radiation; some does not. For tourmaline that does fluoresce, the fluorescent colors differ from specimen to specimen and the intensity of the fluorescent color has

Table 6-4
Tourmaline Fluorescence ("Short Wave" Ultraviolet Radiation)

Specimen	*Fluorescence*	*Reference*
dravite and uvite ("low in iron content"—general)	mustard yellow	Dunn, 1977*b*
elbaite (pink to purplish red, Newry, Maine)	chalky to "very strong" blue	Beesley, 1975
elbaite (pink, Brazil; pink, Mozambique)	none to faint chalky blue	Beesley, 1975
elbaite (colorless, Nuristan, Afghanistan)	bright vivid violet	Dunn, 1974
elbaite (pink, Nuristan, Afghanistan, and Newry, Maine)	weak violet	Dunn, 1974[a] and 1975*a*
elbaite (green, Nuristan, Afghanistan, and Newry, Maine)	chalky blue to weak violet	Dunn, 1974[a] and 1975*a*

[a]Dunn has reported that these specimens exhibit a fluorescence the depth of color of which varies inversely with the depth of color under white light, and that crystals with gradational zoning fluoresce more vividly than those with sharper zoning.

been recorded as ranging from faint to very strong (see Table 6-4). Fluorescence may be added or modified when tourmaline is heat-treated; for example, Beesley (1975) found that some gemstones whose colors had been reduced by heat treatment had their fluorescent color intensified.

The electron paramagnetic resonance studies of Bershov et al. (1969), which led to their concluding that electron-hole centers (i.e., oxygen vacancies with trapped electrons) account for the fluorescence as well as for the color of the pink elbaite they studied, have been commented upon in the section "Causes of Colors." In addition, they found that both the pink color and the fluorescence were lost upon heating to "above 400°C," but that upon subsequent irradiation by either X-rays or γ-rays, both were restored.

Przibam (1929) found tourmaline not to be triboluminescent as a consequence of pressure ("piezochromatic" in his words). However, Krotova and Karasev (1953) found tourmaline, otherwise not identified, to emit light when cleaved in a vacuum; they referred to the phenomenon as "luminescence of mechanical breaking."

With one exception, no tourmaline has been recorded to fluoresce under long-wave ultraviolet, CuKα or MoKα X-rays, β (γ - ϕ), or cathode rays, or to phosphoresce (e.g., Barsanov and Sheveleva, 1952; Beesley, 1975; Dunn, 1975*b*). The exception is some pink elbaite recorded by Bauer (1896) to fluoresce, giving a weak violet glow, under X-rays and also to phosphoresce after the same exposure.

Lind and Bardwell (1923) reported that no thermoluminescence is shown by irradiated green tourmaline during heating to 300°C. Precambrian tourmaline from Chakai, India, was checked for thermoluminescence between 0° and 400°C (heating rate 80°C per minute) and also found to exhibit neither natural thermoluminescence nor thermoluminescence after irradiation of 3.3×10^5R of ^{60}Co (Jain and Mitra, 1977). It will be interesting to check essentially Fe-free tourmalines.

OPTICAL PROPERTIES

Befitting its crystal structure, most tourmaline is uniaxial negative. Like many uniaxial minerals, however, some tourmalines have been found to be biaxial, with optical angles (2V) up to nearly 25°. All but colorless varieties are pleochroic with absorption $\omega > \epsilon$, and both the pleochroism and the differential absorption are macroscopically discernible in some crystals. Recorded indices of refraction for natural tourmalines range from 1.619 to 1.82 for ω and 1.603 to 1.772 for ϵ, and birefringence (Δ) ranges from 0.006 to 0.080 (see Table 6-6). Dispersion is weak, typically ~0.017 for tourmaline gemstones. Optical properties have been correlated with chemical composition and, indirectly, with other physical properties since the late 1800s.

Uniaxial Character

In the late 1800s, Madelung (1883) described tourmaline with 2E ≃ 10°, and Karnojitzky (1894) described and illustrated tourmaline crystals with different domains having 2V values of 0-7°, 10°-20°, and up to 23°20′. Biaxial tourmalines have also been recorded by, for example, DuParc, Wunder, and Sabot (1910) and Kunitz (1929). Takano and Takano (1959) discussed these "optical anomalies" in tourmaline and attributed them to different domains within individual crystals and grains.

More recently, Foord and Mills (1978), again noting that some tourmaline exhibits anomalous biaxiality, expressed by birefringence perpendicular to *c* and 2V values up to 20°, investigated several schorl-elbaite specimens from San Diego County (California), Oxford County (Maine), and Minas Gerais (Brazil), and reported: biaxiality occurs in deformed (e.g., bent) crystals and in strained and cracked zones, and the azimuth of the optic planes is typically irregular (Fig. 6-8). They attributed the biaxiality to abrupt differences in indices of refraction, thermal expansion coefficients, and compressibility constants controlled by compositional differences, and concluded that the biaxiality is an expression of such things as growth imperfections (e.g., screw dislocations), thermal shocks, and/or pressure changes.

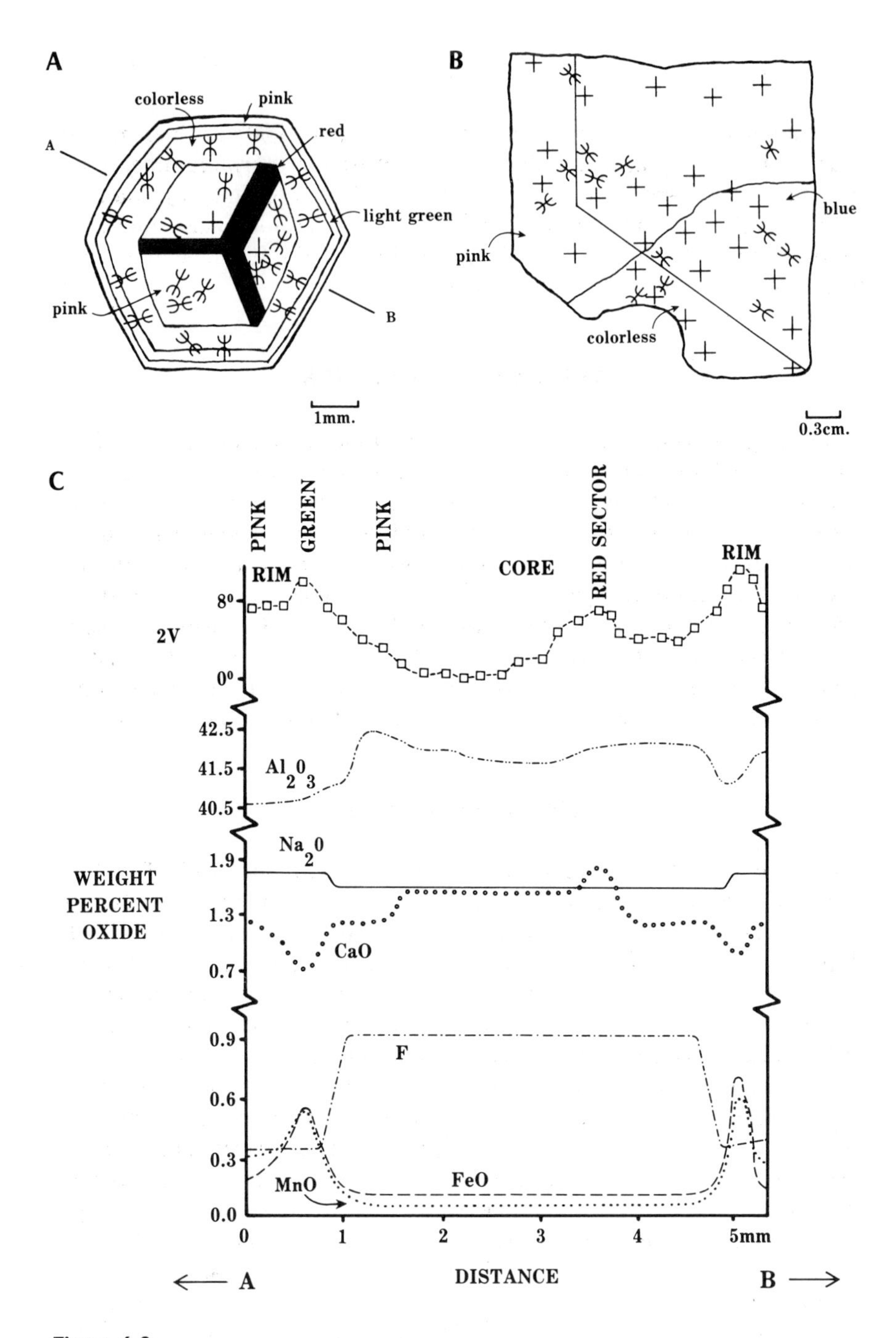

Figure 6-8
Biaxiality in tourmaline crystals. ⋺⋲ indicates azimuth of optic planes on both *A* (section perpendicular to *c*) and *B* (section parallel to *c*); *C*, chemical differences along traverse shown on *A*. (After Foord and Mills, 1978.)

Previously, Foord (1976) had reported that the schorls tend to be uniaxial whereas the elbaites are nearly all biaxial in the pegmatites of the Mesa Grande district of San Diego County, California.

The "Kerez effect" (Mitchell, 1976) whereby some cut and faceted tourmalines exhibit double, triple, or even quadruple index of refraction shadows (i.e., up to eight shadows) on refractometers was first reported by Mitchell in 1967. The phenomenon was subsequently investigated by Schiffmann (1972 and 1975) and by Bank and Berdesinski (1975) who found that careful recutting of a few such stones indicates that the multiple readings are given only by the outer zones of cut and polished stones that apparently were overheated and/or underwent thermal shock during polishing. In addition, they found that the lowest of the multiple shadow values were the correct indices for the original stones.

Absorption and Pleochroism

Absorption, with $\omega > \epsilon$, ranges from essentially no difference for colorless dravite and colorless elbaite to the extreme where ϵ exhibits little absorption compared to ω, which is virtually opaque because absorption is so great—for example, in iron-rich ferridravite (Walenta and Dunn, 1979) and some schorl (Ward, 1931). Common pleochroic colors are given in Table 6-5.

Three tourmaline specimens, other than colorless tourmalines, have been reported as nonpleochroic: a yellow-green V-rich dravite (Bassett, 1956) and two elbaites, one pink and one green (Němec, 1951).

Biaxial tourmalines may be trichroic. Karnojitsky (1894) gave, for example, yellow-mud brown, light yellow-mud brown, and bright green for three directions in the previously mentioned tourmaline with a $23°20'$ optic angle.

Pleochroic halos in tourmaline, around inclusions of, for example, zircon have been recorded (e.g., Michel-Lévy, 1889).

Refractive Indices and Birefringence

The previously mentioned extreme values recorded for natural tourmaline species, though of little significance, are given in Table 6-6. "Common values," as noted in the footnote to the table, can be logically given for elbaites and schorls because most specimens of these species tend to have about the same values; "common values" are not given for dravite because of the relatively large differences among "common" specimens (see, for example, Barsanov and Yakovleva, 1964). As can be seen in Figure 6-9: the recorded values for both refractive indices of dravite/uvite and elbaite/liddicoatite have about the same ranges; the low values for schorl are in the same range;

Table 6-5
Typical Pleochroic Colors for Tourmaline Species

Specimen	ω	ϵ
Buergerite	yellow-brown	very pale yellow
Dravite	yellow	colorless
	orange-yellow	pale yellow
	dark green	olive green
	bluish green	yellowish green
	medium to dark brown	yellowish to light brown
Elbaite	medium pink	light pink or colorless
	green	yellow to olive green
	blue-green	light green to purplish
	blue	colorless to pink to purple
Ferridravite	dark brown to dark olive green	light olive green to light brown
Liddicoatite (like elbaite but type specimen is	dark brown	light brown)
Schorl	blue to greenish blue	yellow, yellow-brown, pale violet, or colorless
	green-brown	rose-yellow
	dark brown	yellow, light brown, or yellowish blue-green
Uvite (like dravite)		
Chromdravite	dark green	yellow-green

values for ferridravite and buergerite are high and apparently definitive; and values for chromian dravites, also high, can be used to distinguish them from, for example, green elbaite. Also, as can be seen on Figure 6-9, birefringence values may be used in much the same way.

A decrease in indices of refraction with increase in wavelength of transmitted light has been reported for dravite from Gouverneur, New York (Rath and Puchelt, 1959); for nine (3 red, 1 yellow, 1 brownish, 2 green, and 2 blue) elbaites from Mozambique, Elba, Brazil, and Namibia (El-Hinnawi and Hofmann, 1966); and for both green and blue elbaite from Brazil (Pinet, 1971); see Table 6-7. The birefringence values were also found to decrease for all but three increases in λ.

As might be expected, different portions of zoned crystals (e.g., Němec, 1955) and detrital tourmalines versus their authigenic overgrowths commonly have different indices of refraction (e.g., Erofeyev, 1871, and Burgelya,

Table 6-6
Refractive Indices and Birefrigence

Species	ω *Range Recorded*	ϵ *Range Recorded*	Δ *Range Recorded*
Buergerite	1.735	1.655-1.670	0.065-0.080
Dravite	1.627-1.675	1.604-1.643	0.016-0.032
Elbaite	1.619-1.655	1.603-1.634	0.013-0.024
Ferridravite	1.80-1.82	1.743	0.057
Liddicoatite (type)	1.637(Na)	1.621(Na)	0.016
Schorl	1.638-1.698	1.620-1.675	0.016-0.046
Uvite	1.638-1.660	1.619-1.639	0.018-0.021
Chromdravite	1.778	1.772	0.006

Note: Common values (see text) are: elbaite $\omega \sim 1.641$; $\epsilon \sim 1.622$; $\Delta \sim 0.019$
schorl $\omega \sim 1.665$; $\epsilon \sim 1.635$; $\Delta \sim 0.030$

1961). The latter relationship serves, in fact, as the basis for recognizing some authigenic tourmaline.

Dispersion

Compared to other relatively commonly used gemstones, tourmaline has dispersion that is comparable to all but diamond and zircon—see Table 6-8.

Correlations Between Optical and Other Physical Properties

Several mineralogists have correlated optical properties with physical properties (e.g., Erofeyev, 1871; Wülfing, 1901; Bank, 1965; Cho, 1974; and Beesley, 1975). All of the correlations are, of course, indirect in that both the optical properties and the other physical properties are based on chemical and/or structural differences.

Four frequently repeated correlations are: (1) refractive indices may be correlated with color (Erofeyev, 1871); (2) refractive indices tend to increase with saturation of body color (Beesley, 1975); (3) refractive indices and birefringence generally vary directly with specific gravity (e.g., Wülfing, 1901; and Kunitz, 1929); and (4) optical properties vary with cell dimensions (Bank, 1965).

On the other hand, Ward (1931), after studying 174 black tourmalines, concluded that (a) there appears to be no close correlation between density

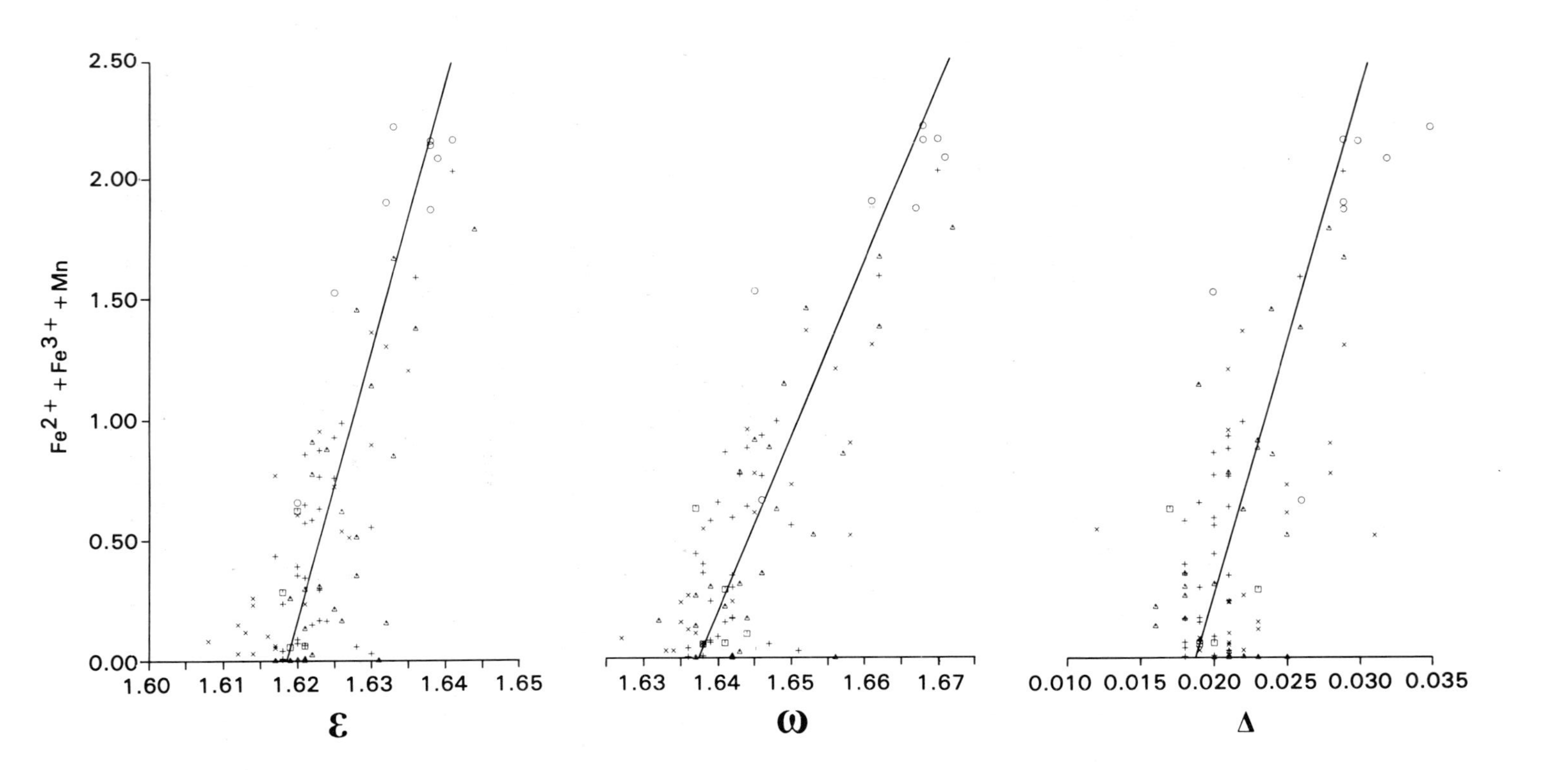

Figure 6-9
Relations between Σ Fe + Mn content and, left to right, ϵ refractive index, ω refractive index, and Δ (birefringence). Symbols same as on Figure 4-2.

Table 6-7
Relationship Between ω and ϵ and λ

	Dravite[a]		*Elbaite*[b]			
			blue		*green*	
λ	ω	ϵ	ω	ϵ	ω	ϵ
420			1.6604	1.6371	1.6567	1.6362
440			1.6577	1.6347	1.6542	1.6343
460			1.6555	1.6325	1.6519	1.6320
480		1.6222	1.6531	1.6307	1.6500	1.6303
500			1.6513	1.6289	1.6482	1.6288
520			1.6495	1.6272	1.6468	1.6274
535	1.6412	1.6199				
540			1.6481	1.6263	1.6453	1.6263
560			1.6467	1.6250	1.6442	1.6251
580			1.6456	1.6240	1.6431	1.6242
587.6	1.6377	1.6173				
589.3	1.6375	1.6171				
600			1.6445	1.6230	1.6421	1.6231
620			1.6434	1.6222	1.6410	1.6220
623.4	1.6338	1.6145				
640			1.6427	1.6213	1.6404	1.6215

[a]from Rath and Puchelt, 1959
[b]from Pinet, 1971

and either index of refraction or birefringence; and (b) although indices of refraction appear to increase, in general, with increases in total iron content, they do not exhibit a similar relation to increases in either Fe^{2+} or Fe^{3+} alone.

Correlations Between Optical Properties and Chemical Composition

In presenting the optical properties of tourmalines from Elba, D'Achiardi (1896) implied a relation between them and their chemical compositions as expressed by color. Wülfing (1901) gave graphs showing apparent relations between birefringence and Fe-content. Schaller (1913) graphed a general correlation of indices of refraction and Al_2O_3-content. In his classical paper on tourmaline chemistry, Kunitz (1929) showed direct correlation between optical and other physical properties and chemical composition. Several investigators (e.g., Quensel, 1939; Slivko, 1955*b*; Bank, 1965; El-Hinnawi and Hofmann, 1966; Cho, 1974; and Dunn, 1977*b*) have studied, and in some cases extended, such correlations. (See Figures 7-1 and 7-2.)

Table 6-8
Dispersion (6870A/4308A) of Some of the Relatively Popular Gemstones

Diamond	0.044
Zircon	0.039
Garnet	0.022-0.028
Corundum	0.018
TOURMALINE	0.017
Topaz	0.014
Beryl	0.0135
Quartz	0.013

Note: Each value given is the mean difference in the visible spectrum; for zircon, corundum, tourmaline, and beryl, it is for the ordinary (ω) ray; for topaz, it is for the intermediate ray (β).

Miscellany

As the result of light absorption measurements on a green elbaite from Minas Gerais, Brazil, Wasastjerna (1922) tabulated data that show that in changing wave lengths from 486 $\mu\mu$ to 656 $\mu\mu$:

η_ω changes regularly from 1.6520 to 1.6450 (cf. Table 6-7);
η_ϵ changes regularly from 1.6317 to 1.6247 (cf. Table 6-7);
β_ω and μ_ω vary directly with wave length;
α_ω and η_ω vary indirectly with wave length; and
α_ϵ, β_ϵ, η_ϵ, and μ_ϵ change irregularly with wave length;

where α = percentage of light intensity absorbed by 1-mm-thick plate; β = percentage of light intensity that passes through 1-mm-thick plate; η = coefficient of absorption; and $\mu = \eta^{1/2}$.

Optical properties have also been correlated with genesis (e.g., McCurry, 1971*b*). Strictly speaking, however, this is merely correlation of chemical composition as reflected by optical parameters.

Chapter 7

Physical Properties

Density, hardness, elasticity, magnetic properties, thermal properties, radioactivity, and electrical properties are treated in this chapter. For additional information about these and other physical properties (e.g., electrooptic coefficients) see the appropriate sections of the American Institute of Physics Handbook (1972), the applicable Landolt-Börnstein tables (1962-   ), and Nye's (1957) tabulations of both the tensors that represent physical properties of crystals and the matrices for equilibrium properties for materials of the *3m* crystal class.

DENSITY

In his classical compendium of specific gravities for minerals, Brisson (1787) included values for 12 specimens identified as schorl (actually, one was apparently axinite, another epidote, and still another possibly staurolite rather than schorl). Most of the described specimens were from the cabinet of Romé de l'Isle. Among the recorded values are 3.1555 for a green tourmaline from Ceylon and 3.3548 for a yellow tourmaline from Tyrol.

Graphic relations between density and Fe-content and between density and birefrigence were reported by Wülfing (1900), and subsequently modified and/or extended by several observers—for example: Schaller (1913) plotted specific gravity versus Al_2O_3-content; Bank (1965) graphed density:index of refraction relations; and Beesley (1975) indicated how density correlates with body color for elbaites. See Figures 7-1 and 7-2.

Actually, recorded specific gravity values for tourmaline species range greatly. Representative values are given in Table 7-1. With a few possible exceptions, the ranges for elbaites and schorls are small and the density does not appear to vary directly with, for example, $FeO + Fe_2O_3 + MnO + TiO_2$-

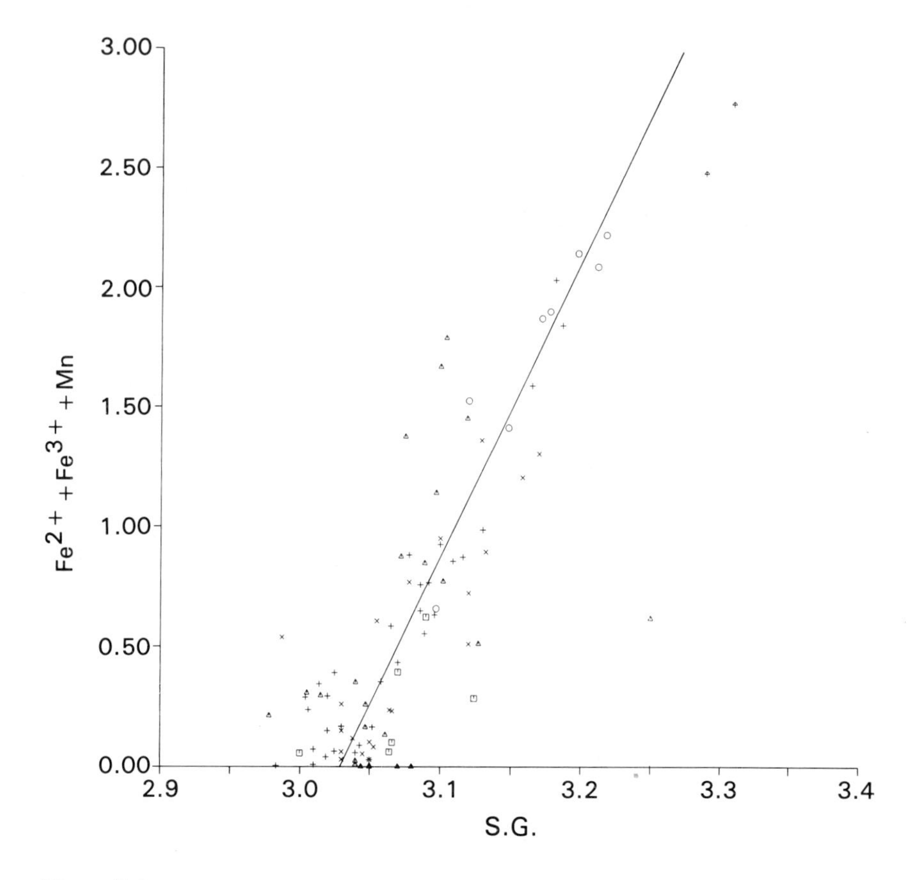

Figure 7-1
Interrelations between specific gravity (S.G.) and ΣFe + Mn content. Symbols same as on Figure 4-2.

content. On the other hand, the range of values for dravite is relatively great and its density does tend to vary directly with FeO + . . .-content (see Barsanov and Yakovleva, 1964, 1965, 1966).

In any case, it must be kept in mind that abundant fluid inclusions may effect density measurements of some tourmalines to the extent that their recorded specific gravity values are quite misleading (Slivko, 1965).

HARDNESS

Tourmaline has been reported to have a Mohs hardness ranging up to 8¾ to 9½ (Breithaupt, 1847). The generally accepted values, however, range from 7 to 7½.

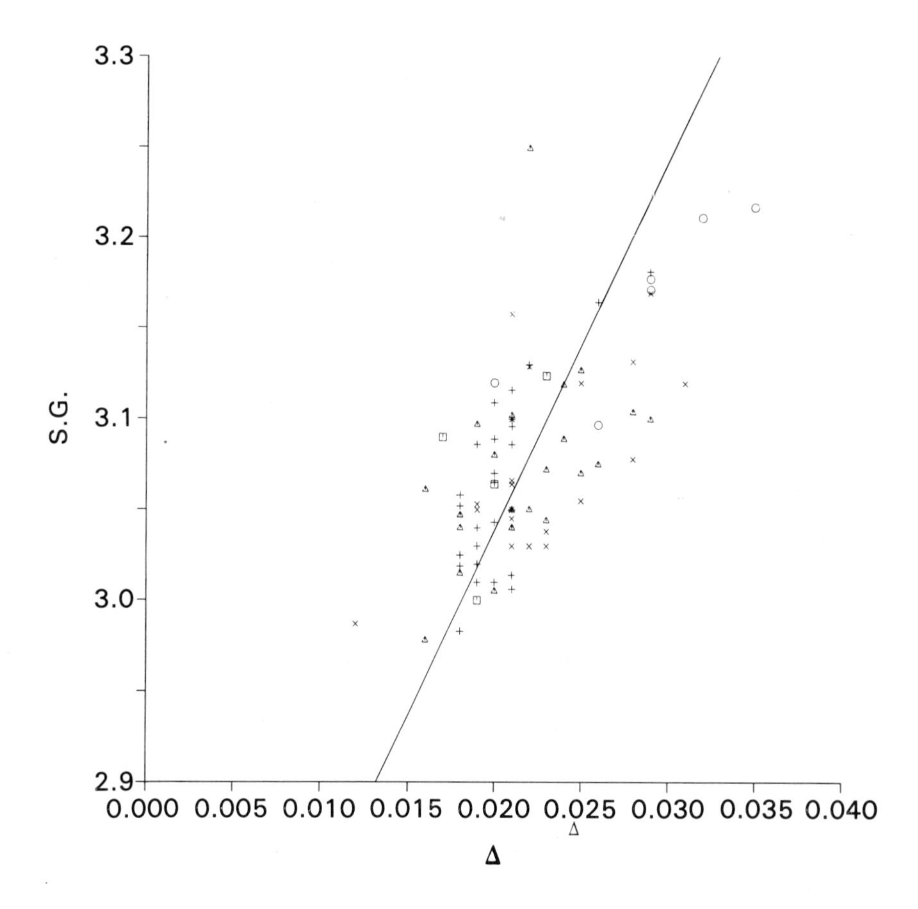

Figure 7-2
Interrelations between specific gravity (S.G.) and birefringence (Δ). Symbols same as on Figure 4-2.

Microhardness values as determined by different investigators range rather widely. For example, Holmquist (1920) gave abrasion hardness values [on (0001)] ranging from 470 to 745 for different specimens, and Young and Millman (1964) gave Vickers hardness (HV in kg/mm^2 with a 100 g load) ranging from 1187 to 1378 (0001) and from 1343 to 1478 (10$\bar{1}$0). These values, according to their mode of measurement, compare to Mohs values as follows:

Vickers 642-933 = Mohs 6
Vickers 1266-1561 = Mohs 7
Vickers 1478-2012 = Mohs 8

Povarennykh and Lebedeva (1970), however, have given the Vickers hardness value of 1133 kg/mm^2—the average of 10 tests, also using a 100 g load, on the

Table 7-1
Specific Gravity Values for Tourmaline Group Minerals

Species	*~Average*[a]	*Range (c = calculated)*	*Reference (for calculated value)*
Buergerite	3.305	3.29-3.32c	Donnay, 1963
Dravite	3.095	2.9-3.29	
Elbaite	3.05	2.84-3.10 (3.069c)	Donnay and Barton, 1972
Ferridravite	3.26	3.18c-3.33c	Frondel, Biedl, and Ito, 1966; Walenta and Dunn, 1979
Liddicoatite	3.02	3.05c	Dunn et al., 1977
Schorl	3.13	2.82-3.244c	Fortier and Donnay, 1975
Uvite	3.04	3.01c-3.089	Dunn, Appleman, and Nelen, 1977
Chromdravite	3.40	3.39-3.41	

[a]For dravite, elbaite, and schorl, the ~average is for several recorded values.

$(10\bar{1}0)$ face of a dravite—and they calculated this value to equal 7.299 on the Mohs scale (cf. Taylor, 1949).

Additional microhardness values recorded for tourmalines include those given in Table 7-2 and values for synthetic tourmalines given by Voskresenskaya and Ivanova (1975). Also, Leoni and Troysi (1975) tabulated values for different faces on four elbaite crystals from Elba and schorl-elbaite from Antisirabé, Madagascar, and Ivanova (1981) tested different faces on a schorl-dravite from Kazakhstan (See Table 7-3).

Von Engelhardt (1942) reported that the abrasive hardness of tourmaline (*inter alia*) depends largely on the type of liquid used—for example, the hardness is greater in nonpolar than in dipolar liquids; he attributed the differences to differences in interfacial energies. As a product of their investigation of the effect of absorbed water on indentation microhardness, as measured with a Vickers diamond indenter, Westbrook and Jorgenson (1968), reported values ranging from 980 to 1210 (wet) and from 1090 to 1320 (dry) for different crystal planes on a single specimen.

ELASTICITY

Volume elasticity, also termed "bulk modulus," is the reciprocal of volume compressibility. Volume compressibility is defined as the fractional decrease in volume per unit volume under the influence of hydrostatic pressure without destruction of crystal structure (linear compressibility is the decrease in length per unit length . . .).

Bridgman (1949) gave the data shown in Table 7-4 for the compressibility

Table 7-2
Microhardness Values for Tourmaline

Specimen	*Average or Mean(m) Value* $kg\ mm^{-2}$	*Number of Tests*	*Remarks*[a]	*Reference*
Elbaite	1237(m)	22	R = 1090-1384	Leoni and Troysi, 1975
Elbaite	1046		R = 966-1126	Voskresenskaya and Ivanova, 1975
Elbaite (pink)	1178	30	Li-1.19, Mg-0.23, Fe-0.42	Ivanova, 1981
Elbaite (pink)	1294	5	Li-1.42, Mg-0.40, Fe-0.64	Ivanova, 1981
Elbaite (green)	1177	26	Li-0.80, Mg-0.27, Fe-5.87	Ivanova, 1981
Elbaite (blue)	1061	30	Li-0.78, Mg-0.31, Fe-7.24	Ivanova, 1981
Dravite	1341(m)		R = 1275-1407	Leoni and Troysi, 1975
Dravite	1058(m)		R = 986-1131	Voskresenskaya and Ivanova, 1975
Dravite	1214	17	Mg-10.12, Fe-0.72	Ivanova, 1981
Uvite	1222	5	Mg-10.01, Fe-9.98	Ivanova, 1981
Dravite-(schorl)	1022	5	Mg-5.88, Fe-9.00	Ivanova, 1981
Dravite-(schorl)	1071	17	Mg-4.29, Fe-9.48	Ivanova, 1981
Schorl-(dravite)	1182	23	Mg-8.96, Fe-9.85	Ivanova, 1981
Schorl-(dravite)	1143	22	Mg-8.81, Fe-9.85	Ivanova, 1981
Schorl	1515(m)		R = 1404-1626	Leoni and Troysi, 1975
Schorl	1146(m)		R = 984-1307	Voskresenskaya and Ivanova, 1975
Schorl	1122	27	Li-0.24, Mg-0.99, Fe-13.67	Ivanova, 1981
Schorl	1105	28	Mg-3.27, Fe-11.72	Ivanoa, 1981

[a]R = range; Li- . . . , Mg- . . . , and Fe- . . . refer to weight percentages of LiO_2, MgO, and FeO + Fe_2O_3, respectively.

Table 7-3
Microhardness Values Obtained on Different Faces of the Dravite-(Schorl) Specimen Listed with the 1071 kg mm^{-2} Value in Table 7-2

Face	*Average kg mm^{-1}*	*Number of Tests*	*Range*
$11\bar{2}0$	1092	10	1025-1126
$10\bar{1}0$	1107	8	1052-1136
$10\bar{1}1$	1037	10	980-1127
$01\bar{1}2$	1074	6	983-1127

Source: Data from Ivanova, 1981.

of "black" tourmaline; he tested two specimens—one (schorl?) from San Diego County, California, the other (uvite?) from Pierrepont, New York. As can be seen in the table, under the pressures checked, these tourmalines are more than two and a half times as compressible parallel to c as they are perpendicular to c. This apparently reflects the fact that bonding within the rings of Si-tetrahedra and the groupings of *Y* and *Z* octahedra, both of which are essentially perpendicular to *c*, is stronger than the bonding between them and between them and the B-triangles and *X*-site cations. This is consistent with the observation of Newnham and Yoon (1973), who directed attention to the fact that tourmaline's stiffness is greater within than perpendicular to the plane of the Si_6O_{18} rings (i.e., $c_{33} = 160$ GPa $< c_{11} = c_{22} =$ 270 GPa). And, as would be expected, other directional properties—for example, appropriate magnetic and electrical properties—are similarly correlative with structure.

In addition, Bridgman (1949) noted: (1) there are small abnormalities in both directions; (2) there is no conventional hysteresis in the parallel direction; (3) there is definite hysteresis in the perpendicular direction; and, (4) the shape of the loop differs somewhat from the canonical, the maximum width of the loop being near 20,000 (kg/cm^2), with a drawn-out tail at the low-pressure end.

Povarennykh (1956) gave values for linear compressibility of 0.49×10^6 cm^{2}kg^{-1} parallel to *c* and 0.17×10^6 cm^{2}kg^{-1} perpendicular to *c* and for volume compressibility of 0.82×10^6 cm^{2}kg^{-1}.

Table 7-5 gives values for stiffness ($c_{\rho\sigma}$ or GPa, with 1GPa $= 10^9$N m^{-2} $= 10^{10}$ dyn cm^{-2}) and for compliances ($s_{\rho\sigma}$ or (TPa)$^{-1}$, with 1(TPa)$^{-1} = 10^{-12}$ m^{2}N$^{-1} = 10^{-13}$ cm^{2}dyn^{-1}). The first six entries in the second and third columns were calculated by Huntington (1958) from the data of Mason (1950) and Voigt (1890);

Birch (1950) gave data for elastic constants, which were determined by resonant frequencies of extensional vibrations, for tourmaline parallel to *c*

Table 7-4
Compression of Tourmaline

Pressure kg/cm²	$\left(\frac{\Delta l}{l_0}\right)_{\parallel c}$	$\left(\frac{\Delta l}{l_0}\right)_{\perp c}$	$\left(\frac{\Delta V}{V_0}\right)$	$(\Delta V/V_0)$ *Calculated Value*[a]
5,000	0.00245	0.00094	0.00431	(0.00433)
10,000	0.00484	0.00185	0.00851	(0.00854)
15,000	0.00715	0.00273	0.01257	(0.01261)
20,000	0.00937	0.00359	0.01647	(0.01655)
25,000	0.01147	0.00444	0.02023	(0.02035)
30,000	0.01336	0.00525	0.02368	(0.02386)
Average Percentage Deviation from Smooth Curve	0.6	0.6		

Source: After P. W. Bridgman, 1949, Linear compressions to 30,000 kg/sq.cm., including relatively incompressible substances, *Am. Acad. Arts. Sci. Proc.* **77**:228.
[a]This value is given for comparison only.

(Table 7-6). Note especially the s_{33} values as compared to those of Mason and of Voigt (Table 7-5).

Ozkan (1980), in an investigation of the effect of thermal neutron irradiation on the elastic moduli and structure of tourmaline, found that

> according to ultrasonic measurements, the elastic constants . . . are little affected by neutron irradiation up to ~8×10^{17} neutrons/cm². Detectable damage . . . starts to appear at a fluence [flux?] of 1×10^{18} neutrons/cm², with a rather rapid decrease of the elastic constants . . . at higher fluences. Above 2×10^{18} neutrons/cm² cracks are generated and tourmalines become rather brittle . . . Rapid decrease of the elastic constants . . . is similar to the observed decrease in the metamict minerals, except that in tourmaline the damage rate is . . . [associated with] neutrons [absorbed] by ^{10}B. [and] Although the positions and intensities of the Bragg peaks and the background scattering are affected by irradiation, in the short range the crystalline nature of tourmalines is preserved up to the observed flux of 7.8×10^{18} neutrons/cm². (from *Chemical Abstracts*-93:105020e).

Mineralogists and petrologists should, nonetheless, keep their eyes out for true metamict tourmalines.

Using the Mason-based values, Chou and Sha (1971) calculated and recorded the energy factor K for a (0001) dislocation and for edge and screw dislocations on the basal plane with Burgers vector in the $(11\bar{2}0)$ direction; computed coefficients, using the same data base; and concluded, *inter alia,* that tourmaline exhibits a much less distinct anisotropic effect than, for example, quartz.

Table 7-5
Stiffness ($c_{\rho\sigma}$ or GPa) and Compliance [$s_{\rho\sigma}$ or $(TPa)^{-1}$] Values (see text)

	Landolt-Boernstein	*Mason Base*	*Voigt Base*
c_{11} $(= c_{22})$	277	272	270
c_{33}	163	165	160.7
c_{44} $(= c_{55})$	64	65	66.7
c_{12}	64.5	40	69.1
c_{13} $(= c_{23})$	31.9	35	8.84
c_{14} $(= -c_{24} = c_{56}/2)$	−6.9	6.8	−7.75
s_{11} $(= s_{22})$	3.92	3.85	3.98
s_{33}	6.46	6.36	6.24
s_{44} $(= s_{55})$	15.7	15.4	15.12
s_{12}	−0.83	−0.48	−0.103
s_{13} $(= s_{23})$	−0.63	−0.71	−0.016
s_{14} $(= -s_{24} = s_{56}/2)$	0.52	−0.45	−0.58

MAGNETIC PROPERTIES

The magnetic behavior and measurements of both the magnetic susceptibility and magnetic anisotropy of diverse tourmalines have been recorded by several mineralogists and physicists. [Magnetic susceptibility is the ratio of intensity of magnetism produced within a substance and the intensity of the magnetic field in which the substance resides. Quantitatively, magnetic anistropy is the difference in magnetization between the easy (greatest intensity) and the hard (least intensity) directions.] Most of the magnetic properties of most tourmalines depend on Fe^{2+} and Fe^{3+} in *Y*-sites; however, Fe in *Z*-sites and Mn and/or Ti in *Y*- and/or *Z*-sites probably contribute to, and perhaps even control, the magnetic properties of some tourmalines.

The fact that diverse tourmalines exhibit different kinds of magnetism serves to nullify general statements made to the effect that pyro- and piezoelectric minerals are also paramagnetic (an idea that apparently evolved from less than complete understanding of the applicability of Maxwell's (1873) equations that relate magnetism and electricity).

The magnetic anisotropy of tourmaline was apparently recognized by at least the middle 1700s (Wilcke, 1766). In any case, it was described as such by Wilson (1920) when he noted that the magnetic susceptibility of both opaque and transparent tourmalines is greater perpendicular to *c* than along *c*.

Later, Sigamony (1944) studied 11 differently colored tourmalines, all of which he observed to be paramagnetic, and tabulated the mass susceptibility and anisotropy of each (see Table 7-7). He also reported that for those

Table 7-6
Elastic Constants for Tourmaline Parallel to c

Tourmaline	*S.G.*	*Velocity (km sec^{-1})*	*Stiffness (GPa)cm*	*Compliance [(TPa)$^{-1}$]*
black	3.091	6.88	c_{33} = 146	S_{33} = 6.84
yellow	3.028	7.40	c_{33} = 166	S_{33} = 6.03
pink	3.031	7.47	c_{33} = 169	S_{33} = 5.91

Source: Data from Birch, 1950.

specimens: (1) susceptibility is greater perpendicular to than parallel to *c* (see above); (2) susceptibility is isotropic in the symmetry plane (it appears that he meant the plane perpendicular to *c*, which would be consistent with his data in that the *c* axis has threefold symmetry); (3) specimens of higher density and greater dichroism show the greater mass susceptibility; and (4) specimens with greater mass susceptibility have, in general, greater anisotropy.

More recently, and again confirming the observation that the magnetic susceptibility is greater perpendicular to than along *c*, Leela (1954) added: (1) magnetic susceptibility is the same at different field strengths; and (2) magnetic susceptibility correlates directly with dielectric constants and refractive indices. However, he also found apparent exceptions to Sigamony's third and fourth conclusions—that is, he found that anisotropy does not vary directly with depth of color for at least some pink elbaites and the anisotropy of some crystals does not vary directly with their mass susceptibility.

On the basis of measurements of magnetic susceptibilities of ten tourmaline specimens from five Russian localities—specimens with Fe-contents of 0.3 to 13.8 weight percent—Kruglyakova (1954) reported magnetic susceptibilities as ranging between 1×10^{-6} and 31.1×10^{-6} cgsm (as compared to the 0.8 to 97.0×10^{-6} cgs range of previously cited values) and noted that they varied directly with Fe-content.

Vinokurov and Zaripov (1960) tabulated magnetic susceptibility values for 28 specimens—11 black, 11 green, and 6 rose colored tourmaline crystals—19 of which were new measurements. Their results also indicate the existence of a direct, albeit general, correlation between magnetic susceptibility and both refractive indices and dielectric constants. In addition, the results show that green tourmalines apparently fall into two groups based on their anisotropy, with the darker ones having the greater anisotropy, whereas pink tourmalines have both low susceptibilities and low anisotropy values. It was also concluded: (1) the shapes of the *Y* octahedra account for magnetic anisotropy, (2) the presence of Fe^{2+} in the $^5D_4$ state leads to large magnetic susceptibilities and high anisotropies, and (3) the presence of $Fe^{3+} + Mn^{2+}$ in the $^6S_{5/2}$ state, which has purely spin magnetism, leads to higher susceptibilities but lower anisotropies.

Table 7-7
Magnetism—Examples of Mass Susceptibility and Anisotropy

Tourmaline Color	*Susceptibility* $\chi m_{\perp}$ 10^{-6}*cgs units*	*Anisotropy* $(\chi_{\perp}-\chi_{\parallel})10^{-6}$*cgs units*	*Reference*
Black	26.99	1.30*	Leela (1954)
	26.96	1.14*	Leela
	26.3	4.4	Vinokurov and Zaripov (1960)
	26.4	3.8	Vinokurov and Zaripov
	24.95	1.20*	Leela
	24.38	1.40*	Leela
	21.9	2.6	Vinokurov and Zaripov
	17.37	3.71#	Vinokurov and Zaripov
	17.34	3.59#	Vinokurov and Zaripov
	17.31	3.50#	Vinokurov and Zaripov
	17.30	3.47#	Vinokurov and Zaripov
	13.4	2.01	Sigamony (1944)
	10.1	0.577	Sigamony
Brown	11.0	0.295	Sigamony
Blue-green	9.3	2.4	Vinokurov and Zaripov
	10.6	0.8	Vinokurov and Zaripov
Dark-green	10.8	1.65	Sigamony
	12.50	2.19	Leela
	12.57	2.40	Leela
	15.2	1.5	Vinokurov and Zaripov
	20.8	0.4	Vinokurov and Zaripov
	21.6	0.3	Vinokurov and Zaripov
Green	11.7	2.0	Vinokurov and Zaripov
	12.0	2.4	Vinokurov and Zaripov
Light green	9.7	0.226	Sigamony
	8.7	0.6	Vinokurov and Zaripov
	3.9	0.7	Vinokurov and Zaripov
Rose	1.18	0.038	Sigamony
	1.1	0.4	Vinokurov and Zaripov
	0.96	0.012	Sigamony
	0.95	0.014	Sigamony
	0.73	0.017	Sigamony
	0.62	0.02	Leela
	0.61	0.07	Leela
	0.58	0.038	Sigamony
	0.4	0.1	Vinokurov and Zaripov
	0.2	0.1	Vinokurov and Zaripov
Colorless	4.96	0.058	Sigamony

*Recorded as having hexagonal cross-sections perpendicular to *c*.
#Recorded as having triangular cross-sections perpendicular to *c*.

As the result of an investigation of the diverse magnetic fields of nine specimens from localities in the United States, Mexico, Madagascar, and England, and the extended study of three of the specimens at temperatures down to less than 10 K, Donnay et al. (1967) concluded: (1) Pierrepont (New York) uvite is paramagnetic both parallel and perpendicular to the *c* axis and is magnetically anisotropic; (2) Mexquitic (Mexico) buergerite exhibits antiferromagnetism both parallel and perpendicular to *c* and is magnetically anisotropic; and (3) Madagascar ferric iron tourmaline exhibits weak antiferromagnetism that is the same both parallel and perpendicular to *c* (i.e., it is magnetically isotropic). Considering the probable Fe^{2+} and/or Fe^{3+} site occupancies in these tourmalines, as indicated by structural studies, these data appear to indicate that the magnetic properties of these tourmalines depend upon differences in the ligands around the iron atoms. Also, as suggested by the authors, it seems that the antiferromagnetic ordering may consist of a trigonal arrangement of spin on Fe ions within the mirror plane. In addition, their data would appear to indicate that the buergerite and ferric iron tourmaline have Curie points and thus would undergo a phase transition at some temperature (however, cf. Tippe and Hamilton, 1971).

Tsang et al. (1971) determined the magnetic susceptibility of three crystals, one for each of the just mentioned localities, between 8 K, and 300 K. They repeated the conclusions of Donnay et al. (1967), presented graphs showing reciprocal magnetic susceptibilities plotted against temperature, and indicated how magnetic exchange interactions of Fe^{2+} and Fe^{3+} can be shown to agree with the data obtained and also how the data are compatible, so far as site locations of magnetic ions, with structural interpretations based on X-ray and neutron diffraction and optical spectroscopic data.

Tippe and Hamilton (1971), in comparing neutron powder diffraction patterns for buergerite at 300 K and 4.2 K, recorded no significant differences and, thus, no evidence for long-range magnetic ordering between those temperatures.

Dobrovol'skaya (1972) gave the range of $\chi = 12.5\text{-}28.5 \times 10^{-6}$ cgs for the magnetic susceptibility of tourmaline and noted that minor amounts of paramagnetic ions other than those of Fe appear to have no effect. Later, along with Kuz'min (Dobrovol'skaya and Kuz'min, 1975), she checked the magnetic susceptibility (χ) of specimens termed paramagnetic tourmaline (tm) from 80 deposits in the U.S.S.R., and found that all fall within a rather small range, with the differences apparently attributable to different Fe-contents and/or Fe^{2+}:Fe^{3+} ratios, and as a result, concluded that the susceptibility is a linear function of Fe-contents according to the following equation:

$$\chi = 423.6 \times 10^{-6} (0.362A - 0.439B)$$

in which A and B are concentrations of FeO and Fe_2O_3, respectively.

THERMAL PROPERTIES

Most thermal data published before the 1950s dealt with "fusion" as determined in blowpipe analysis, determination of the melting temperatures of a few specimens, and/or information gained from attempts to change the color of certain tourmaline gemstones. These data may be summarized briefly as follows.

In 1830, Shepard reported that pink tourmaline, upon being heated to redness before the blowpipe, loses its color, upon increased heating whitens and becomes opaque and swells and fractures, and that microscopic examination of the end product indicated partial vitrification. He also noted that with borax, pink tourmaline dissolves readily with effervescence to form a deep rose colored glass, and with soda, it gives a bluish green opaque glass. In addition, he found that green tourmaline upon heating to redness becomes pale; upon continued heating, becomes milk-white and swells, cracks, and vitrifies; and with borax, it dissolves, giving a transparent glass with a slight "tinge of iron." Furthermore, he noted that blue tourmaline upon heating to low redness retains its color; in higher heat, it swells slightly and turns yellowish gray, vitrifies, and undergoes partial fusion; with borax, it dissolves easily to form a pale green transparent glass; and with salt of phosphorus, it dissolves, leaving a skeleton of "silex," which while hot has a "feeble color of iron."

In the sixth edition of Dana's "System . . . ", E. S. Dana (1892) recorded the following: Mg varieties fuse rather easily to white blebby glass; Fe-Mg varieties fuse with strong heat to a white, greenish, or brownish slag or enamel; Fe varieties fuse with difficulty to a brown, brownish red, gray, or black slag; Fe-Mg-Li varieties fuse with great difficulty to a yellowish, grayish, bluish, or whitish slag or enamel; and Li varieties are infusible. . . .

Melting temperatures are recorded as follows:

Fe-rich tourmaline	1050–1200°C with the melting point inversely proportional to the Fe-content (Frondel, Hurlbut, and Collette, 1947)
Li tourmaline	1725°C, total melting (Frondel, Hurlbut, and Collette, 1947)
Dravite	1100°C, beginning of melting (Kurylenko, 1950)
"Black tourmaline"	1105°C, beginning of melting (Kurylenko, 1950)
"Rubellite"	1350°C, beginning of melting (Kurylenko, 1950)

However, Robbins, Yoder, and Schairer (1959) and Robbins and Yoder (1962) reported that (1) dravite from Mesa Grande, California "appeared to melt incongruently at about 700°C and 2000 bars" and (2) synthetic dravite decomposes or dissociates, as shown on Figure 4-11.

With regard to flame color, Vorob'ev (1955), as the result of the study of the behavior of several minerals in an electric arc at 7600°C, found that tourmaline, otherwise not identified, gave green in the central reducing part of the flame and blue in the outer oxidizing part of the flame.

Thermal Conductivity

Thermal conductivity varies with both temperature and crystallographic direction. Knapp (1943) obtained the data given in Table 7-8 for a green elbaite from Brazil.

Horai (1971), however, recorded the thermal conductivity as 0.0126 cal sec^{-1} cm^{-1} $°C^{-1}$ (= 5.27 W m^{-1} K^{-1}) for a tourmaline (apparently schorl) from Keystone, South Dakota, and Sigrist and Balzer (1977) gave the value of 20×10^{-3} W cm^{-1} K^{-1}, which converts to .00478 cal sec^{-1} cm^{-1} $°C^{-1}$, for a tourmaline not otherwise described.

Differential Thermal Analyses

The earliest recorded DTA data for tourmaline are those of Frondel, Hurlbut and Colette (1947), who noted that for a high-iron tourmaline: 1) there were no thermal peaks below 1050°C, 2) an endothermic peak accompanies melting at 1050°C, and 3) the peak at 1050°C is followed immediately by a sharp exothermic peak "presumably representing partial crystallization of a transient complete melt."

Kauffman and Dilling (1950) presented a DTA curve for pink elbaite from Paris, Maine, that shows a broad shallow endothermic peak at ~560°C and a large relatively sharp endothermic peak at ~1000°C.

Kurylenko (1950, 1953, and 1957) gave DTA curves for diverse tourmaline specimens (see, for example, Fig. 7-3). Along with the beginning of melting temperatures just listed (1100°-1350°C), he stated that B is lost between

Table 7-8
Thermal Conductivity

Mean Temperature °C	*Thermal Conductivity cal sec^{-1} cm^{-1} $°C^{-1}$*
Parallel to *c*	
125	.00695
267	.00765
340	.00770
450	.00838
Perpendicular to *c*	
120	.00708
219	.00826
318	.00916
456	.00996

Source: Data from Knapp, 1943.

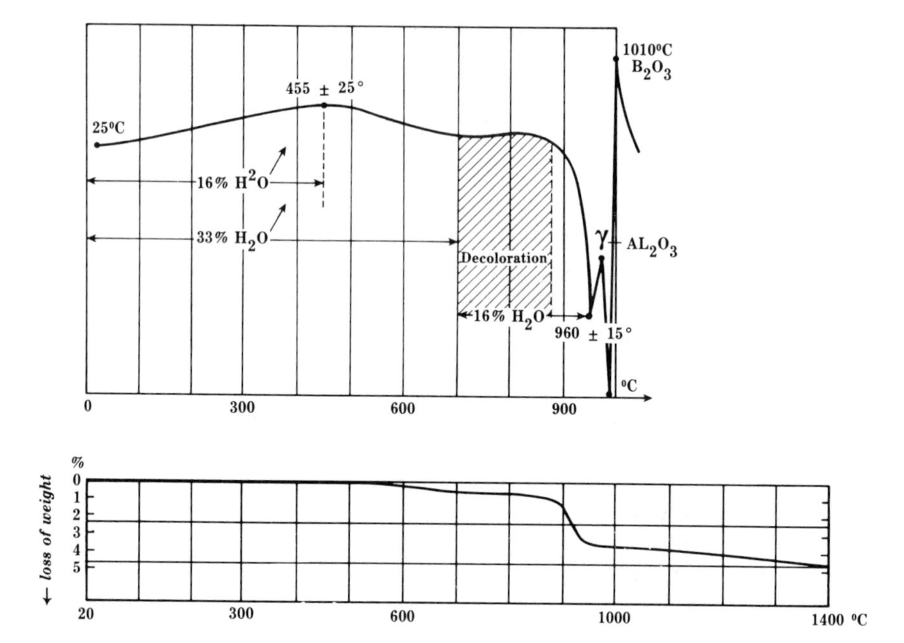

Figure 7-3
Differential thermal analyses curve of type locality dravite (after Kurylenko, 1953), and thermogravimetric curve of "dravite asbestos" from northeast of Chvaletice, eastern Bohemia (after Novak and Zak, 1970).

1000°C and 1100°C, and H_2O is lost in stages at diverse temperatures between 145°C and ~1000°C (e.g., a dravite lost H_2O at ~470°C, ~680°C, and ~960°C—cf. Machatschki, 1941 and Novak and Zak, 1970). He also suggested that the temperatures at which water is lost correlate with species. It appears, however, that more specimens need to be checked before this correlation is proved.

Korzhinskii (1958) published curves for five elbaite-schorl specimens for which he also gave chemical analyses, optical data, and X-ray diffraction data. All of his DTA patterns exhibit a rather sharp endothermic peak between ~935° and 1010°C, which he interpreted to represent almost simultaneous dehydroxylization and beginning of melting, plus a small exothermic peak at ~1100°C, which he related to sintering. (The dehydroxylization products are recorded as predominantly mullite and maghemite.) In addition, he noted that the melting temperature appears to vary indirectly with higher total alkali-, alkali-earth-, and iron-contents, and that Fe^{2+}-rich tourmalines exhibit sharp increases in both their indices of refraction and their birefringences, apparently as the result of an oxidation of the Fe^{2+}, between 600°C and 800°C.

Breskovska and Eskenazi (1959-1960) reported that the strong endothermic reaction occurs between 960°-980°C for Fe-rich tourmalines and at 1020°C for Fe-poor tourmalines. They found the higher temperature products to have a hematite-type structure and a mullite-type structure, respectively.

Wilkins, Farrell, and Naiman (1969) found that heating of a pale blue elbaite from Mt. Mica, Maine, changed the cell dimensions—*a* decreased from 15.908 to 15.838 and *c* increased from 7.123 to 7.135. They concluded that the changes appear to indicate that the included Fe^{2+} was oxidized to Fe^{3+} and that there was a loss of H (i.e., the OH^- was converted to O^{2-}). This apparent change from an elbaite-schorl towards a buergerite, as well as other such possible changes, should be investigated further. (Along this line, it is unfortunate that so little is known about the occurrence of the type buergerite specimen. Could it, for example, have been formed from an original schorl?)

Novak and Zak (1970) interpret their DTA and TG data more or less to support Kurylenko's scheme.

Miscellany

Both linear (λ) and volume (α) coefficients of thermal expansion for tourmaline are reported by Povarennykh (1956): parallel to c, $\lambda = 8 \cdot 10^{-6}$ $°C^{-1}$; perpendicular to c, $\lambda = 6 \cdot 10^{-6}$ $°C^{-1}$; and $\alpha = 20 \cdot 10^{-6}$ $°C^{-1}$.

For the range 193 K to 293 K, Donnay (1977) reported results obtained by Fortier to show the following increases in cell dimensions: $a = 0.020A$ ($= 13 \cdot 10^{-6}$ $°C^{-1}$), $c = 0.015A$ ($= 21 \cdot 10^{-6}$ $°C^{-1}$) and $V = 7A^3$ ($= 46 \cdot 10^{-6}$ $°C^{-1}$).

Sigrist and Balzer (1977) gave: Specific Heat = > 0.2 cal/g · K; Thermal Diffusivity = $7 \cdot 10^{-3}$ cm^2 sec^{-1}; (and, the minimal energy loss for the formation of a track = $16.9(dE/d\xi)_k$ [MeV/(mg/cm^2)]). Hamid (1980) gave: Specific Heat (C_ρ) 0.82 J g^{-1} K^{-1} (= .196 cal/g K). Hoover (1983), using the "assumed value" of 0.2 cal/g°C for elbaite, gave: Thermal Diffusivity = 0.0202 cm^2/sec^{-1} and "Thermal Inertia" – 0.0889 cal/cm^{-2} C m $sec^{1/2}$ for elbaite from Keystone, South Dakota.

Michel-Lévy and Kurylenko (1952) added an interesting note to the effect that tourmaline grains heated to temperatures greater than about 1000°C may develop a fibrous character with fibers perpendicular to the *c* axis of the original grains. As previously mentioned, they observed the relation in tourmaline from a xenolith within trachyte from Mont Dore (France), as well as in artificially heated tourmalines.

Martin and Witten (1961) investigated the temperature dependence of microwave phonon attenuation in green tourmaline, and found that

> the exponent (n) in the T^n dependence of the attenuation . . . varied with temperature. At very low attenuation values (below 0.1 dB/cm) . . . n = 7.5 + 1.6 . . . at ~6 K. The magnitude of n gradually diminished with increasing temperature and attenuation. . . . At 15 dB/cm . . . n = 0.8 + 0.1 for tourmaline at 90 K. The T^7 dependence . . . at low attenuation levels seems characteristic of scattering predominantly through four phonon processes. However, it may result from nearly colinear three-phonon, scattering processes. . . (p. 447).

RADIOACTIVITY

The use of tourmaline in geologic dating and the checking of the transparency of tourmaline by X-rays are described in this section. Other aspects of radioactivity and tourmaline are described in other sections as follows: X-ray diffraction patterns in Chapter 3 and effect of radiation on color and on luminescence in Chapter 6.

Dating

Much tourmaline contains small percentages of potassium and is argon-retentive. Consequently, tourmaline has been used for K/Ar dating. Damon and Kulp (1958), however, suggested that tourmaline should not be so used because of possible excess Ar error; Fitch and Miller (1972) disagreed.

Gerling, Shukolyukov, and Matveeva (1962) found that tourmaline, along with beryl and quartz, give Rb/Sr ages that differ markedly from the K/Ar ages of micas from the same rocks; they suggested that the $^{87}Sr/^{88}Sr$ ratio of tourmaline, which deviates some 64 percent from the "normal," may reflect capture of Sr with anomalous isotopic composition during crystallization.

More recently, the dating method involving total degassing $^{40}Ar/^{39}Ar$ has been used for tourmaline with apparent success (Fitch and Miller, 1972; and Allen et al., 1972; and Allen, Sutton, and Watson, 1974).

On the basis of their investigation of fission track etching and annealing of tourmaline, Lal, Parchad, and Nagpaul (1977) concluded that in 1000 million years all tracks would be erased if the host tourmaline was held at 220°C or higher temperatures but that no annealing would occur if it was held at 150°C or lower temperatures.

Transparency to X-rays

Many different gemstones have been submitted to X-rays in order to determine their relative transparencies to that radiation. Doelter (1915) found the tourmaline that he checked to be essentially opaque. Notes for the colored stone course offered by the Gemological Institute of America indicate the gem varieties of tourmaline to be semitransparent to X-rays. These differences very likely depend on the fact that transparency to X-rays depends upon the composition of the crystal tested and also upon both the wavelength and intensity of the radiation.

ELECTRICAL PROPERTIES

Tourmaline occupied the center of the scientific stage during much of the eighteenth century because of its pyroelectric properties. In this century, the piezoelectric properties of tourmaline have been investigated extensively. Both of these and other electrical properties are described in this section.

Electrical Conductivity

In 1881, Thompson reported that he found tourmaline to conduct electricity better "*across* the long axis of the crystal than *along* that axis." The fact that his "long axis" corresponds to crystallographic c is indicated by his remarks about the light absorption of the crystal. Indeed, a difference does exist; however, subsequent measurements show the electrical conductivity to be greater parallel to c than perpendicular to c. For example, Gomm (1973) recorded the electrical conductivity of "black" tourmaline as 5.5×10^{-10} Ω^{-1} cm^{-1} along c and 1.1×10^{-10} Ω^{-1} cm^{-1} perpendicular to c, both at 25°C and with a temperature dependency of 0.25 eV for all directions.

Directionally dependent dielectric constants (permittivity) of tourmalines have been reported by several workers (e.g., Curie, 1889; Schmidt, 1902 and 1903; Fellinger, 1919; Pulou, 1947; Rao, 1949; Rozhkova and Proskurovskii, 1957; Wappler, 1965; Koja and Zheku, 1968). These data show, just as would be expected for an optically negative uniaxial mineral, that tourmaline has constants that are greater for the $\perp^{r}$ direction than for the $||^{l}$ direction. Also, as expected, the constants vary with composition as well as with frequency. The values reported in the literature range from 5.43 ($||$) and 6.50 ($\perp$) at a frequency of 7 MHz (Wappler) to 6.5 ($||$) and 8.0 ($\perp$) at 1.6 MHz (Rao). Schmidt's values of 5.05 ($||$) and 6.75 ($\perp$) at 400 MHz are also noteworthy. In addition, it appears from the data of Rao (Table 7-9) that the dielectric constants may vary inversely with specific gravity for the elbaites (?) and directly with the specific gravity for the schorls (!).

Rao (1950) directed attention to the following relations:

direction of higher dielectric constant
= direction of smaller modulus of elasticity
= direction of greater bond strength within the structure.

Leela (1954) added:

= direction of greater index of refraction
= direction of greater magnetic susceptibility.

Pyroelectricity and Piezoelectricity—An Introduction

Tourmaline, like other minerals that possess only a single polar axis of symmetry, exhibits both pyroelectric and piezoelectric properties. Born (1790) and Melon (1930), however, have reported black tourmalines from

Table 7-9
Dielectric Constants for Tourmaline

Specimen	Specific Gravity	Dielectric Constant ∥ 11	⊥ r
light yellow	3.06	6.5	7.9
red	3.05	6.0	7.4
green	3.16	5.9	6.8
black	3.18	6.4	8.0
black	3.14	6.3	7.9
black	3.11	6.1	7.6
black	3.11	5.9	7.5
black	3.09	5.8	—

Source: Data from Rao, 1949.
Note: Frequency is 1.6 MHz.

Sri Lanka (formerly Ceylon) and Bas-Uele, Zaire, (formerly Belgian Congo), respectively, that exhibit no pyroelectric properties. In fact, most schorl that has been checked exhibits only weak—in some cases hardly detectable—pyroelectricity.

Pyroelectricity, the generation of electricity and the development of electrical polarity on the surfaces of dielectric substances as the result of temperature change (either heating or cooling), was possibly recognized in tourmaline by Theophrastus (~315 B.C.). In any case, it was definitely observed by 1703 when tourmaline was termed as "aschentrekker" (Schmidt, 1707). (That name, meaning ash-drawer, was given because gem tourmaline brought to Holland from Ceylon, now Sri Lanka, was found to attract ashes when heated.)

Piezoelectricity, essentially the analog of pyroelectricity that is caused by subjecting dielectric substances to mechanical stress, was first reported for tourmaline by Jacques and Pierre Curie (1880). They found that "La tourmaline donne les effets trés-énergiques suivant son axe principal" as follows: upon compression along c, the crystals develop electrical polarity; upon decompression, the polarity is reversed; and the polarity developed during compression is the same as that developed during cooling, whereas that developed during decompression is the same as that developed during heating.

Tourmaline also exhibits converse (often termed indirect) pyroelectricity and piezoelectricity. That is, if tourmaline crystals are appropriately oriented within an electric field, they will undergo temperature changes and develop stress.

The history of discovery, already mentioned, and of early demonstrations and investigations of electrical properties of tourmaline are reviewed by, for example, Home (1976), Lang (1974), and Schedtler (1886).

To summarize briefly, after its discovery, Lemery (1717) demonstrated pyroelectric properties of a polished piece of tourmaline to the French Academy of Sciences in Paris; he attributed the properties to some sort of magnetism. In a series of papers, Æpinus (1756 and 1762) reported systematic experiments made to discover the nature of pyroelectric effects in tourmaline to the Berlin Academy of Sciences; his results and conclusions, although objected to in part by de Noya (1759), were checked and for the most part confirmed by several others—for example, Bergmann (see Fig. 7-4), Coulomb, Franklin, Priestley, and Wilson (see Wilson, 1759).

In the 1800s, the work of Rose (e.g., 1838) is particularly noteworthy. He observed, for example, that intensity of electrical polarity appears to differ with color. On the basis of checking 25 tourmaline specimens, he concluded that it is weak for black, strong to weak for brown, very strong for green, and strong for red tourmalines.

Pyroelectricity

During heating, the analogous end of a tourmaline crystal, or any fragment thereof, becomes charged positively while the antilogous end becomes charged negatively. During cooling, after the charges developed during heating have been "removed," the analogous end becomes charged negatively while the antilogous end becomes charged positively. Furthermore, when an electric field is applied along the *c* axis, heating occurs when the current is directed from the analogous end toward the antilogous end, and cooling occurs if the field is directed in the opposite direction.

Both true and false pyroelectricity have been described for crystalline materials. True pyroelectricity can develop only in tourmaline and other crystalline substances having a single polar axis; on the other hand, false pyroelectricity can develop in any crystalline substance that lacks a center of symmetry (e.g., quartz). (False pyroelectricity is, in essence, piezoelectricity developed in response to strains caused by heating and cooling.) Unfortunately, true and false pyroelectricity cannot be distinguished easily, so the exhibition of a pyroelectric effect can be taken only to indicate the lack of a center of symmetry, not the presence of a polar axis. Also, it must be kept in mind that some minerals that have polar axes do not readily exhibit pyroelectric effects (e.g., some schorl). Nonetheless, checking for the pyroelectric properties of minerals of unknown identity or symmetry has been used for untold decades to determine the presence or absence of the center of symmetry. In fact, Wooster once (1938) stated "This diagnostic test is almost the only practical application of the knowledge about pyroelectricity." Several methods have been devised to detect, demonstrate, or measure this effect. A few of the more frequently employed qualitative tests,

Figure 7-4
Sketch of appratuses used by Bergmann (1976).

used particularly for demonstrations, plus the results given by crystals (or fragments or grains) that exhibit pyroelectricity effects, follow.

The procedure of Kundt (1883):

Heat crystals to about 200°C.

Pass crystal through a flame (to remove surface charges).

Shake a mixture of powdered sulfur and red lead through a brass sieve onto the crystal.

Result: The red lead will be attracted to the negatively charged end of the crystal and the yellow sulfur to the positively charged end.

The procedure of Bleekrode (1903):
Cool crystal in liquid air.
Suspend it on a glass filament in room conditions.
Result: Ice particles will grow, as the result of condensation and freezing of water vapor in the air, along lines of force.

The procedure of Maurice (1930):
Heat crystal to about 200°C.
Pass crystal through a flame (to remove surface charges).
Burn some magnesium ribbon under a belljar.
While full of smoke, place belljar over crystal.
Result: Magnesium oxide "hairs" will form a pattern outlining the lines of force around the crystal.

The procedure of Martin (1931):
Attach crystal to a glass fiber.
Suspend it in liquid air
Move it close to a metal plate previously immersed in liquid air.
Result: The crystal will be attracted to the plate.

If a prismatic crystal or crystal fragment that is elongate parallel to c is available, the procedure first used by Gaugain (1856) may be used. This procedure, using a "flicking" electrometer, can be monitored to measure, for example, quantitative differences in charges developed by crystals of small versus large cross-sections (see, for example, Wooster and Breton, 1970).

In addition, as Thompson (1881) reported, *Lichtenberg's figures* can be produced on, for example, glass or shellac by using powdered and warmed tourmaline.

Donnay (1977) has concluded that the pyroelectric effect in tourmaline is largely, if not completely, dependent upon the asymmetric aharmonic vibration in the z direction of O(1), the oxygen atom with point symmetry $3m$ that has a polar environment. (The O(1) site was occupied by $F_{0.60}\,OH_{0.20}\,O_{0.20}$ in the elbaite in question.) X-site and O(2) atoms are thought also possibly to contribute to the pyroelectric effect.

From a practical standpoint, it was difficult to make quantitative measurements of pyroelectric effects until relatively recently. Among other things, the electrical charges escape readily. Some of the earlier employed methods were devised by Gaugain (e.g., 1856) and the Curies (1880). More recently devised methods are described by, for example, Gladkii and Zheludev (1965).

An early, widely quoted result is that of Lange (1905) who reported a temperature change of 1/500°C for a tourmaline grain put into a field measuring 30,000 volts/cm. Another frequently quoted value is 1.2 c.g.s. units/cm^2 · °C for the pyroelectric constant (e.g., Ackermann, 1915). According to Cady (1946), this value is a good figure for the sum of primary and secondary effects. It can also be expressed in m.k.s. units as $p = 1.2(\frac{1}{3} \times 10^{-5}) = 4.0\,\mu C\, m^{-2}\, K^{-1}$. Therefore, as Nye (1957) noted, by using a series of assumptions and calculations, it may be shown that the field strength along the c axis = 740 volts cm. More recently, Meissner and Bechmann (1928) have given 1.18 . . . and Sil'vestrova and Sil'vestrov (1958) have reported 1.26-1.28 cgs units/cm^2/°C (see Table 7-10A & B and Fig. 7-5); these values translate to 3.93 and 4.2-4.27 μC m^{-2} K^{-1}, respectively.

Drozhdin et al. (1974) investigated the effect of temperature on the pyroelectric coefficient of tourmalines of six different colors and reported that the coefficient approaches zero at temperatures near 0 K. They found that spontaneous pyroelectric polarization decreases as temperature is increased and that the coefficient passes through a sign reversal at 18 ± 2 K; they suggested that a dependence on the thermal expansion coefficient accounts best for this behavior, at least at lower temperatures.

For most temperature ranges in which tourmaline would likely be used, results of quantitative measurements have led to general conclusions such as the following: (1) over a given temperature range, the same quantity of charge is developed by either heating or cooling; (2) the rate of temperature change does not effect the amount of charge developed; (3) the developed charge is proportional to the cross-sectional area perpendicular

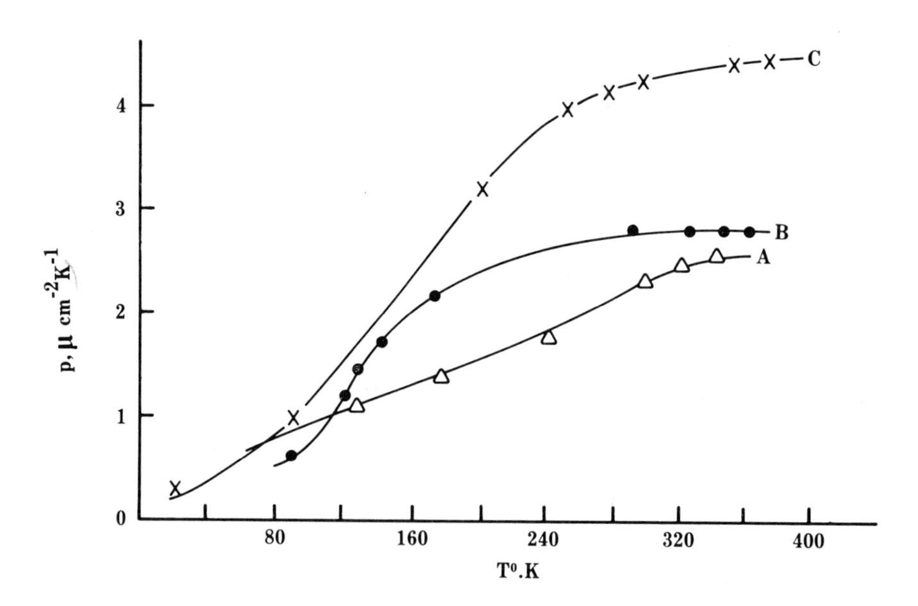

Figure 7-5
Temperature dependence of pyroelectric constant, p, for single crystals of black tourmaline *(A)*, yellow-green tourmaline *(B)*, and another yellow-green tourmaline *(C)* (cf. Table 7-11). (After Gladkii and Zheludev, 1965.)

Table 7-10A
Pyroelectric Coefficient Data

Type	*Temperature* K	*Pyroelectric coefficient,* $\mu C\ m^{-2}K^{-1}$	*Temperature Range Studied*	*Method*[a]	*Reference (cited by Lang)*
green	293	3.4	20-313	S	Röntgen, 1914
yellow-green	293	4.3	23-648	S	Ackermann, 1915
rose-red	293	4.4	23-648	S	Ackermann, 1915
blue-green	293	3.5	23-648	S	Ackermann, 1915
—	294	4.23	—	D	Sil'vestrova and Sil'vestrov, 1958
green	293	3.9	150-410	S	Gavrilova, 1965
black	298	2.1	60-360	S	Gladkii and Zheludev, 1965
yellow-green	298	2.7	80-400	S	Gladkii and Zheludev, 1965
green	298	3.6	—	D	Fabel and Henisch, 1971
rose	473	6.	473-873	S	Karyakina, Novik, and Cavrilova, 1971

Source: After S. B. Lang, 1974, *Sourcebook of Pyroelectricity,* Gordon and Breach, p. 77.
[a]S = static method (i.e., stress-free) and D = a dynamic method referred to as the "pyroelectric current method" (see Lang, 1974).

Table 7-10B
Temperature Dependence of Pyroelectric Coefficients, $\mu C\ m^{-2}K^{-1}$

	Temperature K											
Type	*23*	*88*	*198*	*253*	*274*	*293*	*352*	*372*	*408*	*483*	*578*	*648*
pink	0.27	1.0	3.3	4.1	4.2	4.4	4.7	4.8	4.9	5.1	5.8	6.5
yellow-green	0.27	0.97	3.2	4.0	4.1	4.3	4.4	4.5	4.6	5.0	5.6	6.2
blue-green	0.13	0.47	2.2	3.1	3.4	3.5	3.9	4.0	4.1	4.2	4.6	5.1

Source: Data from Ackermann, 1915.

to the electric axis but independent of the thickness (i.e., the length parallel to the electric axis).

Potential uses of tourmaline based on its pyroelectric properties are given in Chapter 9, "Uses . . .".

Piezoelectricity

As mentioned in the introduction to this section, when crystals, or fragments thereof, of tourmaline are subjected to pressure changes, they exhibit piezoelectric effects. For tourmaline: upon compression along *c*, the antilogous end develops positive electric charges while the analogous end becomes negatively charged; under tension or decompression, opposite charges are developed at the ends; and, conversely, when an electric field is directed along the *c* axis, stress develops.

Each of the effects is thought to be a manifestation of the phenomenon whereby uniform stress produces a separation of originally coincident planes of ions thus producing a dipole moment (see, for example, Mason, 1950). According to Wooster (1946), the separation is of the same order as the unit cell contraction (i.e., $\sim 10^{-19}$ cm).

Unlike pyroelectricity, piezoelectricity has been a seldom used aid to determining symmetry characteristics of crystalline materials. To outline demonstrations and tests would, therefore, be superfluous. On the other hand, also unlike pyroelectricity, piezoelectricity has found an ever increasing number of applications in both science and industry, and consequently the piezoelectric properties of tourmaline have received a great deal of attention.

In the literature, descriptions of the piezoelectric properties of tourmaline are usually referenced to the orthogonal axis system XYZ in which X = *a*, Z = *c*, and Y is perpendicular to both X and Z and thus coincident with $[10\bar{1}0]$. Cady (1946) summarized pertinent measurements and calculations of earlier workers (e.g., Riecke and Voigt, 1892, and Röntgen, 1914) and concluded that the differences in recorded values are not surprising when one considers the variability in the composition of tourmaline. There are, of course, additional contingencies that must be kept in mind both by investigators and potential utilizers of tourmaline as a piezoelectric material. Representative values are given in Table 7-11.

Using a dynamic method with the hydrostatic pressure caused by acoustic waves around a California schorl, Cook (1940) determined 6.7×10^{-8} as the value for the coefficient $(d_{31} + d_{32} + d_{33})$ $[= 2d_{31} + d_{33}]$, which is the expression for polarization produced by hydrostatic pressure (Cady, 1946). This value corresponds to 7.3×10^{-8}, calculated $(2d_{31} + d_{33})$ from both Röntgen's and Riecke and Voigt's values, and 8.0×10^{-8} given by Koch (1906), for Brazilian tourmalines.

Table 7-11
Piezoelectric Constants

	Static Method (all values × 10^{-8})[a]			*Stress Constants (all values × 10^{-4})*	
	Röntgen and Voigt per Cady, 1946	*Mason, 1950*	*Riecke and Voigt per Cady, 1946*	*Cady, 1946*[b]	
d_{15} ($=d_{24}$)	11.0	−10.9	11.0	e_{15}	7.40
d_{22} ($=-d_{21}=-d_{16}$)	−0.94	−1.0	−0.69	e_{22}	−0.53
d_{31} ($=d_{32}$)	0.96	−1.53	0.74	e_{31}	3.09
d_{33}	5.4	−5.5	5.8	e_{33}	9.60

[a]Some of the signs differ because of different choices of the $+Z$ axis.
[b]Cady used Riecke and Voigt's piezoelectric constants.

Also, Voigt (1910) reported that Lissauer found d_{33} to be constant over the temperature range −192°-+19°C.

Mason (1950) measured the temperature coefficients for the piezoelectric and elastic constants on plates with different orientations and concluded that "the temperature coefficients for all the modes of motions are negative and hence no possibilities exist for zero temperature coefficient cuts" (p. 209). He also directed attention to the fact that "This necessitates a different set of crystal orientations to obtain all the elastic and piezoelectric constants, since, for example, a crystal cut normal to the x-axis cannot be driven in a longitudinal mode" (see Figure 7-6).

Zheludev and Tagieva (1963) investigated the electrical polarization of tourmaline during hydrostatic tension and reported the piezomodulus of hydrostatic compression to be $d_{\text{hydr}} = 12 \times 10^{8}$ (average of 25 measurements on several crystals) as compared to $d_{\text{hydr}} = 8 \times 10^{8}$, which they calculated on the basis of published piezoelectric coefficients. [The values should be . . . × 10^{-8}.] With respect to this difference in values, they again directed attention to the fact that the piezoelectric properties of tourmalines range "considerably" from specimen to specimen, and also noted that the coefficients used in their calculations were determined under directed rather than hydrostatic pressure.

In an attempt to determine the usefulness of tourmaline in diverse transducers, Hearst, Irani, and Geesman (1965) studied the piezoelectric response of Z-cut tourmaline to shocks of up to 21 kbar.

As a result of their investigation of the piezoelectric behavior of tourmaline (*inter alia*) "beams" bent in the cantilever mode, Williams et al. (1975) reported: 1) the relations between the measured voltage and the applied force is linear; 2) the sign of measured voltage reverses when the beams are turned end-for-end in the clamping device; and 3) the magnitude of measured voltage does not vary with electrode location along the length of the beam. The second result, although apparently not expected, is in accord

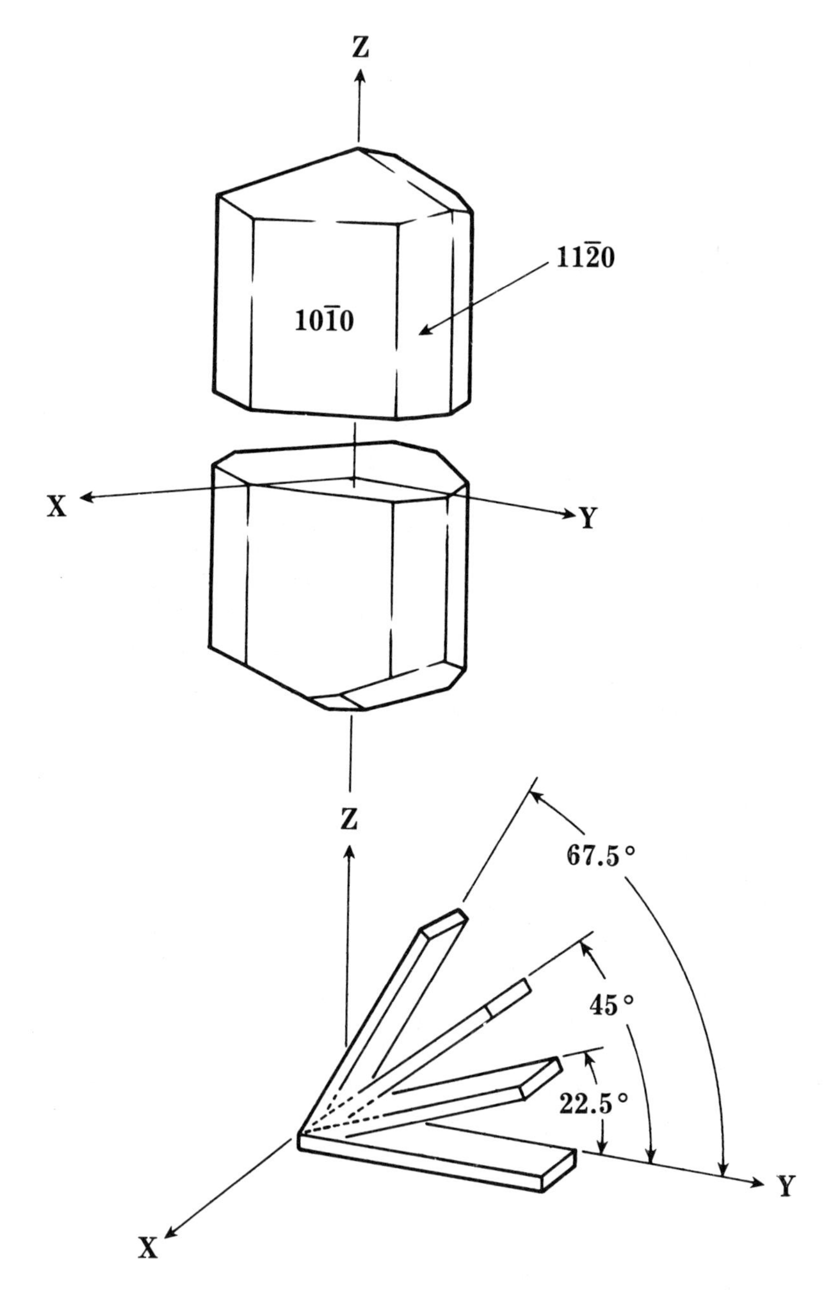

Figure 7-6
Plates used for measuring piezoelectric constants of tourmaline (after Mason, 1950).

with the rules of tensor transformation for this geometry (see, for example, Nye, 1954).

By using Lycopodium nodal patterns, luminous effects, and frequency measurements, Petrzilka (1932*a*, 1932*b*, and 1937) studied diverse vibrational modes of Z-cut tourmaline. The patterns were studied primarily to check Love's theory of radial and circumferential waves. Koga (1933), by the way, has derived equations for thickness vibrations in plates of tourmaline. Much additional information is given about this subject in the papers of Petrzilka and a few others cited by Cady (1946).

Frondel (1948) gave "11.08 micromicrocoulombs per lb. per sq. in." as the best value for the electrostatic charge developed per unit area of a Z-cut plate per unit change of pressure; the value is for somewhat zoned, chiefly green tourmalines from Madagascar and Brazil (Frondel, 1984, personal communication). Poynting and Thomson (1914) gave the piezoelectric charge produced by 100 atmospheres as 5.4 e.s.u. cm^{-2} (= 16.2×10^9 Coulombs) for adiabatic pressure, which is only 0.18 different from the value 5.32 e.s.u. cm^{-2} (= 15.96×10^9 Coulombs) for the isothermal constant. Keys (1923) stated that the difference between the isothermal and adiabatic piezoelectric constants of tourmaline is so small—he gave a figure of < 0.33 percent—that the error resulting in calibrating crystals isothermally and using them adiabatically is essentially negligible.

(The compiler (Dietrich, 1982) posed the question as to whether anyone has investigated the possibility that pre-earthquake anomalous activities of, for example, chickens might be responses triggered by the presence of tourmaline or some other piezoelectric material in the animals' systems and/or concentrated in the ground material beneath them(?). The question, posed despite the fact that most biophysicists believe that the reactions are probably triggered by a release of radon and consequent ionization, was prompted by the knowledge that tiny particles of magnetite are present in several animals—for example, certain bacteria (Blakemore, 1975), chitons' teeth (Kirschvink and Lowenstam, 1979), pigeons' heads (Walcott, Gould, and Kirschvink, 1979), and honey bees (Gould, Kirschvink, and Deffeyes 1978)—*and* appear to influence certain activities; for example, in pigeons, the particles are thought to influence, if not control, homing. The compiler did not know, at the time, that there has been significant research of both piezo- and pyroelectric effects in bones, tendons, cartilege, and nerves of certain vertebrates (see, for example, Shamos and Levine, 1967, and Lang, 1969). Nonetheless, the question remains.)

Chapter 8

Synthesis

Tourmaline synthesis has been investigated with one or both of two objectives—to produce crystals suitable for use in, for example, piezoelectric gauges and/or to determine stability ranges.

SYNTHESIS

In 1882, in broaching the subject of tourmaline synthesis, Fouqué and Michel-Lévy noted that (1) no proved tourmaline had yet been synthesized, and (2) in their opinion, tourmaline is formed in nature in the presence of volatile halides. Sixty-five years later, synthesis of undoubted tourmaline was recorded by Frondel, Hurlbut, and Collette (1947). Subsequently, several syntheses—using diverse feeds plus or minus seeds, in hydrothermal apparatus or autoclaves, at temperatures ranging from 350° to 900°C and pressures up to 3 kbar—have been successful (see Table 8-1).

To elaborate on the information upon which Table 8-1 is based: Frondel, Hurlbut, and Collette (1947) melted schorl in contact with water solutions of Mg and alkali borates during all of their runs. Slender prismatic crystals about 0.5 microns long were obtained. Some of these crystals were described and illustrated on the basis of electron microscopy by Barnes (1950).

Smith (1949) reported the stability field of his tourmaline to be in weakly alkaline to weakly acid conditions and, along with albite(!), in alkaline solutions (pH 8-8.6) containing less than ~45 percent H_2O.

The growth of tourmaline upon heating sericitic shale in the presence of a K-borate solution by Michel Lévy (1949) is particularly interesting. Her other syntheses (1949 and 1953) resulted in the formation of sheaflike crystals and rosettes of Fe-tourmalines. She concluded, *inter alia,* that tourmaline must form above the critical temperature of water and that the presence of fluorine is not essential to tourmaline synthesis.

Table 8-1
Tourmaline Syntheses

Feed	*Conditions*	*Reference*
glass (melted schorl)	400°-500°C	Frondel, Hurlbut, and Collette, 1947
component oxides + water	400°-450°C, >~1 kbar	Smith, 1949
shale + K-boarate	400°C, 0.45 kbar	Michel-Lévy, 1949
diverse[a]	400°-500°C, 0.2-4 kbar	Michel-Lévy, 1953
minerals[b] + NaCl and H_3BO_3	350°C, 2 kbar to 550°C, 0.7 kbar	Frondel and Collette, 1957
powdered glasses of B_2O_3 and other constituents	(see Fig. 4.1)	Robbins, Yoder, and Schairer, 1959
oxides or minerals[c] in H_2O sols of H_3BO_3 + NaCl or NaF	350°-600°C, 0.4-3 kbar	Emel'yanova and Zigareva, 1960
		Darragh, Gaskin, and Taylor, 1967
oxides ± carbonates	400°-600°C	Taylor and Terrell, 1967
see below[d]	[d]600°-700°C, 0.7-2 kbar	Tomisaka, 1968
mixture with dravite composition	400°-900°C, 0.2-1 kbar[e]	Ushio and Sumiyoshi, 1971
diverse mixtures, some with minerals[f]	400°-750±°C, 1-7 kbar	Voskresenskaya (see text)
like above in diverse, high-concentration, chloride media	650°-750°C, 2-6 kbar	Kovyzhenko, 1974
oxides in gels and H_3BO_3	650°-830°C, 4-7 kbar	Werding and Schreyer, 1977
oxide system	hydrothermal	Rosenberg and Foit, 1979

Note: Durations of reported runs range from 6 hours to 30 days, the latter reported to be without the attainment of equilibrium.
[a] SiO_2, Al_2O_3, $Na_2B_4O_7 \cdot 10H_2O$, K_2SiF_6 (or Na_2SiF_6), Li_2CO_3, and Fe ± Mg, Mn and/or Ca oxides or carbonates.
[b] Several mineral mixtures that would supply the required Si, Al, Fe were used, for example magnetite, quartz, and corundum.
[c] Minerals were aluminosilicates such as topaz, pyrophyllite, and kaolinite.
[d] H_2BO_3, $NaHCO_3$ or Na_2CO_3, MgO, $FeCO_3$, $MnCO_3$, Li_2SiO_4 or $LiNO_4$, γAl_2O_3, cristobalite (etc.) + an excess of H_2O.
[e] Formation was at ≤750°C at 1 kbar and ≤700°C at 0.5 kbar.
[f] For example, in the work reported in 1965, she used topaz, kyanite, sillimanite, with an aqueous soluton of H_3BO_3 and NaF plus small quantities of LiF and CaF_2.

Frondel and Collette (1957) reported that

In general tourmaline is not formed if the added minerals contribute alkalies to produce strongly alkaline solutions. [And] A high content of Ca or a high ratio of Ca, Mg and Fe relative to (Al, Si) is also unfavorable. [That is,] tourmaline is obtained, depending on temperature and other factors, in weakly alkaline to moderately acid solutions even though alkalies may be present in amounts considerably over the requirements of the formula. [and] The best results are obtained in acid solutions (pp. 754, 755).

In addition, they suggested that the pH control may really be "through an influence on the type of borate anions present in the solution, with (BO_3) groups, such as in the structure of tourmaline, present in the acid range and $[B(OH)_4]$ or other tetrahedral or polynuclear complexes present in the strongly alkaline range" (Frondel and Collette, 1957, pp. 755, 756).

Dravite prisms up to 0.3 mm long and uvite spherulites ~0.1 mm in diameter were synthesized during the investigation of the stability ranges of tourmaline by Robbins, Yoder, and Schairer (1959) (see Fig. 4-11).

As a result of their experiments, Emel'yanova and Zigareva (1960) recorded crystals up to 0.6×0.1 mm and black tourmaline layers up to 1 mm thick on natural "rubellite" (elbaite) seed crystals.

Darragh, Gaskin, and Taylor (1967) synthesized tourmalines with K and Ca (but not Na) in *X*-sites and V, Cr, Mn, Co, Ni, Cu, and Zn in "Mg or Fe [*Y*-]-sites."

Among the tourmalines produced by Taylor and Terrell (1967) are those characterized, according to *X*- and *Y*-sites, as NaFe, CaFe, LiFe, KFe, NaCo, KCo, NaNi, LiNi, KNi, CaNi, NaCu, NaZn, NaMn, KMn, CaMn, NaCr, NaV, NaMg, CaMg, and LiMg tourmalines and also an alkali-deficient Fe-tourmaline that contains only 0.12 atoms (0.1 Na and 0.02 K) in the *X*-site and $Fe_{2.2}$ $Al_{0.9}$ in the *Y*-site. It is not clear from their paper what the probable *X*-site occupancy is in the LiFe, LiMg, and LiNi tourmalines.

Tomisaka (1968) synthesized dravite, schorl, "tsilaisite," and two elbaites in sealed gold tubes in a hydrothermal apparatus.

Ushio and Sumiyoshi (1971) synthesized a fluorohydroxyl dravite ($NaMg_3$ $Al_6B_3Si_6O_{27}F_x$ $(OH)_{4-x}$) and found it to be stable in aqueous solutions of low alkalinity or low acidity but unstable in solutions that were strongly alkaline (cf. Smith, 1949). In addition, they found that the F-content increased with increased temperatures and pressures and that x was in the range 0.60-0.85.

Kovyzhenko (1974) reported the c and a cell dimensions of synthetic colorless dravite to be less than those of similarly constituted natural dravites. He also noted that the synthetic crystals tend to have fewer crystal faces than their natural analogs.

Voskresenskaya, alone and with various coworkers (e.g., 1965, 1967, 1968, 1971, 1973, and 1975), has recorded the following results and observations *(inter alia)* relating to tourmaline synthesis.

The Fe-content, as manifested by black color, has been derived by leaching from the walls of the autoclaves because of corrosion by F and/or Cl ions. (This is cited as support for the idea that tourmaline is an "iron-absorber".)

Conditions favorable to, charges required for, and speed of growth during synthesis of several diverse tourmalines—for example, Al and Ga analogs of elbaite, bright cherry-red Co-tourmalines, bright green Ni-, dark green Cr-, and colorless Mn-tourmalines (cf. "Miscellany" subsection in Chapter 6, "Color . . .")—were determined. It is concluded for instance, that a small amount of Ca seems to be essential for the formation of elbaite (Voskresenskaya and Barsukova, 1971) and that homogeneous colorless crystals can be synthesized at 750°C and pressures in the 1 kbar range by using highly concentrated chloride media (Voskresenskaya and Shternberg, 1973).

Runs yielding tourmaline alone and runs that gave tourmaline plus one or more other phases—for example, chiolite, cryolithionite, cryolite, fayalite, hercynite, and mullite—are recorded. The charges and conditions yielding each product are given but apparently no stability relations were determined.

Preliminary measurements indicate synthetic colorless elbaite crystals to have superior piezoelectric properties to those of natural achroites (elbaites?).

Werding and Schreyer (1977) found tourmaline, with no Na, Ca, or K, as a possible phase in the system SiO_2-Al_2O_3-MgO-B_2O_3-H_2O.

Rosenberg and Foit (1979) synthesized what they termed alkali-free tourmalines in the system MgO-Al_2O_3-SiO_2-B_2O_3-H_2O in the presence of excess B_2O_3 and H_2O at 1 kbar. The recorded formulae (and temperatures of synthesis) are: $\square_{0.92}\ Mg_{0.03}\ (Mg_{1.98}\ Al_{1.02})\ Al_6\ (Si_{5.59}\ Al_{0.41})\ (BO_3)_3\ O_{17.74}\ (OH)_{4.26}$- (600°C); and $\square_{0.04}\ Mg_{0.96}\ (Mg_{1.52}\ Al_{1.48})\ Al_6\ (Si_{5.23}\ Al_{0.77})\ (BO_3)_3\ O_{19.60}\ (OH)_{2.40}$- (450°C). They explained these tourmaline formulae by invoking the substitution $Mg^{2+} + Na^{+} \rightarrow Al^{3+} + \square$, where $\square$ indicates vacancy. They also stated that the tourmaline produced at 600°C is an "Al-rich, alkali-defective dravite" (cf. the so-called magnodravite of Wang and Hsu, 1966, alluded to in the chapter on Chemistry).

STABILITY

Results of the experiments of Frondel and Collette (1957), if sorted out, may bear on stability relations and consequently on the petrogenesis of tourmaline-bearing suites. This is true because the feed for each experiment was one or more natural minerals—for example: corundum and quartz;

topaz; staurolite; montmorillonite; and zoisite—in addition to a water solution of NaCl (0.3N), H_3BO_3 (3.ON), and magnetite, and the temperature and pressure conditions ranged between 250° and 550°C and 250 and 2000 bars.

Michel-Lévy (1953) submitted diverse mixtures with compositions of biotite plus borax plus or minus other components to temperatures of 450°-500°C in order to investigate the apparent incompatibility of biotite and tourmaline (see, for example, Termier, Owodenko, and Agard, 1950). As shown in Table 8-2, she did not obtain tourmaline *and* biotite from any of her mixtures.

As mentioned in the section about thermal properties, Robbins, Yoder, and Schairer (1959) and Robbins and Yoder (1962) have investigated the stability of dravite. They reported that dravite from Mesa Grande, California, appears to melt incongruently at ~700°C and 2 kbar and that synthetic dravite decomposes or dissociates as indicated on Figure 4-11.

MISCELLANY

Mason, Donnay, and Hardie (1964) reported that Barton synthesized tourmaline with cell dimensions similar to those of buergerite. The syntheses were accomplished at 700°C and 2 kbar in runs ranging from 166 to 333 days; the feeds were diverse mixtures of oxides plus, for example, $NaHCO_3$, H_3BO_3, and H_2O (Barton, 1982, personal communication, and Appendix to Barton, 1968). In addition, Barton noted that E. H. Chown (1960, unpublished) also synthesized tourmaline in the Johns Hopkins University laboratory, apparently within the stability field of magnetite.

Slivko and Voskresenskaya (1966) have compared certain growth features in synthetic crystals with similar features in natural crystals. Among other things, they directed attention to polar growth on spherical seeds (Fig. 8-1), tubular cavities above solid inclusions (Fig. 5-1) and multi-headed growth.

An unexpected and unexplained result—well-terminated acicular crystals of tourmaline, up to 0.3 mm long and 10μm thick—was obtained by reacting

Table 8-2
Investigations of Tourmaline-biotite "Incompatibility"

Feed	*Temperature*	*Product*
biotite + borax (~10%)	500°C	biotite
biotite + borax (~10%)	450°C	tourmaline + white mica
biotite + borax + SiO_2 + Al_2O_3	450° and 550°C	tourmaline + white mica
biotite + tourmaline + borax	450° and 550°C	tourmaline + white(?) mica

Source: Data from Michel-Lévy (1953).

Figure 8-1
Polar growth on a spherical seed (~4X); growth is on analogous end (after Voskresenskaya and Barsukova, 1971).

sulfur with natural almandine ($Alm_{53.5}$ $Pyr_{20.4}$ $Grs_{21.0}$ $And_{3.8}$ $Sps_{1.3}$) in the presence of H_2O in sealed gold tubes in hydrothermal apparatus at 700°C and 1 kbar (MacRae and Kullerud, 1971). The products were a Mg-tourmaline (1% or less), Mg-cordierite (80-85%), Fe-sulfide (10-15%), and anhydrite (3-4%). (The boron required for the tourmaline was probably an impurity in the garnet.)

Chapter 9

Uses and Recovery

USES

Actual and suggested potential uses for tourmaline may be considered in five categories: (1) in geologic investigations, (2) as a standard test material, (3) as a component of scientific and industrial instruments, (4) as a gemstone and in the decorative arts, and (5) for miscellaneous purposes.

The utilization of tourmaline in geologic, especially petrologic, investigations is treated in the "Synthesis" and "Geneses . . ." chapters. Use as a gemstone and in the decorative arts is the theme of the next chapter. Uses in the other three categories are dealt with in this chapter.

Tourmaline as a Standard Test Material

Tourmaline has been used rather frequently as a standard for calibration of, for example, piezoelectric manometers, and for testing, for example, the possible applicability of a device or procedure for use on materials having complex structures and/or compositions. In addition, it has been used as an experimental control substance because of its being a boron-containing compound that is essentially inert under nearly all conditions known to obtain at or near the surface of the earth.

The following examples are listed in order of publication dates of the cited reports.

For well over a half century, tourmaline has been used as a standard in tests made to check possible effects of water-soluble boron in mixed fertilizers (e.g., Bartlett, 1923). This use is based on tourmaline's being a boron-containing mineral that is so chemically nonreactive (i.e., a nonreleaser of boron). (However, cf. Brammall and Leech, 1937.)

In 1958, Vishnevskii and Klimovskaya used tourmaline as a check in their study of absorption spectra of microcrystals.

In 1966, Kupkova used tourmaline as one of the minerals in the study of the effect of the structure of silicates on the spectral emission of their main components (e.g., Fe, Mg, Al, and Si). Correlations were made between arc temperature and $\Delta Y_{Mg/Mo}$ values. Among other things, Kupkova stated that

> from the practical pont of view the effect of the structure, expressed quantitatively by the ΔY_{Mg} values, can be useful for deciding whether the element analyzed can be determined from a single calibration curve constructed by means of one or more minerals . . . [and] This criterion can also be used for eliminating the effect of the structure in spectral analysis, since its size is known (p. 4056).

Diehl (1970), after reviewing the physical constants of tourmaline and quartz, the construction of piezoelectric pressure gauges, and the possible errors of measurement, showed how quartz and tourmaline can be used to evaluate the accuracy of static calibration of such gauges.

Pretorius, Odendaal, and Peisach (1972) chose tourmaline, because "its chemical composition is typical of most geological samples," to check the applicability of deuteron activation analysis for geologic samples. Their main conclusion was that deuteron activation analysis can be used in conjunction with neutron activation analysis on a routine basis, and that large numbers of samples can be analyzed for Mg, Al, Mn, Li, Na, Fe, and B so long as they are present in quantities of greater than about 0.1 weight percent.

Another example of these kinds of uses depends on the already mentioned unexpected result of an investigation of Iijima, Cowley, and Donnay (1973). They discovered that electron irradiation used in high-resolution imaging of tourmaline's structure apparently caused structural changes. Consequently, this possibility must be kept in mind when any mineral is so studied.

Tourmaline in Instruments

As noted in the following subsections, tourmaline has been used or suggested for use as a component in several scientific and other instruments and devices because of its optical or electrical properties. In addition, Spottswood (1911) has alluded to its use for jewel bearings.

Optical Property Use

Deeply colored tourmaline plane-polarizes light with an effect that is readily visible macroscopically. Because of this and because of its availability,

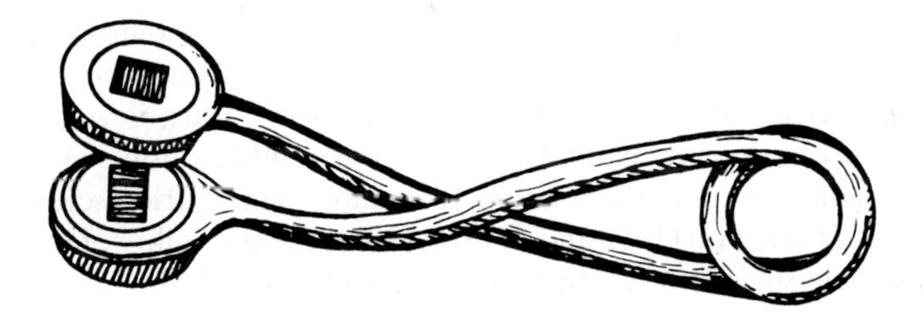

Figure 9-1
Tourmaline tongs (after a sketch in Haidinger, 1845).

tourmaline may have found use, along with cordierite, as a "sunstone" to aid the Vikings in their navigation in the north Atlantic (Ramiskou, 1969).

According to duBois-Reymond and Schaefer (1908), Thomas Johann Seebeck first discovered the polarizing quality of tourmaline in 1813 (a year before he detected it in biotite!), and Karl Michael Marx first constructed tourmaline tongs in 1827. Tourmaline tongs have found widespread use, especially by jewelers, as an easily portable polariscope to distinguish isotropic from anisotropic materials that resemble each other (e.g., red spinel from ruby). The "tongs" consist of two thin slices of colored tourmaline cut parallel to *c* and mounted so that they polarize and analyze light (Fig. 9-1).

Electrical Property Uses

Although tourmaline is both pyro- and piezoelectric, to the present, the latter, especially the converse piezoelectric effect, has received most of the consideration so far as possible application. Nonetheless, pyroelectric effects may find use in the future.

Piezoelectric properties of certain tourmalines have led to its being considered for utilization in several instruments and devices. Waesche (1949) has given a résumé of uses of piezoelectric materials known or predicted up to that time. For some of the possible applications, tourmaline has important advantages over other piezoelectric substances (e.g., quartz and Rochelle salt) because tourmaline is mechanically strong and chemically insoluble in acid, has a high frequency:thickness constant and hydrostatic sensitivity, and exhibits linear piezoelectric effects. On the other hand, "no known slice of tourmaline, cut with specific relations to the crystallographic axes, has the so-called zero-coefficient frequency-temperature characteristic of certain quartz 'cuts' " (Waesche, 1949, p. 13).

Kurtz and Perry (1968) investigated the second-harmonic generation of a powder (75-100μ) of green tourmaline during their evaluation of approximately 100 nonlinear optical materials, and stated that the tourmaline tested falls in "Class C: nonlinear coefficients greater than crystalline quartz; not phase-matchable for second-harmonic generation"; thus, it would not be suitable for use in, for example, efficient second-harmonic generators, broadband electro-optic modulators, or tunable parametric oscillators.

Lewis and Patterson (1972 and 1973) assessed the suitability of tourmaline as an acoustic-surface-wave-delay medium and concluded

> the advantages of tourmaline relative to the most commonly used crystal, viz. quartz [are]: (1) Taking 1 μm as a practical lower limit to the finger widths/gaps of an IDT, quartz can be used up to 800 MHz, tourmaline up to 1100 MHz. (2) The attenuation at 1 GHz is of the order of 4 dB/μsec for quartz, but probably less than 1 dB/μsec for tourmaline. (3) The optimum fractional bandwidth is 5% for quartz and 6% for tourmaline. [However,] One disadvantage of tourmaline is the absence of cuts with the zero-temperature coefficient of delay (Lewis and Patterson, 1972).

Examples of possible uses are in instruments to detect submarines and other manmade and also natural underwater obstructions (see, for example, Cady, 1921), depth-sounding apparatuses, transducers to detect and measure sudden large hydrostatic pressure differences produced by explosions either in air or under water and including such diverse phenomena as nuclear blasts and the firing of internal combustion engines (e.g., Keys, 1921; Arons et al., 1947; Frondel, 1948; Arons and Cole, 1950; Sinani, 1957; Hearst, Irani, and Geesaman, 1965; and Hofmann, 1970), motors for converting electrical into mechanical energy, generators for turning mechanical energy into electrical energy, capacitors for storing electrical energy, frequency control devices (especially for communication equipment such as radio transmitters), gas-detecting sensors (Yamaguchi, 1983), and piezoelectric electric lighters (i.e., spark generators) for gas torches, gas stoves, bunsen burners (etc.). (See Mason, 1950, for other former uses.)

[It is of passing interest that the felt need for tourmaline for pressure gauges during and immediately after World War II led to most of the early attempts to synthesize tourmaline and also to a widespread search for natural tourmaline crystals from which suitable plates could be cut.—In 1944, the latter search is reported to have involved Martin Ehrmann in some cloak-and-dagger work: as a member of the U.S. Intelligence Service, he was to go "into German-occupied France and bring out several tons of Madagascar tourmaline urgently needed for piezoelectric plates for making pressure gauges." (Switzer, 1974). He did, by the way, succeed (Frondel, 1984, personal communication).

Pyroelectric properties of tourmaline have been suggested for use as highly sensitive and accurate pyroelectric thermometers for large temperature ranges, especially at high temperatures (Lang, 1974, and Hamid, 1980) and also as pyroelectric infrared radiation detectors (Hamid, 1980). Both uses, according to Hamid, are possible because tourmaline is not ferroelectric and does not undergo any phase change up to relatively high temperatures.

Miscellaneous Uses

The following additional uses—some realized, others only suggested as potential—are noteworthy:

1. the manufacture of Al_2O_3 (Kamiko, 1944);
2. the manufacture of borax (Kang and Pak, 1971);
3. the manufacture of boric acid, H_3BO_3 (e.g., Kamiko, 1944; Karashima et al., 1946; Hatoya, 1948; and Nagai, 1944);
4. the source of H_3BO_3 for the manufacture of ampule glass—also containing Al_2O_3 in excess of 10 percent and typically green in color because of Fe-content (Kaneko, Yasuhara, and Toshio, 1943);
5. the manufacture of a B-fertilizer after fusion with limestone at 1000°C for two hours and powdering of the resulting sintered mass (Katalymov, 1941; cf. Polyak and Devyatovskaya, 1957);
6. a component of certain specialty concretes for nuclear (neutron) reactor shields (see, for example, Bourgeois, Ertand, and Jacquesson, 1953, and Rastoin, 1966);
7. a welding flux (Okuda, Tamita, and Tanaka, 1973);
8. a flux in blast furnace production of low-S pig iron (Zagyaneskĭi, 1958);
9. a laser detector;
10. a track detector; that is, for differentiation of heavier ions and energies (e.g., Balzer and Sigrist, 1972, and Sigrist and Balzer, 1977); and
11. a thermal dosimeter to measure the intensity of radium (Aitkens, 1932) and/or neutron (Vargas and Tupinambá, 1962; cf. Moon, 1948) irradiation.

Other Data Relating to Possible Uses

Keys (1923) stated that the difference between the isothermal and adiabatic piezoelectric constants of tourmaline is so small (< 0.33 percent) that the error resulting in calibrating crystals isothermally and using them adiabatically is essentially neglibible.

Thuraisingham and Stephens (1975) investigated the effects of gamma-irradiation and annealing on the attenuation of hypersonic compressional waves in tourmaline, and measured changes in the ultrasonic attenuation of a crystal that was submitted to successive doses of irradiation and annealing that apparently altered the concentration of the crystal's imperfections (defects). Their results appear to support the hypothesis of Silverman (1968) that holds that irradiation can cuase imperfections that can act as scattering centers, thus lowering the thermal phonon relaxation time (frequency) and thereby influencing ultrasonic attenuation.

The combination of its high Debye temperature (550-800 K) and its large number of atoms per unit cell indicate that tourmaline should have a very low thermal phonon attenuation (Oliver and Slack, 1966). In fact, Lewis and Patterson (1973), upon measuring the attenuation of longitudinal waves propagated along *c* for a dravite, found the attenuation to be low enough to make tourmaline of possible use in instruments for measuring ultrasonic absorption in the gigahertz range.

Later, Helme and King (1977) made microwave acoustic absorption measurements on dark green schorl (of the schorl-elbaite series) at 580 MHz and 1.03 GHz as a function of temperature between 1.5K and 300K and reported:

> In addition to thermal-phonon attenuation, two large low temperature peaks are found. These peaks [at ~11K and 45K] are most pronounced for transverse wave propagation and are found together or individually, depending on the mode propagated. . . . It is tentatively suggested that the peaks may be caused by Fe^{2+} ions on distorted octahedral sites (p. 1535).

In addition, they also noted that (1) in some cases, the peaks extend to room temperature, which would adversely affect the performance of tourmaline as a low loss instrument material; and (2) in tourmalines having little or no Fe^{2+}, the only significant attenuation—among the lowest observed for any crystalline material—is that due to phonon-phonon interactions.

RECOVERY

Processes used in the recovery of tourmaline, both as described in the literature and as observed by the writer, are summarized briefly in this section.

In a study of the relative floatability of silicate minerals, Patek (1934) presented data on the order of floatability and the average contact angles between each of the minerals and liquid drops. He found black tourmaline to be easily separable from all minerals studied except topaz, almandine, and zircon and to assume a contact angle of 47° with the liquid drops. This contrasts with the contact angle of 80-88° for drops of water on elbaite recorded by Nassau and Schonhorn (1978).

Methods yielding up to 98 percent of total B-content of tourmaline-bearing rock are discussed by Stoicovici, Ghergaru, and Motiu (1957); flotation methods for tourmaline in general are described by Naifonov, Pol'kin, and Shafeev (1963) and Gladkikh and Pol'kin (1963); study of diverse methods of separating tourmaline from pegmatites, hornfels, and aplite show that flotation and electrostatic methods are rather unsatisfactory whereas high-intensity electromagnetic separation methods give good results (Potuzah, 1965); selective flotation of tourmaline, beryl, and garnet are described by Chipanin and Kozhukhovskaya (1967).

Other mineral dressers or extractive metallurgists have discussed the separation of, for example, cassiterite from tourmaline and other associated minerals. Some of them noted the yield of tourmaline-rich fractions (e.g., Robert, 1968, and Berger, Hoberg, and Schneider, 1980).

The mining observed by the writer has had as its objective the recovery of gem-quality tourmaline and/or mineral specimens. It consists of following pegmatite dikes and lenses, by both surface and subsurface methods, opening the tourmaline-bearing pockets by careful blasting, and hand excavation of the pockets to remove the clay (etc.) while recovering unaltered tourmaline (etc.).

As an example of production, Dillon (1983) gives the following figures for the Himalaya Mine of San Diego County, California: "The average yield for one month is 100 to 200 pounds of all qualities. Approximately 80% is bead material, 10% fine bead, 8% cabochon, and 2% suitable for faceting" (p. 61). The "yield," of course, includes only the usable material.

Chapter 10

Tourmaline as a Gemstone and in the Decorative Arts

It seems likely that tourmaline was used as a decorative stone by at least the time of the Roman Empire (~27 B.C. to 395 A.D.). This conjecture is based on the fact that tourmaline occurred associated with many of the gems and decorative stones that were taken from deposits in the Orient during that period. In any case, it is known that tourmaline, for the most part elbaite but also dravite/uvite and schorl, has been used for gems and small carvings for the last three centuries (see, for example, Brückmann, 1773).

During the early 1900s, the value of the annual production of tourmaline in the United States averaged about $100,000 (Spottswood, 1911). More recently, tourmaline has become the most popular of all colored stones, and currently its annual market value runs into the millions of dollars.

Another "use" that appeals chiefly to man's aesthetic sense is the inclusion of tourmaline specimens as showpieces in mineral collections. Many mineral collectors consider it almost imperative that they have at least one showpiece specimen of tourmaline. Both free crystals and crystals attached to their original matrices are prized. This practice dates back to at least the late 1700s (see, for example, Wagner, 1818; Lévy, 1838; and Giesecke, 1832); also, according to Greene and Burke (1978) "In 1780 Archduchess Marie-Anne of Austria, who had formed an extensive cabinet, presented several beautiful specimens of the tourmaline of Tyrol to Duke Charles of Lorraine [her future spouse] to enrich his collection" (p. 8) [ref. Birembaut, A., 1957, "La Minéralogie," in *Historie de la Science* M. Daumas, ed., Paris]; these specimens, however, may have been epidote. According to another interesting report (Mawe, 1818): at a Devonshire locality, there was

> tourmaline in crystals of from two to four inches in diameter [for which] the owner asked a gentleman a few shillings for one of them, who, after selecting as many as he could well carry, replied, "Farmer, they are neither gold nor silver," and gave him a

shilling . . . [In addition, the owner used the tourmaline-containing rock] to build a wall on his farm, and disliking its "magpie" appearance, had it *whitewashed* (p.369).

Fairly recently, a single specimen, the "Joninha" (Plate III) from Minas Gerias, Brazil, was sold for $1.3 million (Keith Proctor, 1984, personal communication).

Perhaps the most august station for tourmaline to the present, however, is its inclusion in the sarcophagus of the Duke of Wellington (1779-1852). According to Bonney (1877), this sarcophagus, which is in St. Paul's in London, is made up of a tourmaline-rich rock, tourmalinite, from Tregonning in Cornwall.

LORE ASSOCIATED WITH TOURMALINE AS A GEMSTONE

Unlike many popular gemstones, tourmaline has little associated lore, particularly in Western cultures (e.g., Kunz, 1913). This lack stems from the fact that tourmaline was not recognized as a distinct mineral (i.e., a gemstone) until relatively recently (see DeBoodt, 1609). A few exceptions follow:

If the *Lyngorium* of Theophrastus (~315 B.C.) was tourmaline, rather than amber or zircon, as sometimes interpreted, it is noteworthy that Theophrastus wrote the following: lyngorium is "formed by the lynx' urine, which the animal, as soon as it parts with it, hides by scraping the earth over it. . . [and] the stones vary according to the sex and disposition of the animal."

Sinkankas (1971) has noted:

Dreaming of tourmaline is supposed to insure success through superior knowledge but there seems to be no evidence that the residents of tourmaline mining areas in Maine or California, where such dreams are particularly common, are any better off as a result (p. 44).

(In corroboration of Sinkankas' overall conclusion, the compiler's name should probably be added to those of the residents.)

Adrian (1962) has reported the beliefs that (1) tourmaline, if worn by a woman, makes her more desirable to men; (2) tourmaline facilitates (promotes) friendship; (3) tourmaline guarantees chastity (purity) and/or good health; and (4) tourmaline stimulates imagination. His source for these beliefs is not given.

Benesch (1981), apparently recording beliefs held by "Christengemeinschaft," a protestant denomination in Switzerland and Germany, which is closely related to the secular anthroposophic movement, has included several mysticial thoughts relating to the "spiritual content" (virtues) of tourmaline. He noted in particular how tourmaline's composition, colors, and color zoning epitomize the essence of flowering plants and animals' (e.g., butterflies')

PLATE I

Plate I
Pink elbaite from Pala Chief mine, San Diego County, California. The size and current location of the specimen is not known for sure, but is probably American Museum of Natural History (N.Y.C.) #42474, which is 8 inches (~20 cm) high. The original water-color painting (8½ × 6½ inches [~21.5 × 16.5 cm]) by an unknown artist, possibly Mrs. W. T. Schaller, was recently found in W. T. Schaller's papers in the U.S. National Museum of Natural History. *(Courtesy of the Smithsonian Institution)*

PLATE II

Plate II. A
Orange, golden yellow, and green elbaite crystals on quartz crystal from the vicinity of Coronel Murta, Minas Gerais, Brazil. The larger tourmaline crystal, 5 inches (~12.5 cm) long, is in the Keith Proctor collection, Colorado Springs, Colorado. Photograph by Harold and Erica Van Pelt. *(Courtesy of K. Proctor)*

Plate II. B
Green elbaite crystals on quartz crystal from the Cruzeiro mine, Minas Gerais, Brazil. The top elbaite crystal, 6 inches (~15 cm) long, is in the Keith Proctor collection, Colorado Springs, Colorado. Photograph by Harold and Erica Van Pelt. *(Courtesy of K. Proctor)*

PLATE III

Plate III. A
Blue elbaite with quartz from the Manoel Mutuca mine, near Virgem da Lapa, Minas Gerais, Brazil. The elbaite crystal, 4 inches (~10 cm) long, is in the Keith Proctor collection, Colorado Springs, Colorado. Photograph © 1983 Harold and Erica Van Pelt. *(Courtesy of Harold and Erica Van Pelt)*

Plate III. B
"Joninha"—cranberry-purple elbaite with quartz and albite (var. cleavelandite) from Itatiaia mine, Conselheiro Pena, Minas Gerais, Brazil. It is approximately 18 inches (~45 cm) across. The specimen was recently sold and is believed now to be in the United States. *(Courtesy METAMIG—Metais de Minas Gerais S/A)*

PLATE IV

Plate IV. A

"Candelabra"—red, white, and blue, color-zoned elbaite with quartz, albite, and lepidolite from the Tourmaline Queen mine, near Pala, San Diego County, California. The overall size is 10 × 9 × 6 inches (~25 × 23 × 15 cm). NMNH #132377. Photograph by Dane Penland. *(Courtesy of the Smithsonian Institution)*

Plate IV. B

"Steamboat"—red and green, color-zoned elbaite with quartz and albite (var. cleavelandite) from the Tourmaline King mine, near Pala, San Diego County, California. The overall height of the specimen is 11 inches (~28 cm). NMNH #R-51. Photograph © 1979 by Harold and Erica Van Pelt.

PLATE V

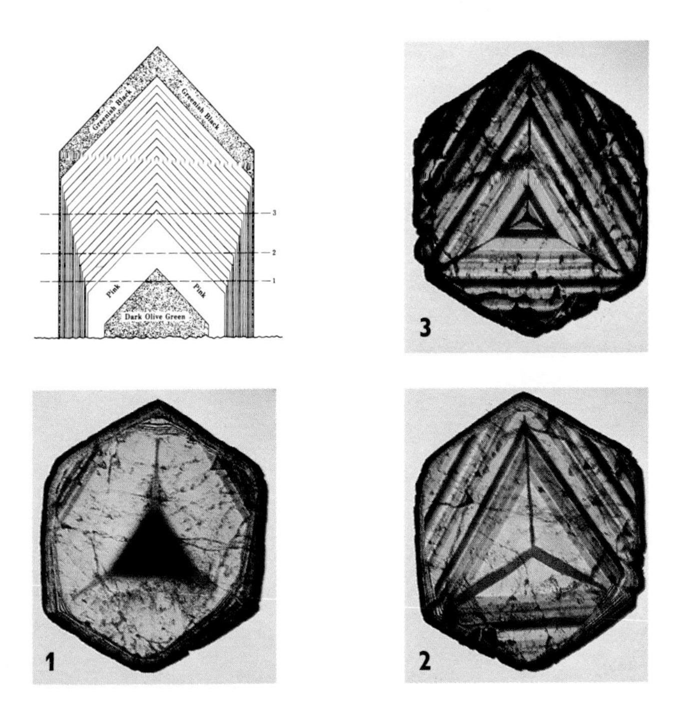

Plate V. A

Three slices of color-zoned elbaite crystal from Anjanabonoi na, Madagascar. The diagram indicates approximate positions from which the slices were cut. Each slice is 6¼ inches (~16 cm) across. The sketch is redrawn and modified after Althaus (1979). Photographs by Werner Lieber. *(Courtesy of W. Lieber)*

Plate V. B

Selection of slices of color-zoned tourmaline, cut perpendicular to *c*, from Madagascar. They range from ¾ to 3¼ inches (~2–8 cm) across. They are in a private collection in West Germany. Photograph by Eugene E. Foord. *(Courtesy of E. E. Foord)*

PLATE VI

Plate VI. A
Cat's-eye elbaite from Brazil, 1⅛ inches (~2.8 cm) long. NMNH #G-5700. Photograph by Dane Penland. *(Courtesy of the Smithsonian Institution)*

Plate VI. B
Faceted tourmalines from world-wide localities with sizes ranging from 8 to 20 carats. Alexander Blythe collection, Paso Robles, California. Photograph © 1983 by Harold and Erica Van Pelt.

PLATE VII

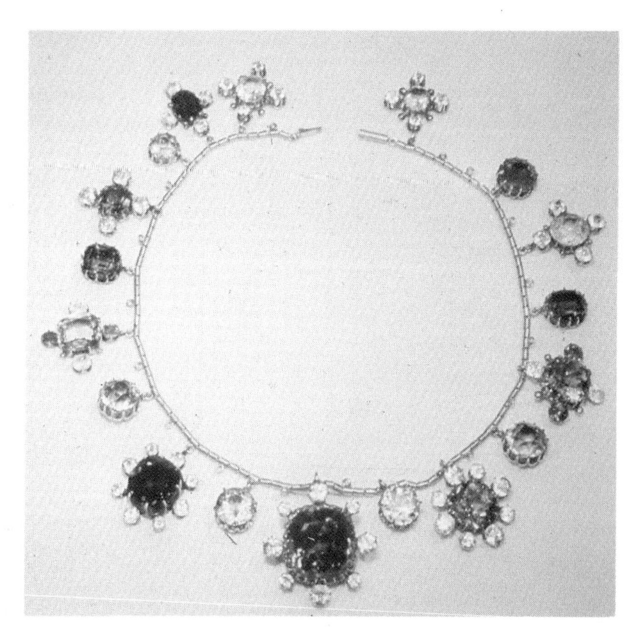

Plate VII. A
Hamlin necklace—all stones are elbaites taken from Mt. Mica, Maine, in the 1800s. Unnumbered piece in Harvard University Mineralogical Museum. For further description, see text. *(Courtesy of the State of Maine Museum)*

Plate VII. B
Cross-shaped pendant or brooch—all stones are tourmalines, from world-wide locations. Piece, 2½ inches (~6.3 cm) high, is in the Alexander Blythe collection, Paso Robles, California. Photograph © 1984 by Harold and Erica Van Pelt.

PLATE VIII

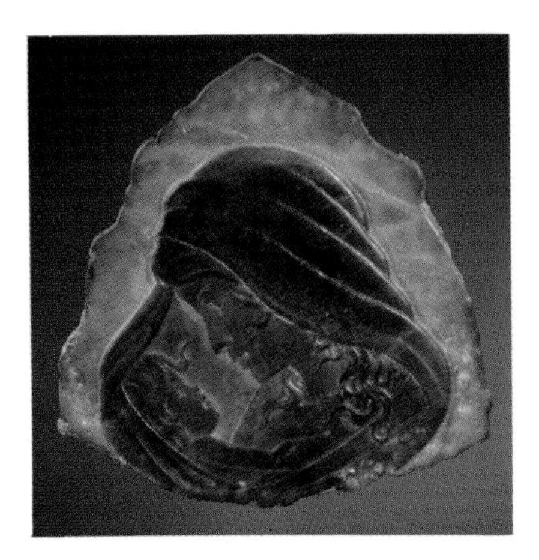

Plate VIII. A
Carving—engraved on blue cap of an elbaite crystal from the Tourmaline Queen Mine, near Pala, San Diego County, California. The crystal, 2⅝ inches (~6.5 cm) across, is in the Gerhard Becker collection, Idar Oberstein, West Germany. Photograph © 1980 by Harold and Erica Van Pelt.

Plate VIII. B
Color-zoned elbaite crystals with quartz and albite from Queen mine, Pala district, San Diego County, California. Overall specimen, 8 × 5 × 5 inches (~20 × 12.7 × 12.7 cm), is Harvard University Mineralogical Museum #87922 (Holden Collection). The original water-color painting by an unknown artist, possibly Mrs. W. T. Schaller, was recently found in W. T. Schaller's papers in the U.S. National Museum of Natural History. *(Courtesy of the Smithsonian Institution)*

colorations, and incarnate Man's soul, psyche, . . . In addition, the color hues are related to the resurrection of Christ, the eternal life of Man, and seasons of the year. Furthermore, it is suggested that tourmaline specimens should be placed between plants and candles on one's table at Easter time.

In his classic *Bunte Steine und Erzählungen,* Stifter (1853) included a story "Turmalin" in which tourmaline is seen as providing a warning against base instincts, especially ill-founded passion. The "warning" consists, in essence, of an analogy to a broken marriage that resulted in "punishment" of related innocents as well as of the unfaithful wife.

In the Orient, tourmaline has long had many uses that may be characterized as tradition-bound. For example, in the past, tourmaline was used with jade in necklaces worn only by Chinese aristocracy, and large, round buttons of red tourmaline were one of the marks distinguishing members of the Mandarin (Magistrate) class (Spottswood, 1911). (It was reported on page 14, Section 2 of the October 1983 issue of "The Stanford Observer" that the large Chinese market was essentially closed when "in 1912, the Dowager Empress of China died, and with news of her death [there was] spread a rumor that tourmaline carried a curse.")

In Brazil, green tourmaline ("Brazilian emerald") has been emblematic of the priesthood and thus worn as a ringstone by many of that calling (Bauer, 1896).

Today, tourmaline is widely accepted as the alternate birthstone (to opal) for the month of October, and, according to a brochure published by the Americn Gem Society, "it is said to exhilarate the spirit." However, as Sinkankas (1971) has noted, "there is little traditional or historical justification for this."

COLOR OF GEM TOURMALINE

According to Hamlin (1873), the French jeweler Barbot made the following statement about tourmaline in allusion to the way its colors resemble the colors of, for example, emerald, ruby, and sapphire: "It seems as if Nature had wished to prove to man that she could imitate in a degree almost perfect even that which she had created most perfect" (pp. 92, 93).

Indeed, gem-quality tourmaline is available in nearly all readily distinguishable color hues and saturations. "Gem-quality," in the case of tourmaline, means transparent with the exception of cat's-eyes and lustrous black tourmalines that were used formerly in mourning jewelry. The required transparency is emphasized by the designations "precious tourmaline" and "common tourmaline," which are used, particularly in the jewelry trade, to refer to transparent colored tourmaline and essentially opaque black to nearly black tourmaline, respectively.

Some of the most desired colors are described as pink, fine ruby- to violet- or cranberry-red, emerald-green, bluish green, yellowish green, indigo-blue, pure brown, golden brown, "champagne," greenish brown, and reddish brown (see Plate VI). In addition, there is always a demand for good so-called particolored stones and rare alexandritelike tourmalines. The particolored stones are those cut to exhibit two or three colors; the pink and green ones, which are the most desired, are widely referred to as "watermelon tourmalines." Most of the alexandritelike stones appear brownish or yellowish green in sunlight and orangy red in artificial light. (Cossa and Arzruni, 1882, in recording a chromium-bearing alexandritelike tourmaline from Schabry in the Urals, proposed that it be called deuterolite; fortunately this term has not found acceptance.)

The colors of tourmalines have been discussed in Chapter 6. The importance of considering dichroism when cutting tourmaline is dealt with in a succeeding section of this chapter. It is perhaps worth repeating here, however, that the green tourmalines are generally classified as Cr- and/or V-bearing or Cr- and/or V-free and, according to the literature, Cr-bearing tourmalines are more common than V-bearing tourmalines. Nonetheless, at least some of the gem-quality green tourmalines from, for example, Kenya and Tanzania are V-, rather than Cr-, rich dravites and uvites (Schmetzer and Bank, 1979); two localities are Gerevi Hills, Tanzania (see also Bassett, 1953), and the Umba Valley, Tanzania (see also Zwaan, 1974).

As noted in the section on the "Effect of Temperature and Irradiation on Color," several attempts have been made to enhance the color of gem tourmalines (see also Nassau, 1984). To the present, although some methods have been found to be somewhat unpredictable, generally the results are fairly consistent for tourmalines from individual localities (cf. Kálmán, 1941). Consequently, heating is used rather widely to lighten blue and green stones, in particular, and also to make some brownish red stones pink. In addition, irradiation is used in some cases to intensify colors, for example, to darken too light pink stones.

A good example of the use of heat treatment to lighten the natural color is the following, which is used by the Plumbago Mining Company of Rumford, Maine. According to Dean McCrillis (1983, personal communication) brownish burgundy colored stones are heated slowly to ~500°C, held there for 10 minutes, and slowly cooled to produce rose colored stones; or, they may be heated to ~505°C, held there for 15 minutes, and slowly cooled to produce pink stones. The "about right" speed of heating and cooling is two to two and a half hours each. The heating is done in a crucible in which the stones are surrounded by sand-sized grains of pegmatite material. Stones are generally treated after, rather than before, cutting and polishing because lapidaries have indicated that the treatment makes some stones become more brittle (i.e., more fragile) and thus more difficult to work.

Treatments to enhance the colors of tourmaline gemstones, as reported by Nassau (1984), appear to fit into one of three main categories: (1) an increase in red (e.g., pale colors to pink or red, blue to purple, and yellow to peach) as a result of irradiation; (2) an increase in yellow (e.g., pale colors to yellow, pink to orange, and blue-green to yellow-green) as a result of irradiation; and (3) a reduction in red (e.g., red to colorless, deep brownish red to pink, and purple to blue) as a result of heating. There are a few observed changes, however, that do not appear to be so easily categorized (e.g., dark blue to lighter blue-green or yellow-green as a result of heating). In addition, it is noteworthy that some treatments yield stable results whereas others yield changes that do not persist. A few stones revert to or toward their natural color even after only several weeks exposure to sunlight.

CAT'S-EYE TOURMALINES

Members of the staff of the Gemological Institute of America have examined several tourmaline cat's-eye stones (Fryer, 1982). The most common colors seen were diverse shades of pink, red, green, and blue. In addition, there are yellowish to greenish brown stones that resemble "precious" cat's-eyes (chrysoberyl), blue-green stones (Plate VI), and even stones that are essentially colorless. As previously mentioned, the chatoyancy in most of these stones is due to the alignment of hollow tubes or needle-like inclusions, including fluid inclusions. In most stones the alignments are more or less parallel to *c*. In a few, however, the chatoyancy is dependent upon striations on the base of the cut stone.

A few of these stones have been cut so the eye appears to float between two colors, such as green and blue, or between zones exhibiting different color saturation. For example, John White (1984, personal communication) has noted that the cat's-eye shown in Plate VI, when held "where a row of incandescent light are on my left and fluorescent light on my right, I get two eyes that are well separated. One is blue and one is green. As you might guess, it is the incandescent response that is green and the fluorescent that is blue."

(Tourmaline crystals that only appear to be chatoyant because of closely spaced striae do not exhibit the cat's-eye effect unless, as just mentioned, the stones are cut so the striae are still present on the relatively flat base.)

TOURMALINATED QUARTZ

Quartz with numerous inclusions of elongate tourmaline crystals has been used for faceted stones, cabochons, spheres, and other decorative items (Fig.

5-2). Some tourmalinated quartz has been incorrectly identified as rutilated quartz and called "Venus hair." As early as 1902, Kunz listed three United States localities—Mesa Grande, San Diego County, California; Haddam Neck, Middlesex County, Connecticut; and east of Butte, in Jefferson County, Montana—from which tourmalinated quartz has been recovered.

NAMES APPLIED

The names of the tourmaline species are not used as much as they should be in the gemstone trade. Instead, tourmaline gems are frequently referred to by: (1) the term tourmaline, either unmodified or described by, for example, a color modifier; (2) the color indicative terms rubellite, verdelite, and indicolite (or indigolite); or (3) terms such as "Brazilian chrysotile," "Brazilian emerald," "Brazilian peridot," "Brazilian sapphire," "Ceylon chrysolite," "Ceylon(ese) peridot," "Emeralite," "Siberian ruby," "siberite," and "Transvaal tourmaline" (a name sometimes used for green tourmaline no matter what its source). Only the first two nomenclatures are at all acceptable and the first is preferable (see Chapter 1).

The already alluded to particolored stones, tourmaline cat's-eyes, and alexandritelike tourmalines are generally so-named.

OCCURRENCES OF GEM-QUALITY TOURMALINE

Nearly all gem-quality tourmalines are elbaites from pegmatites or placers derived from pegmatities. In addition, there are gem-quality dravites and uvites, most of which have been found in carbonate rocks that have undergone contact metamorphism. (According to Dunn, 1977*b*, "all gem-grade facetable brown tourmalines examined . . . are uvite" (p. 307). If his sample is representative, gem-quality dravites are even less common than generally believed.)

Appendix III includes the better known localities from which gem-quality tourmaline has been produced.

FASHIONING

Some tourmaline crystals are clear and free from unsightly flaws. Others, though of fine color, are so highly flawed that they are of no or little use as rough material for gemstones. Still others, for the most part flawed, contain subspherical zones of clear material of the quality most desired by lapidaries. These zones are the nodules described in Chapter 2 (see Fig. 2-23). Suitable crystals and nodules have been faceted, cut en cabochon, made into beads, carved, or tumbled.

Most fine gem-quality tourmalines are faceted, and most of the faceted stones are table-cut, low step-cut, mixed-cut (e.g., a modified brilliant-cut top and a step-cut bottom), trap-cut, or cut free-form (e.g., "flashing eyes"). In order to reduce waste, elongate stones such as baguettes are frequently cut with their lengths parallel to the lengths of the original crystals.

As noted in Chapter 7, in the section on optical properties, nearly all tourmaline exhibits differential absorption. Thus, for many stones, moderately thick plates cut perpendicualr to *c* may appear macroscopically opaque whereas plates of equal thickness cut parallel to *c* may appear transparent. Also, as previously noted, most tourmaline is dichroic with its different colors or color saturations such that the mineral appears darker perpendicular to *c* than parallel to *c*. As a consequence, the knowledgeable lapidary can exercise noteworthy control over the appearance of gemstones cut from many tourmaline crystals merely by choosing the orientation of, for example, the table facet or even the long dimension of a rectangular table facet.

Two general "rules of thumb" of lapidaries emphasize these facts: (1) Gems cut from crystals with great depth of color should be cut with their tables parallel to *c* (e.g., parallel to a prism face of the crystal); and (2) gems cut from crystals with only pale color may exhibit acceptable color if cut with their tables perpendicualr to *c* (i.e., perpendicular to prism faces).

Some of the red tourmalines from Newry, Maine, serve as good examples in that they are a pleasing burgundy hue when cut perpendicular to *c* and a weak diffuse pink when cut in some directions parallel to *c*. However, either color might be the choice of a given individual, so this must also be taken into consideration.

A few color-zoned liddicoatites have been cut with their tables parallel to color banding in order to superimpose colors and thus produce some desired third color (John S. White, Jr., 1983, personal communication).

In most cases, rectangular stones are cut with the longer dimension of their table parallel to *c* and with steep pavilion facets on the ends to minimize the darker dichroic color. A crown angle of 43° and a pavilion angle of 39° have been found to be suitable (GIA—Colored Stones course notes). The critical angle for $n = 1.610$ is $38°23'52''$.

In any case, because of dichroism, the two colors that could be exhibited should be determined before deciding the direction a stone is to be cut. In addition, some lapidaries also take into account the claim that the luster of stones cut parallel to *c* is better than that of those cut perpendicular to *c;* to date, however, this claim appears to be a moot point.

Fischer (1979) suggested that *for general purposes* blue and green tourmalines exhibit their best color when the table is cut parallel to *c* whereas red tourmalines exhibit their best color when the table is cut perpendicular to *c;* this suggestion notwithstanding, he also restated the already mentioned "rules of thumb." In addition, he summarized opinions as to the best abrasives to use.

Especially in the past, foils were frequently used to brighten the appearance of dark-colored tourmalines.

Tourmaline cat's-eyes and most tourmalinated quartz used in jewelry are cut en cabochon (Plate VI). The chatoyant stones with hollow channels must have their tubes sealed off during polishing so the polishing compounds will not penetrate the channels and discolor the stones. The sealing can be accomplished by gently heating the stone and dipping it in shellac, which is drawn into the tubes, and then removing the shellac by heating the polished stone in alcohol (GIA-Colored stones course notes). According to Nassau (1984), Crowningshield has observed that cat's-eyes with dark-colored tube fillings have sometimes been treated with acid to remove the dark-colored material(s) and then filled with "wax (or plastic?)" to inhibit subsequent refilling by dirt. In addition, a few of the chatoyant tourmalines used for cat's-eye stones have their tubes or inclusions so highly and sporadically concentrated that they are cut with the concentrations at or near their bottoms, below relatively clear and transparent zones. As with nearly all such materials, the size, shape, and proportions tend to be dictated by the dimensions of the raw material.

Both solid color and multicolored tourmaline crystals have been carved (Plate VIII). Prior to the 1912-1913 Revolution, much of the carving was done in China. Carvings with the shapes of leaves, fish, butterflies, and bunches of grapes were especially common during the nineteenth century. A few of the carvings were fashioned rather interestingly from the black-topped "mohrenkopf" crystals so the black portions became, for example, the hair on the heads of miniature busts. Multicolored tourmaline crystals are sometimes carved into small birds (e.g., humming birds, parrots, and birds of paradise) or animals that are given such appurtenances as ruby eyes and gold beaks and then mounted on mineral or rock bases. A few of these figurines are carved from single crystals; others are "glued" composites. In the writer's opinion, such pieces range from choice to trash.

Tourmaline that is clear and of good color but has too many flaws to be used in faceted stones is often tumbled or cut and used for beads, bracelets, or other costume jewelry. Small tumbled pieces have been used for "Ming trees."

Both carved and tumbled pieces with small cracks have frequently had those cracks filled with wax or plastic.

FAMOUS STONES

A few fashioned tourmaline gems have been named or have especially interesting histories. These include:

"emerald green tourmalines were taken (from Brazil) to Portugal to be cut and mounted as emeralds for the crown of Nossa Senhora de

Penha in the early 1700's. Years later it was discovered that they were tourmaline . . ." (Crowley, 1984);

a grape-shaped, 255-carat (~4 × 2.7 × 2.3 cm), red pendant given to Catherine II of Russia by Gustav III of Sweden in 1777 (see Donova, 1972);

a heart-shaped, 50.59 carat, blue-green stone fashioned from a Mount Mica crystal (Graves, 1967);

the Rozek tourmaline, a 25-carat, octagon step-cut stone also fashioned from a Mount Mica (Maine) tourmaline (Sinkankas, 1976).

There also are some well-known tourmaline necklaces. Perhaps the most famous has been described as follows in an item in the *Lewiston (Maine) Journal* (from an undated clipping):

MISS ELINOR CUTTING HAMLIN, a Boston young lady of Bay State Road, a granddaughter of Lieut. Col. Dr. Augustus C. Hamlin of Bangor, and a great-niece of the late Ex-Vice-President Hannibal Hamlin, is the proud possessor of the most valuable necklace in America, . . . It is a necklace that is the envy of all the queens and princesses of Europe. This famous necklace is entirely made up from the finest gems ever taken from the famous Hamlin mine at Mt. Mica, Paris, Maine, and includes the first green tourmaline gem ever found there ninety years ago. The necklace is composed of all the vast range of colors found in the tourmaline; white, yellow, green, black, blue and red. Attached to a chain of gold are 17 large tourmalines, ranging from three karats up to thirty karats, arranged so that they can be detached and others of different shades and sizes can be put in their places. In the set are thirty additional stones, many of them being mounted with white tourmalines and beryls. . . .

This necklace, now on display at the Harvard Museum, is shown on Plate VII. Another necklace of note is one made for wearing by the Governor of Maine or spouse. This necklace, the "State of Maine necklace," was commissioned by the Maine Retail Jewelers Association and presented to the State in 1975. It consists of stones from the Newry find of 1972 and gold nuggets from the Swift River in Byron, Maine. The center stone of 24 carats was donated by the Plumbago Mining Corporation of Rumford, Maine.

It has been suggested that "the monster red gem which hangs as a pendant to the jade necklace which belonged to the Chinese emperor, and which was captured by the French in the sack of Peking" and is now the property of the French Government, is tourmaline (Hamlin, 1873). Furthermore, Hamlin also suggested that "the magnificent ruby in the Russian crown of the Empress Anne Ivanova" may be a tourmaline (var. siberite). An inquiry about the latter, addressed to the U.S.S.R. Embassy in Washington, D.C., received no reply.

The 509-carat tourmaline fashioned from material from the Tourmaline Queen pegmatite, near Pala, San Diego County, California, is apparently the largest faceted tourmaline recorded to the present. It is described by Sinkankas (1976) as "lightly flawed . . . [and] dark pink in color."

WELL-KNOWN SPECIMENS

Tourmaline crystals, as previously mentioned, are prized for exhibits by collectors and museum curators alike (see Plates I, II, III, IV, and VIII). Several tourmaline specimens are widely known, and a few have been given names. Examples of the latter are: The *Cloven Hoof* and the *Jolly Green Giant* from Newry, Maine; the *Candelabra,* and *Crowning Glory,* from the Tourmaline Queen mine of San Diego County, California; *Steamboat* from the Tourmaline King mine of San Diego County; and *Flor de Liz, Foguete* (= "The Rocket"), *Joninha,* and the *Star of Itatiaia,* from the Jonas mine of Minas Gerais, Brazil.

Richard V. Gaines (1984, personal communication) has said that the "Joninha" specimen (Plate III) is perhaps the single most valuable mineral specimen in the world.

DISTINGUISHING TOURMALINE GEMSTONES

Although tourmaline, with its great range of colors, may resemble many other gemstones, with few exceptions it can be distinguished readily from each of the other stones by nondestructive means. One or more of the following properties are usually used: refractive indices, birefringence, dichroism, lack of "fire" (dispersion), and specific gravity. (Absorption spectra are less than satisfactory for distinguishing most tourmalines from gems they resemble.) Tests frequently conducted by jewelers include:

Red and pink tourmaline floats in pure methylene iodide whereas ruby (corundum), ruby spinel, and rose topaz sink. Pink tourmaline and kunzite (spodumeme) may be distinguished by the fact that the higher index of refraction of tourmaline is constant for each individual stone whereas that of kunzite varies as the stone is rotated on its table on the refractometer hemisphere. Pink tourmaline has higher indices of refraction and a higher specific gravity than morganite (beryl).

Yellow and yellowish brown tourmalines have much greater birefringences and lower specific gravities than topaz.

Green tourmaline has higher indices of refraction and a higher specific gravity than emerald. Green tourmaline is strongly dichroic whereas demantoid (garnet), peridot (olivine), and apatite are not. Green tourmaline has lower indices of refraction than peridot.

Blue tourmaline floats in methylene iodide whereas sapphire (corundum) sinks. Blue tourmaline has higher indices of refraction and a higher specific gravity than aquamarine (beryl).

Colorless tourmaline, seldom used as a gemstone, can be distinguished from other colorless stones on the basis of its specific gravity. For example, tourmaline has a higher specific gravity than phenakite, colorless beryl, and quartz and a lower specific gravity than diamond, colorless topaz, corundum, and zircon.

Black tourmaline is anisotropic whereas black garnet, jet, and both natural and manufactured black glass are isotropic.

Tourmaline and chrysoberyl cat's-eyes of similar color may be distinguished by tourmaline's lower refractive indices, higher birefringence, and lower specific gravity.

Tourmalinated quartz can generally be distinguished from rutilated quartz because of tourmaline's typical rounded triangular cross-sections and dark green, dark blue, or black color.

Liddicoat (1975) has, in addition, noted

> Green tourmaline . . . is characterized by long, irregular, threadlike liquid and gas inclusions, evenly distributed in abundance throughtout. . . . The numerous tiny liquid inclusions . . . have an appearance unlike other gems. . . . [and] rubellite has these same capillary-size liquid inclusions, but seldom in the abundance common in green tourmaline (p. 91).

SYNTHETICS AND SIMULANTS

To the present, there appears to be no reason for using synthetic tourmaline (even if it were available in suitable size) in lieu of natural tourmaline for gemstones.

Synthetic spinel has been used to simulate green tourmaline. It is easily distinguished from tourmaline because spinel is isotropic whereas tourmaline is anisotropic. Aquamarine has been used to simulate greenish blue tourmaline (Anderson, 1980). As just noted, refractive indices or specific gravity determinations may be used to distinguish between these two gemstones.

In the opposite sense, Cr-tourmaline has apparently been used as a simulant for transparent grossular garnet (Bank, 1974*b*). Again, tourmaline is anisotropic whereas grossular is isotropic.

MISCELLANY

Most jewelers do not display tourmaline jewelry in show-windows that get heated by, for example, the sun. This is true because tourmaline attracts dust as a consequence of its pyroelectric property.

Chapter 11

Occurrences and Geneses

Some descriptions of tourmaline occurrences and associations clearly support one versus other geneses. Other descriptions are unclear. Still others are clear but the relations appear to be amenable to alternative interpretations. Whatever, one or more of the tourmaline species have been recorded as occurring in rocks or geological deposits of nearly all common origins, of all geologic ages, and in essentially all regions on Earth. None, however, has been found in extraterrestrial rocks. Briefly (and alphabetically):

Buergerite, to date, has been found at only one locality: Mexquitic, San Luis Potosi, Mexico. It occurs coating fractures in a rhyolite and is thought to be of pneumatolytic origin (Mason, Donnay, and Hardie, 1964).

Dravite and uvite (see the last paragraph of these summary statements) typically occur in metamorphosed Mg- ± Ca-rich rocks. Dravite also constitutes authigenic overgrowths in many sedimentary rocks (see, for example, Krynine, 1946), and occurs, albeit less commonly, in several other kinds of rocks (see, for example Barsanov and Yakovleva, 1964).

Elbaite and probably liddicoatite, originate nearly exclusively in complex Li-bearing pegmatites. Elbaite-schorl, however, is also recorded from, for example, schists with no apparently associated pegmatites (e.g., Hutton, 1939; Pichamuthu, 1944; and Cooper, 1971).

Ferridravite is known from only one locality: the San Francisco mine, near Villa Tunari (Alto Chapare), Bolivia. It occurs as subparallel aggregates of intergrown crystals coating a K-feldspar-bearing muscovite-alkali amphibole schist (Walenta and Dunn, 1979).

Liddicoatite, as noted under elbaite, probably originates in complex Li-rich pegmatites. The type material was found in "detrital soils" of Antsirabe, Madagascar (Dunn, Appleman, and Nelen, 1977).

Schorl, frequently referred to as "common tourmaline," is ubiquitous (e.g., Ward, 1931, and Barsanov and Yakovleva, 1965).

"Uvite" is a common mineral . . . In general, brown tourmalines from metamorphosed limestones are . . . uvite. In terms of attractive mineral specimens, dravite is considerably less abundant than uvite . . . *most* brown tourmaline associated with calcium-rich minerals is uvite, and *most* brown tourmaline associated with schist and non-pegmatite micas, is dravite." (Dunn et al., 1977, p. 107).

Chromdravite occurs in a Cr-rich V-bearing micaceous metasomatite in the Onezhkii depression of central Karelia, U.S.S.R. (Rumantseva, 1983).

In addition, because of their widespread occurrence and their chemical and physical stability under essentially all conditions of weathering and erosion, tourmalines are relatively common in both unconsolidated and consolidated residual and transported deposits, for example, in soils (e.g., Dakshinamurti, Satyanarayana, and Jingh, 1955), laterites (e.g., Sirotin, 1966), and detrital sedimentary rocks (e.g., Houser, 1979).

Kinds of occurrences recorded for tourmaline group minerals are *briefly* summarized in this chapter. Complete coverage of this subject would more than treble the size of this book. For most types of occurrences, only one or a few representative references are cited; that is, nearly all of the references in this chapter should be considered as being preceded by "e.g." In addition, both physical and chemical characteristics that may indicate one versus other geneses and ways whereby tourmaline may be utilized to aid petrogenetic and/or geologic interpretations are noted.

IGNEOUS AND PYROCLASTIC ROCKS

Tourmaline, most descriptions of which indicate schorl, has been recorded as occurring as an accessory mineral in many intrusive igneous rocks (including some that may have been formed as a result of metamorphic processes (e.g., granitization), a few extrusive igneous rocks, and also in a few pyroclastic rocks. Examples are: granite (Messina, 1940), alaskite (Mirgabitov, 1970), charnockite (Sobolev et al., 1964), aplite (Tilley, 1919), and quartz porphyry (Eliseeva, 1958); quartz monzodiorite (Constantinescu, 1976), tonalite (Tsukanov and Esipchuk, 1971), quartz diorite (Gnativ, 1959), and diorite (Bazhenov and Poluektova, 1969); gabbro (Karpova, 1965) and diabase (Tomlinson, 1945); peridotite (Smith, 1971) and kimberlite (Shumaro, 1971); kyschtymite (Papastamatios, 1939) and alkalic (e.g., sodalite) rocks (Mawson and Dallwitz, 1945); and rhyolite (Demay, 1938). In addition, Scherillo (1941) has recorded sanidinite fragments that contain both tourmaline and danburite among volcanics in the vicinity of Cimina, Italy, and Gapon and Gapeeva (1969) have recorded wind-blown pyroclastic tourmaline in the Dogaldynsk Valley of the Bodaibo Basin of central Siberia.

Within these diverse rocks, the tourmaline occurs as disseminated grains, aggregated with quartz in "nodules" (Edwards, 1936) or "spots" (Němec, 1975), as the predominant constituent of one or more shells of some orbicules (Oba and Ishikawa, 1959), and as crystals in miarolitic cavities (Wagner, 1973).

Several of the recorded occurrences of tourmaline within these rocks may represent only spatial association; that is, some of these tourmalines may not be of igneous origin or even consanguineous with their surrounding rocks. Instead, they may have originated during, for example, another, genetically unassociated period of metasomatism.

Actually, for the bulk of the tourmaline that is usually designated igneous, the weight of available evidence appears to support best the contention that although some of it may have crystallized directly from magma (i.e., a silicate melt), much more of it probably had some other origin. This contention is supported by several facts. For example: (1) even the tourmaline that is disseminated in large igneous masses tends to be most common in zones that appear to have been subjected to late stage (epimagmatic) alteration (e.g., Wasserstein, 1951); (2) tourmaline crystals are common in miarolitic cavities (e.g., Brammall and Harwood, 1925, and Stoicovici and von Gliszczynski, 1944); and (3) tourmaline is common in greisens, pegmatites, and metasomatically deposted veins. In addition, it seems noteworthy that even some of the geologists (e.g., Ivanov, 1970) who support formation of "igneous" tourmaline by direct crystallization from magma have described textural relations that appear more likely to indicate its late-stage epimagmatic and/or later hydrothermal origin.

Experimental Data

Experiments dealing with apparent effects of boron on water-saturated "haplogranitic" magma at 1 kbar vapor pressure (Chorlton and Martin, 1978; and Pichavant, 1981*a* and 1981*b*) are possibly pertinent to the origin of igneous tourmalines and to tourmalines in greisens, pegmatites, veins, and other cavity fillings. The runs indicate, *inter alia,* that the presence of boron appears to: (1) increase H_2O-solubility in the melt; (2) lower the temperature of the solidus, and possibly the liquidus; and (3) cause the resulting end-stage vapor phase to be Na- and Si-rich and K- and Al-depleted.

Trace Element Studies

Power (1966*a,* 1966*b,* and 1968) reported differences in trace element concentrations between tourmalines of two generations, termed igneous and secondary, within rocks of southwestern England. He recorded higher contents of F, Fe, Mn and lower contents of Mg, Ca, and possibly of Sr, Sn, V, Cr, Ni, and Sc in the disseminated "igneous" tourmaline grains than in the secondary "hydrothermal" tourmalines.

Lyakhovich (1963) reported a rather ambiguous "fact" that Cs tends to accumulate preferentially in tourmaline-bearing granites but that there appears to be no Cs:B correlation.

Neiva (1974) reported relations and trends in the trace element contents of schorls in apparently consanguineous granites, aplites, and pegmatites of northern Portugal. She stated that

> In the schorlite from comagmatic granites and composite aplite-pegmatite veins, B, Cl, F, Ge, Ga, Nb, Zn, Li, Cu, Er tend to increase in the sequence granite-aplite-pegmatite while In, La and Au tend to decrease. A less clearly defined increase in $Fe^{3+} + Fe^{2+}$, coupled with a decrease in Ti, Mg, V, Ni, Sc, Dy, Tb, Gd, Sm, Nd, Pr, Ce, Pb, was observed in the same sequence . . . The chemical evidence is interpreted as indicating the influence of fractionation in late stage 'magmatic' fluids . . . [Also,] The ratios Fe^{2+}/Mg, $Fe_t/(Fe_t + Mg)$, Mn/Fe^{2+}, F/K, F/Li, Li/Mg show a general increase and K/Tl, Ni/Fe^{2+}, Co/Fe^{2+}, Co/Mg, Sc/Fe^{2+}, Sc/Mg a general decrease from the schorlites of the granite to those of the associated aplite and pegmatite, but generally, for each ratio, the trend of differentiation is not so clear between the schorlite of the aplite and pegmatite coexisting in the same vein . . . [And, when the schorls from all of the granites are compared to the schorls from all of the aplites and pegmatites of the region,] it is found that Tb, Gd, Pr, Ce are concentrated in the schorlites from the granite, while in the aplites and pegmatites they generally occur below the limit of sensitivity . . . . [Further,] schorlite from granites contains more Ti, ΣCe(Ce, Pr, Nd, Sm), ΣY(Gd, Tb, Dy, Er), RE and less Mn than the schorlites from aplites and pegmatites (pp. 1308, 1310, 1311).

In addition, she notes that in these rocks the schorls that coexist with biotite and muscovite contain more Ti, Mg, La, and ΣCe and less ΣFe than do those with only muscovite.

Unfortunately, when all the recorded trace element data for tourmalines from "igneous" and other origins are considered, even on a species by species basis, their applications seem to be rather limited. It appears that at best they may: indicate relative stages of evolution within a group of consanguineous units; serve to distinguish among spatially associated, similar units of different kinships; and/or aid distinction among tourmalines of diverse origins (e.g., different generations within individual rocks). Also, trace elements do not appear to be present in absolute precentages that are significant so far as resolving broad-scale petrogenetic problems. However, it is known that, for example, granites that contain tourmaline tend to have higher Sn-contents—here and there to the point of being cassiterite-rich—than granites with no tourmaline (see, for example, Lyakhovich, 1965). And, as implied in the section on isotopes in Chapter 4, it appears that isotopic studies may prove fruitful.

SOILS, SEDIMENTS, AND SEDIMENTARY ROCKS

Although tourmaline is typically resistant to both physical and chemical weathering, alteration due to weathering has been recorded (see section on alteration in Chap. 4) An especially interesting, apparently exceptional, case involving the breakdown of tourmaline as a consequence of weathering is reported by Dubinina and Kornilovich (1968): Within a zone of supergene enrichment, dark-colored tourmaline (schorl?) was partially decomposed to a powdery crust of colorless dravite plus plumbojarosite and beudantite. This occurrence could, of course, have involved low-temperature hydrothermal activity.

In general, however, tourmaline is a widely occurring, rather common constituent of residual soils, transported detritus, and their lithified products.

Residual Soils, Crusts, and Laterites

Tourmalines, for the most part not identified as to species, have been recorded from residual soils (Adams and Matelski, 1955), "weathering crusts" (Dodatko. 1969), laterites (Adams and Matelski, 1955, and Khan, 1960), and bauxites (Göksu, 1953), as well as from the just mentioned supergene zone of an ore vein.

Detrital Sediments and Sedimentary Rocks

Minerals of the tourmaline group are relatively widespread and common "heavy mineral" constituents of detrital sediments and their lithified equivalents. One or more species have been recorded as occurring in essentially all kinds of unconsolidated sediments and sedimentary rocks of this general category. Examples are: aeolian sands (Wyo and Reitz, 1944) and loess (Codarcea, 1977); alluvium (Cuppels and White, 1973), including placers (Barbosa, 1974); lacustrine sands (Vetrunov, 1962) and glacial deposits (Tsvetkova-Goleva, 1964); sediments in transitional environments, including deltas (deCarvalho, 1955), estuary deposits (Dropsy, 1943), and beach deposits (Foxworth et al., 1962); and marine deposits (Cazeau, 1962); also, conglomerate (Fieremans, 1955), "grit" (Muir, 1964), sandstone (Gault, 1939), siltstone (Dromasko, Lukashev, and Matveeva, 1966), shale (Blatt and Sutherland, 1969), and clastic limestone (Crampton, 1958).

As might be expected, several of the sediments, and in some cases the included tourmaline, have undergone more than one cycle; for example, the above alluded-to estuarian deposit contains detritus that has undergone aeolian transport (Dropsy, 1943).

One somewhat exceptional occurrence includes detrital tourmaline as euhedral clasts, which lack any evidence of abrasion, within sediments off the southeastern coast of Spain. Berthois (1939) has interpreted these euhedra as detrital on the basis of their similarities with tourmaline crystals in the nearby igneous rocks that appear to have been a source for the sediments.

Thiel (1941) has illustrated and tabulated abrasion effects and development of characteristic degrees of sphericity for tourmaline during specified time periods. Alling (1944), as the result of ball-mill runs to determine the rates of reduction in size and changes in sphericity and roundness for quartz, microcline, garnet, and tourmaline, found tourmaline to undergo the least change during any given elapsed time. Dryden and Dryden (1946), on the basis of their investigation of zircon, staurolite, hornblende, kyanite, monazite, chloritoid, sillimanite, and tourmaline in weathered Wissahickon Schist of the eastern United States, concluded that only zircon is less affected than tourmaline during weathering. Previously, Scheidhauer (1940) had recorded a tourmaline:zircon ratio of >1 for the 0.1-0.05 mm fraction of coarse sandstone but of <1 for fine sandstone in some Cretaceous sandstones in the vicinity of Dresden, Germany.

In his well-known paper "The tourmaline group in sediments," P. D. Krynine (1946) stated that tourmaline, along with zircon, is the most abundant and most "frequently found" heavy mineral in both sediments and sedimentary rocks of all kinds and of all ages (cf. Topkaya, 1950). He then showed how color, grain shape, and identity and/or abundance of inclusions can be used to establish identities of diverse tourmalines and thus aid the resolution of diverse stratigraphic problems. (By the way, he reported 21 varieties of tourmalines in one sediment!) On the basis of a somewhat simplistic view, he described five kinds of tourmalines: (1) granitic, (2) pegmatitic, (3) "tourmaline from pegmatized injected metamorphic terranes," (4) authigenic, and (5) reworked tourmaline from older sediments or sedimentary rocks. He also outlined investigations that illustrate how the study of sedimentary tourmaline may be used, for example: (1) to establish the petrology of both the immediate and ultimate source areas; (2) to determine the general climatic and diastrophic history of both the basin of sedimentation and the source rock area; (3) to establish paleogeographic relations during the sedimentation periods; and (4) to differentiate within and correlate among stratigraphic units. The provenance studies must, of course, be based on only the detrital (i.e., not include authigenic overgrowth) portions of grains.

Currently, zircon:tourmaline, zircon:garnet:tourmaline, and zircon:tourmaline:rutile ratios are used as, for example, maturity indices (e.g., Houser, 1979; Sato, 1969; and Hubert, 1962; respectively). Each of these indices tends to increase with maturity.

Evaporites

The tourmaline in evaporite deposits may, of course, be of windblown or other detrital origin. Hiller and Keller (1965) have recorded tourmaline in small quantities (~2% of the >40 clay fraction, which in turn makes up <1% of the total rock) in the salt deposits of Freiburg, Germany. Previously, Braitsch (1959) had noted that the tourmaline grains of these deposits are euhedral, and in a discussion following Braitsch's paper it is recorded that similar tourmaline grains occur in fetid shales of the same stratigraphic sequence. Popov and Sadykov (1962) consider the idiomorphic tourmaline crystals in the salt deposits of Khodzha Mumyn (Central Asia) possibly to represent direct precipitation from B-bearing seawater. Alternatively, it would appear that they may represent diagenesis.

DIAGENETIC ROCKS

Some tourmaline in both sedimentary and diagenetic rocks is of diagenetic origin. However, some tourmaline grains within diagenetic rocks—for example, in certain dolostones (Bessonova, 1970) and coals (Beising, 1972)—are not of diagenetic origin; rather, they are detrital grains inherited from the precursor sediments. Furthermore, many of the grains mentioned in this section are composite; that is, part of each grain is detrital and part is diagenetic. For example, tourmaline grains concentrated along stylolite seams commonly consist of detrital grains with authigenic overgrowths (W. D. Lowry, 1983, personal communication; cf. Heald, 1955).

For some reason or other, there appears to have been a preoccupation with who first recorded authigenic tourmaline. Although Derby (1898) reported tourmaline with authigenic overgrowths some 27-36 years earlier, there are statements in the literature that variously claim or attribute the "first report" of authigenic tourmaline to Dévérin (see remarks in his note of 1934), Spencer (1925), and even to Stow (1932). This despite the fact that Brammall (1921) had also recorded authigenic tourmaline before any of this trio.

In any case, authigenic tourmaline has been recorded in such diverse rocks as sandstones (Stow, 1932), limestones (Geletsyan, 1967), and rock salt (Popov and Sadykov, 1962). The authigenic overgrowth character of tourmaline is usually recognized on the basis of color differences between the detrital substrate and the overgrowth.

To the present, only a few investigators (e.g., Gautier, 1979) have determined the chemistry—that is, the species identity—of the two components of these composite grains. Several of the better descriptions, however, do indicate that authigenic tourmaline is commonly of the dravite(uvite)-schorl series whereas diverse tourmalines constitute the detrital substrate grains. Rajulu and Nagaraja (1969), however, have reported authigenic schorl, elbaite, and "Cr-tourmaline" in sandstone from Mysore state (India).

Substrate grains commonly exhibit pitting and etching at contacts with their authigenic overgrowths. The overgrowths are typically in optical continuity with their substrates, and growth tends to be greater, if not exclusively, on the antilogous ends of original grains (e.g., Alty, 1933). Although overgrowths tend to be small, some have been recorded up to 0.25 mm across and to comprise volumes of up to two times that of their substrate grains (Krynine, 1946). Most authigenic tourmaline is inclusion-free, but some contains resistant materials such a magnetite (Derby, 1898) and coal (Perozio, 1959), which were apparently engulfed by the authigenic growth.

Krynine (1946) noted that tourmaline overgrowths are typically colorless or "very very pale blue," and this statement has been repeated several times in the literature. There are many exceptions to this color-genesis correlation. Overgrowths with identical properties, though in some cases different nuclei, have been reported to be characteristic of individual formations (Krynine, 1946; Gokhale and Bagchi, 1959), and where present can be used as stratigraphic markers.

Authingenic tourmaline was found in 18 percent of 211 samples from diverse Swiss localities (Topkaya, 1950).

Some idiomorphic grains of tourmaline within sedimentary rocks (e.g., Tikhomirova, 1957, and Popov and Sadykov, 1962) along with tourmaline concretions and cement (Vetrunov, 1955) and tourmaline crystals in vugs (Valeton, 1955, and Rickets, 1978) have also been considered to be of diagenetic origin. Most of the crystals are so interpreted because of their association with grains that exhibit authigenic overgrowths and/or with tourmaline that is acting as a cement. Postdepositional lateral secretion is the usually hypothesized genesis.(As already mentioned, however, idiomorphic tourmaline grains in, for example, some salt deposits have been alternatively interpreted to represent direct precipitation from sea water.)

The constituents of diagenetic tourmalines, including authigenic overgrowths, have been attributed diverse derivations. Four suggested sources involve (1) local derivation as a result of, for example, "pressure solution" of detrital tourmaline grains (e.g., Mader, 1978); (2) relatively local derivation of boron that was: (a) absorbed by the iron in primary Fe-hydroxides (Zubova and Rezapova, 1964), (b) adsorbed by clay particles (Pieruccini, 1950), or (c) in precursor boron-bearing minerals within the parent sediment (Slack et al., 1984); (3) direct crystallization from a B-rich interstitial brine within the sediment (Ethier and Campbell, 1977); or (4) formation from hydrothermal solutions derived from nearby volcanic activity (Ricketts, 1978, and Gautier, 1979).

The fact that diagenesis grades into metamorphism is well illustrated by the occurrence of diverse tourmaline overgrowths. For example, whereas tourmaline overgrowths on rounded tourmaline grains in itacolumite (generally considered to be a metamorphic rock) of Minas Gerais are interpreted as authigenic (Derby, 1898), tourmaline overgrowths on detrital tourmaline grains in low-grade metasedimentary rocks of Cambrian through Middle Ordovician age in southern Quebec are interpreted to be of metamorphic origin (Rickard, 1964). (This latter interpretation is, by the way, based in part on the fact that tourmaline overgrowths within the Quebec rocks contain inclusions that are parallel to slaty cleavage in the host rocks; however, Frey, 1969, thinks otherwise.) Another way that tourmaline overgrowths might be identified as metamorphic versus diagenetic would be to investigate the presence versus absence of chemical equilibrium relations (e.g., partition coefficients) between the overgrowths and other mineral constituents of the containing rocks.

(Since preparation of the manuscript for this book, it has been learned that Henry and Guidotti, 1985, have indeed found that "chemical equilibrium has been attained among the rims of the tourmaline grains and the matrix minerals" in staurolite-grade metapelites of northwestern Maine, which, of course, indicates those overgrowths to be metamorphic.)

METAMORPHIC ROCKS

Tourmaline occurs in essentially all kinds of metamorphic rocks. This is a reflection of its widespread occurrence in precursor rocks, its apparent ease of formation under a large range of conditions, and its stability under most conditions of metamorphism.

Tourmaline occurs in both exomorphic and endomorphic parts of contact metamorphic zones (e.g., Brammall and Harwood, 1925), in shear zones (e.g., Goroshnikov, 1956), and in metamorphic rocks of terrains that are representative of all grades of "regional" metamorphism.

Just for the record, among the metamorphic rocks within which tourmaline has been recorded are: gneiss (Casasopra, 1939), schist (Khasnova, 1960), amphibolite (DuRietz, 1935), and slate (Gillot, 1955); meta-argillite (Aoki, Konno, and Abe, 1975), marble (Jan, Kempe, and Symes, 1972), metaquartzite (Jain, 1972), and granulite (Deprat, 1909); mylonite (Ladurner, 1951); serpentinite (DuParc and Sigg, 1914) and eclogite (Smith, 1971); hornfels (Mawson and Dallwitz, 1945) and skarn (Otroshchenko and Zanin, 1965); "magnesite rocks" (Bouška, Povondra, and Lisý, 1973); diverse tourmalinites; meta-evaporite (Moine, 1979) and meta-chert (Irving, 1937). In addition, tourmaline has been recorded in some rare metamorphic rocks—for example, in a Cr-tourmaline + pink rutile + talc rock from the Barberton district of South Africa (Hall, 1918) and in tourmaline + phlogopite + zoisite ± muscovite rocks from Loh-Oeolo, Java (Koomans, 1938). Also, as might be expected, tourmaline has been listed as a constituent of diverse suites that include just about every group of minerals described for metamorphic rocks.

Unfortunately, it is not possible, on the basis of descriptions in the literature, to classify many of these tourmalines as of pre- versus syn- versus post-metamorphic origin. And, many of the reported tourmalines have not been identified to species. It is true, however, that of those that have been identified or characterized chemically, most have compositions that correlate with the compositions of their containing rocks—for example, uvite in a fassaite-, serendibite-, clinozoisite-bearing calcite tactite (Hutcheon, Gunter, and Lecheminant, 1977); chromian dravite in fuchsite schist (Lum, 1972); manganian dravite in a braunite-, hexagonite- (Mn-bearing tremolite), quartz-bearing meta-evaporite (Ayuso and Brown, 1984); and vanadian dravite in a Ba- and V-rich muscovitic graphite schist (Snetsinger, 1966).

Among "metamorphic" tourmalines, there are: (1) tourmaline porphyroblasts in, for example, schists; (2) tourmaline grains within rocks in contact metamorphic zones (i.e., in hornfelses and tactites, including the diverse rocks termed skarns); (3) tourmaline in zones of greisenization; and (4) tourmaline as a predominant constituent in masses called tourmalinites.

Porphyroblastic Tourmaline

Tourmaline grains, some of which are well-defined euhedral or subhedral porphyroblasts, have formed in many diverse metamorphic rocks. The porphyroblasts range from those with no readily apparent preferred orientation [e.g., the "pseudomonoclinic" prisms (plus $11\bar{2}0$) of dravite, which resemble hornblende, in chlorite schist near Hidrolandia, Goias State, Brazil (DeCamargo and Souza, 1970)] to those, which are more common, that exhibit both foliation and lineation (e.g., those in gneisses of the Molare regions, Ticino, Switzerland [Thakur, 1972]).

Tourmaline-bearing Rocks of Contact Metamorphic Zones

Several tourmaline-bearing hornfelses and tactites, many of which are termed skarns (e.g., Shabynin, 1974), are described in the literature. One especially significant, common association in these rocks is that of tourmaline, commonly a colorless or nearly colorless dravite, and cassiterite (e.g., in the hornfels in the vicinity of the bismuth and tin mines at Tasna, Nor-Chickas Province, Bolivia [Spencer, 1907]). On the other hand, several tourmaline-bearing rocks of contact metamorphic zones do not contain cassiterite; in fact, the compositions of these tourmaline-bearing, thermally metamorphosed rocks are extremely diverse.

Also, although tourmaline is rather widely recognized as the characteristic mineral of so-called B-mineralized contact zones, several other B-bearing minerals have also been found to occur in these zones. As an example, Dorofeev (1974) lists some 14 B-bearing minerals from dolostone and limestone skarns within Paleozoic rocks intruded by Early Cretaceous granites in Siberia.

Greisens

Brown to black tourmaline is a constituent of many tin deposits, especially cassiterite-bearing greisens (e.g., Ferguson and Bateman, 1912). Quartz, fluorite, wolframite, muscovite, topaz, and diverse sulfides are commonly associated minerals in these altered granites. Well described relations indicate that the tourmaline within greisens typically replaces feldspar grains, and that it, along with many of the other minerals, can be interpreted to have been formed from magmatically derived pneumatolytic or possibly hydrothermal fluids.

Tourmalines from some of these masses have been recorded as Sn-bearing. In the compiler's opinion, however, additional analyses should be made to see if the Sn is actually part of the tourmaline rather than reflecting the presence of minute inclusions of, for example, cassiterite.

Greisenization involving tourmaline and tourmalinization in general, including the formation of certain tourmaline-bearing veins, are commonly related and in many cases probably consanguineous.

Tourmalinization and "Tourmalinites"

Tourmalinization, typically involving replacement by tourmalines of the schorl-dravite series, is widespread. Its overall effects range from disseminated tourmaline grains to predominantly tourmaline rocks. In many places, tourmalinization appears to have been selective (e.g., Chauris, 1965) and to have involved migration of B ($\pm$ Al, etc.) into rocks already containing the required *Y*-site cations $\pm$ Si (see, for example, DuRietz, 1935). Several rocks—for example, granitic igneous rocks, pegmatites, migmatites, serpentinites, gneisses, schists, amphibolites, tectonic and sedimentary breccias, conglomerates, sandstones, and shales (see, for example, Irving, 1937, and Coetzee, 1966)—have been recorded as having undergone tourmalinization; several minerals—ranging from oxides (e.g., chromite—Mukherjee, 1968) to diverse silicates (e.g., both K- and plagioclase feldspars—e.g., Kulikov et al., 1976)—are recorded as having been replaced by tourmaline.

One of the more interesting examples of tourmalinization described in the literature is "a diatreme, blown by boron-bearing gases" that is now expressed by an aureole ~900 feet (~270 m) across, in the Bloemfontein district, Orange Free State (Coetzee, 1966). At that locality, the Karroo System sandstone contains about 25 percent tourmaline and the shale contains about 50 percent tourmaline (schorl$_{66}$ dravite$_{34}$).

Predominantly tourmaline rocks are usually termed tourmalinites or luxullianites. Quartz is the other main constituent in nearly all of these rocks. Other names applied to tourmaline-quartz rocks include schörlischiefer, schörlquarzit, schörlfels, turmalinschiefer, hyalotourmalite, tourmalite, and carvoeira (see Johannsen, 1932). According to Johannsen, C.F.G. Freiesleben (in *Geognostische Arbeiten,* VI, pp. 1-16) subdivided these rocks into granular, schistose, and dense varieties.

Some predominantly tourmaline rocks comprise zones along or near contacts rich in *Y*-site cation-bearing rocks; others occur as apparently isolated tabular vein- or dike-like masses; still others are stratabound. (Several occurrences are tabulated by Slack, 1982.) Most tourmalinites have been attributed to metamorphism, either metasomatism or treptomorphism (essentially isochemical metamorphism), with some of the latter apparently representing stratabound masses in which the tourmaline appears to be of early diagenetic origin; a few have been interpreted as igneous (for example, Hatch et al. [1949] have suggested possible formation from a magma "drastically enriched in fugitives").

Much recent work has been concerned with stratiform tourmalinites; for a summary, see Slack et al. (1984). According to those authors: Most of these masses, which occur worldwide (North America, Europe, Africa, Asia, and Australia), are in Proterozoic and lower Paleozoic metamorphosed, predominantly clastic, sedimentary sequences, many of which also include other relatively uncommon lithologies such as manganese- and iron-rich rocks. Some of these tourmalinites, which are typically layers less than a meter thick, exhibit striking sedimentary features such as cross-lamination and rip-up clasts. The typical tourmaline is a dravite-schorl. In the authors' opinion, all data is best interpreted to indicate formation of the tourmaline during sedimentation or early diagenesis, within either a deep-water submarine environment or a shallow-water basin in which evaporite sequences prevail. In addition, they think that the boron must have come from subaqueous fumaroles or nearby hot springs. They further note that the unusual chemical and paleo-enviromental conditions reflected by these tourmalinites were, in many places, suitable for the formation of precursors of stratabound mineral deposits, in which there were accumulations of minerals rich in cobalt, gold, tin, and/or tungsten, along with base metal sulfides. Consequently, they suggest that the presence of stratiform tourmalinites and spatially associated Mg-rich tourmalines may constitute a guide for prospecting for such deposits.

Origin of Metamorphic Tourmalines

Some of the tourmalines of metamorphic rocks appear to represent detrital tourmaline grains that were in the precursor sediments. Where obviously relict clastic grains occur, this origin is well supported; where such evidence is lacking, other derivations seem equally possible.

For metamorphic tourmalines of unproved derivation, considerations of the alternative sources of the constituents lead to two main questions: Were the constituent elements derived locally or introduced? [and] If any were introduced, from where did those components move?

Some tourmaline-bearing metamorphic rocks have been interpreted to represent essentially isochemical metamorphism; that is, the constituents of the metamorphic rocks, including those of the tourmaline, are thought to have been in the precursor rocks (see, for example, Serdyuchenko, 1960; McCurry, 1971*a;* Aoki, Konno, and Abe, 1975; and Wadhawan and Roonwal, 1977). The data of Goldschmidt and Peters (1932), Landergren (1945), and Pieruccini (1950) support this possibility, at least for tourmalines in rocks that were originally carbonaceous marine clays. The boron in the clays is generally thought to have come from sea water and to have been adsorbed by clay particles. In some of the rocks, however, it may have been present in boron-bearing minerals such as reedmergnerite and searlesite; for example, Milton (1971) found such minerals in Green River Shale that had undergone marked diagenesis, and Eugster (1980) found searlesite in two Recent California and Nevada lacustrine evaporites. (Perhaps trace element and/or isotope data will provide a way to detect such precursors.) Alternatively, the boron within the sedimentary precursors may have been added by submarine volcanic exhalations during sedimentation (e.g., Abraham, Mielke, and Povondra, 1972; Slack, 1982; and Plimer, 1983). With regard to this last possibility, however, White and Waring (1963) have stated that elemental boron has not been found in volcanic gases and that condensates around fumaroles idicate a ratio of H_3BO_3:HCl that generally ranges from 1:1,000 to 1:20.

On the other hand, many tourmalines of metamorphic rocks are widely believed to have been formed as the result of metasomatism; that is, their formation is thought to have required the introduction of boron and perhaps some of the other constituents by pneumatolytic and/or hydrothermal fluids of magmatic derivation (e.g., Hutton, 1939, and Bouška, Povondra, and Lisý, 1973). The evidences usually cited as supporting this origin are mineralogic and/or geologic associations. (In several cases, in fact, the presence of tourmaline has been interpreted, conversely, to indicate metasomatism and, in turn, the existence of a parent magma (e.g., Van Houten, 1971); in fact, in some cases, where no nearby magmatic rock was known, a source magma was postulated as, for example, "below" (Gillot, 1955). Extrapolations like this, which invoke former magmatic activities on no better basis than the presence of tourmaline, should be avoided!)

Although both acidic (e.g., Agrell, 1941) and basic (e.g., LaCroix, 1914) magmas have been suggested as having been the source for the magmatically derived B(etc.)-containing fluids responsible for tourmalinization, current consensus holds that most such tourmaline-forming fluids have come from granitic magmas. Actually, later investigations have turned up possible acid magma sources or alternative explanations for some of the occurrences where basic magmas were first favored; for example, in discussing the occurrence of the tourmaline-bearing skarn associated with Triassic diabase near Lambertville, New Jersey, Grant (1964) directed attention to the fact that the Triassic rocks of eastern North America may constitute a boron province.

In any case, in discussing the fact that their synthesis experiments indicate that tourmaline forms under acid or weakly alkaline conditions but not under strongly alkaline conditions, Frondel and Collette (1957) noted that most, if not all, boron in meta-argillites is in tourmaline, whereas the boron in metamorphosed calcic rocks is typically in several different minerals—for example, in axinite, danburite, ludwigite, serendibite, sinhalite, and warwickite, as well as in tourmaline. This, they suggested, may support the contention that a high Ca-content or a high Ca:Al ratio may favor formation of some B-bearing mineral other than tourmaline.

Uses of Tourmalines in Metamorphic Petrology

So far as utilizing tourmaline as an aid to interpreting the geology of tourmaline-bearing rocks and the genesis of a tourmaline-bearing metamorphic host rock, the following facts and fancies appear to warrant further study and/or consideration and, in some cases, broader application.

The identity and/or distribution of detrital tourmaline grains have been used as "markers" to help decipher folding and faulting patterns in metamorphosed, as well as unmetamorphosed, sedimentary rocks (e.g., Jain, 1972).

The volume ratios of metamorphic overgrowth:original clastic grain increase with increasing grade of metamorphism in rocks of the higher greenschist and lower amphibolite facies of the Juras (Frey, 1969).

"Stretched" crystals of black tourmaline and deformed quartz pebbles, both with their long axes parallel to axial trends of folds involving their host rocks, have been used to study the finite strain in rocks from the Molare region, Ticino, Switzerland (Thakur, 1972).

Tourmaline grains commonly have a preferred orientation in deformed rocks. In many of the rocks, the orientation is discernible macroscopically; in others, it is usually established by petrofabric analyses (e.g., Sahama, 1945); it also has been investigated by X-ray Universal Stage and X-ray texture goniometry (Mons and Paulitsch, 1972). As might be expected, Mons and Paulitsch found, among other things, that there is a preferred orientation of the prism forms of tourmaline to the foliation (*ab*-plane) in rocks from several widespread localities.

It may be possible to correlate effects (or lack of effects) of pressure and/or temperature on both metamorphic and authigenic overgrowths on the antilogous versus analogous ends of tourmaline grains with, for example, tectonic directions.

Devore (1966) has developed a model to account

> for mineral recrystallization under directed stress that results in a minimum Gibbs and maximum Helmholtz free energy for the equilibrium orientation. . . Correlations between calculated orientations, and the observed preferred orientations are established for . . . tourmaline [*inter alia*] . . . Shear strain parallel to tectonic A is responsible for determining the fabric of the rock but apparently is a relatively unimportant strain in determining the final mineral recrystallization . . . The possibility that strain energy and thermodynamic equilibrium throughout a stressed system may be used quantitatively to describe relative strain differences in different parts of the deformed system is indicated (p. 19).

In classifying the minerals considered, Devore characterized tourmaline, along with calcite, corundum, hematite, and quartz, as follows: "tectonic B can be perpendicular to crystallographic *a* or at an angle to crystallographic *a* but at a fixed angle to crystallographic *c*" (p. 30), he gave the tectonic B-to-crystallographic *c* angle as 75° for tourmaline.

It has been suggested that under equilibrium conditions, where sufficient boron is available, tourmaline and biotite are incompatible—that is, only tourmaline will be formed (e.g., Park, 1928). There are, however, many rocks recorded as containing both minerals, and some of the rocks have been interpreted to represent equilibrium assemblages; an especially interesting example is a garnet-clinopyroxene-orthopyroxene-clinoamphibole-biotite-tourmaline-rutile eclogite from southwestern Norway (Smith, 1971).

MIGMATITES

As might be predicted on the basis of its presence in both metamorphic and plutonic igneous rocks, tourmaline occurs in either one or both constituents of many of these composite rocks. As in metamorphic rocks, the tourmaline may have been inherited from precursor rocks or have been formed during migmatization.

There are, however, only a few papers dealing with migmatites in which tourmaline is treated as anything other than a minor constituent. DuRietz (1935) has recorded a "tourmaline migmatite-schist" at Muruhatten, Sweden. He used that designation because he interpreted the rock as a predominantly amphibolite sequence that had undergone pneumatolysis. His extended use of the term migmatite, though not unprecedented, is unwarranted (see, for example, Dietrich, 1974). At the southwestern tip of the Isle of Ouessant (northwest of France), Jérémine and Sandréa (1957) recorded banded tourmaline-oligoclase (~50:50) rocks that exhibit ptygmas, which are typically associated with migmatization. The recorded relations are, however, too vaguely described to assume that this occurrence is another ptygma-migma enigma.

PEGMATITES

Tourmalines in pegmatites, like pegmatites themselves, appear to be of diverse origins (see, for example, Cameron et al., 1949). The literature is rife with interpretations ranging from "authigenic pegmatites" with the boron of the tourmaline being derived from surrounding country rocks, through hydrothermal pegmatites with some of the tourmaline constituents being autochthonous and the others carried in by hydrothermal solutions of one or another derivation, to magmatic pegmatites with tourmaline formed by crystallization from so-called pegmatite magmas. In addition, late and post-pegmatite formation replacement processes are attributed negligible to large roles in the formation of the tourmalines (see, for example, Quensel, 1957).

Quite realistically, it seems likely that each, or even some combination, of the suggested origins may have obtained for tourmalines in diverse pegmatite masses. Therefore, critical evidence must be sought for each occurrence. Field relations and chemical studies, including those of fluid inclusions, may yield definitive results, but structural analyses of the tourmalines may also be required. And, in some cases, it may be that none of these studies will be definitive.

Nonetheless, a few of the recorded generalizations appear to be especially noteworthy.

Schorl is the typical tourmaline in simple granitic pegmatites and also occurs, commonly as the predominant tourmaline, in many complex pegmatites. It is so recorded from alkalic (e.g., sodalite-bearing) syenite pegmatites (Il'inskii and Shinkarev, 1964) as well as from granitic pegmatites.

Elbaite is characteristic of central zones (commonly in "pockets") of complex, typically Li-bearing pegmatites, and liddicoatite is apparently of the same general occurrence. (Descriptions of some so-called elbaite-bearing veins indicate that they, too, are pegmatitic.)

Dravite has been recorded from a few pegmatite masses, for example, those in Mg-rich metasedimentary rocks of the Turkestan Ridge (Ginzburg, Kruglova, and Moleva, 1951) and those in magnesite marbles of the basement of Central Asia (Melent'ev, Sharybkin, and Mart'yanov, 1971). It also has been recorded from surrounding country rocks into which boron has apparently been introduced—for example, from the norite of the Mesa Grande district, San Diego County, California (Foord, 1976).

Tourmaline, typically schorl, is common in many of the aplites that are spatially, and apparently genetically, related to pegmatites (e.g., Beus, 1950).

Several investigators have given detailed descriptions of tourmaline-bearing pegmatites (see, for example, Jahns and Wright, 1951, and Foord, 1976). A few of them have directed attention to particular features and/or relations that involve tourmaline. Some of these are also especially noteworthy.

Green and pink tourmaline-bearing pegmatites tend to be farther from their parent intrusives than consanguineous black tourmaline-bearing pegmatites are (Melent'ev et al., 1967).

The Li-content of tourmalines is greater in albitized than nonalbitized pegmatites, and Rb and Cs, as well as Li, are apparently sequentially enriched in later generations (Konovalenka and Bozin, 1976).

Li-bearing tourmalines appear to be most common in core borders whereas schorls tend to occur most commonly in wall and intermediate zones. Both species, however, have been recorded from other zones in one or more well-described pegmatite masses.

Although some gem-quality elbaite crystals are attached to quartz and/or other minerals that protrude into cavities or "pockets," several of these crystals lie free in loose clay within the pockets (Jahns and Wright, 1951). Broken fragments of crystals, with their originally adjacent segments, or segments and their roots, separated by up to several centimeters, occur in some pockets (John Sinkankas, 1983, personal communication), and some such fragments exhibit abrasion, a few to the extent that they appear tumbled. (Possible origins for the corraded fragments are given in Chap. 2.)

Staatz, Murata, and Glass (1955) have recorded a systematic variation in color, composition, and dependent physical properties of tourmaline that correlates with position within the Li-bearing Brown Derby pegmatites #1 and #3 of the Quartz Creek district, Gunnison County, Colorado. Concentrations of K, Ca, Pb, Be, and Li increase, whereas concentrations of Fe, Mg, Ti, and Na decrease toward the cores; and Rb and Cs are concentrated in pink elbaite within the core, whereas Co, Ni, Cr, and V occur only in black tourmaline near the wall. They interpret the relations to reflect crystallization during fractionation within a closed system.

Shmakin (1972) has reported five generations of tourmaline—each with distinctive Ba, Rb, Cs, Tl, and Be contents—in muscovite pegmatites of eastern Siberia.

Foord (1976 and 1977) found oxide variations similar to those recorded by Staatz, Murata, and Glass (1955), and also found a correlation between crystal habit and apparent period of growth for tourmalines of the Mesa Grande district of San Diego County, California.

Karskii et al. (1977) found the morphology, composition, and composition of fluid inclusions in tourmalines to differ within consanguineous aplites, typical pegmatites, and pegmatites having graphic textures. In the last, for example, the tourmaline occurs as short prismatic crystals.They related the differences to temperature of crystallization.

Correlations of tourmaline species and/or color varieties with overall pegmatite compositions have been recorded; for example, Ginzburg (1954) concluded: (1) black tourmaline indicates lack of rare elements in the host pegmatite, (2) blue and green tourmalines indicate the probable presence of Sn-Nb-Ta minerals, and (3) red tourmalines indicate a likelihood of accompanying Cs-Li-Rb-Ta minerals. However, except for pegmatites that contain only black tourmaline, his first conclusion is invalid. In addition: Okulov (1973*a*) found a direct correlation between the Li, Rb, and Cs contents of the tourmalines and the associated minerals other than microcline; Glebov et al. (1975) concluded that tourmaline composition may reflect the kind of rare-metal ore present within a pegmatite; Kulikov et al. (1976) concluded that the elbaites of rare-earth pegmatites contain more Al, Mn, Na and Li and less Si, Fe, and Mg than elbaites of other heavy mineral associations; and Čech, Litomiský, and Novotný (1965) reported that Mn-bearing schorls do not occur in Li-bearing pegmatites in Czechoslovakia.

Němec (1951), as a result of his spectrographic analyses of tourmalines, *inter alia,* from 36 western Moravian pegmatites of diverse origins, reported: (1) in general, schorls of simple pegmatites contain fewer trace elements than schorls of complex pegmatites, aplites, and low temperature veins; (2) Pb is lacking in schorls from simple pegmatites; and, as a result of later even more extensive studies (Němec, 1973), (3) Sn-content varies directly with F-content and is characteristic of elbaites from complex pegmatites (and of schorls from cassiterite-bearing rocks).

Results and interpretations dealing with fluid inclusions in tourmalines of pegmatites are diverse, and in some cases, apparently conflicting. For example, Weiss (1953) found no consistent trend in the temperatures indicated for tourmalines from different zones of pegmatites of the Black Hills of South Dakota, whereas Slivko (e.g., 1974) reported rather consistent trends both within individual crystals and for crystals within different zones of individual pegmatites. In any case, nearly all investigators have interpreted homogenization temperatures (etc.) to indicate formation of tourmalines at temperatures between ~200°C and 350°C, at pressures between ~800 bars and 1500 bars, and Slivko (1969*b*) has concluded that the homogenization, typically in the liquid state, is "direct proof of their hydrothermal origin." It also appears that, in general, temperatures of formation were higher for schorl than for elbaite within at least some individual masses (see, for example, Taylor, Foord, and Friedrichsen, 1979). Furthermore, investigations of fluid inclusions also indicate that the tourmalines were formed from CO-rich fluids in which halogens, especially F and Cl, played major roles.

Somewhat at odds with general conclusions, Pomârleanu and Movileanu (1971), as the result of an investigation of fluid inclusions in pegmatite minerals from the Carpathians of Romania, described primary, "pseudo-secondary," and secondary monophase, biphase, and triphase inclusions in tourmaline *(inter alia)*, and concluded that the inclusions indicate formation from CO_2-saturated solutions and/or hydrothermal solutions in the temperature range 300-530°C, which they termed the pegmatite phases, *and* in the range 100-320°C for their hydrothermal replacement phase.

VEINS AND OTHER CAVITY FILLINGS

Tourmaline of one or another species is recorded as occurring in veins, some of which are identified as joint fillings (e.g., Simpson, 1931); in "vugs", some of which are likely miarolitic cavities (e.g., Tomlinson, 1945); in tectonic fracture zones as, for example, breccia cement (Simboli, 1959) or the cryptocrystalline matrix of fault gouge (Kramer and Allen, 1954); as cement for clastic sediments (Vetrunov, 1955); in agglomerates that represent diatremes (Coetzee, 1966); and in "breccia pipes," whatever their origin (see, for example, Sillitoe and Sawkins, 1971), and Fletcher, 1977). Nearly all of these tourmalines that have been described by species are schorls or dravite/schorls, but a few are apparently elbaites.

Much cavity-filling tourmaline is fibrous, typically in either radial groups (e.g., Simpson, 1931) or resembling cross-fiber asbestos (e.g., Simpson, 1931, and Novak and Zak, 1970).

Juxtaposed rocks are of nearly all kinds and origins, and commonly they are partially tourmalinized. Spatially associated minerals include essentially all common and several uncommon "vein minerals."

The most frequently suggested origin for tourmalines in veins is deposition from hydrothermal solutions, with the source of the solutions usually inferred to have been magmatic. For some occurrences, however, the solutions appear to have arisen rather locally, some during diagenesis, others during metamorphism. Pneumatolysis has also been suggested as responsible for some vein tourmaline. One group of tourmaline-rich veins in Devonshire (England) has been interpreted as magmatic (Chaudhry and Howie, 1976); the apparently discrepant vein, rather than dike, designation for these masses is confusing.

Tourmaline (dravite/schorl) has been found to increase in Fe-content, specific gravity, intensity of color, indices of refraction, birefringence, and size with depth within veins of an unidentifed Be-district in Siberia (Getmanskaya et al., 1970). Rudakova (1972) has reported that for the tourmaline (schorl) of Transbaikalian cassiterite-bearing rocks and veins the Mn:Fe ratio decreases while the Fe^{2+}:Fe^{2+} + Fe^{3+} ratio increases with depth of formation. According to, for example, Radkevich et al. (1972), tourmaline compositions commonly vary along with other compositional variations within many zoned ore deposits and districts.

Sillitoe, Halls, and Grant (1975), as a result of their interpretation of "porphyry tin deposits" in Bolivia, have propsed a model whereby the porphyries were intruded, brecciated, veined and otherwise mineralized (including tourmalinization), and underwent wall-rock alteration during a single sequence of cogenetic events. This hypothesis serves to direct attention to the widespread spatial and possibly genetic relations between magmatic activity *per se* and the fluids responsible for the tourmaline in at least some veins and replacement masses, a few of which constitute ore deposits.

ORE DEPOSITS

Most, if not all, tourmalines that are within or spatially (and very likely genetically) associated with ore deposits are of replacement (e.g., in greisens), cavity-filling (e.g., in veins and breccia pipes), or metamorphic (including both allochemical and isochemical metamorphic) origin. Typically, tourmaline is a gangue mineral. Each of these kinds of occurrences is treated in the appropriate, preceding sections of this chapter.

Kuz'min, Dobrovol'skaya, and Solntseva (1979) have published a rather extensive review (plus 374 bibliographic entries) deaing with this subject in their book *Tourmaline and Its Use in Prospecting-Evaluation Studies* (cf. Otroshchenko et al., 1972).

Slack (1982) has summarized the literature relating to tourmaline in both "sedimentary-hosted . . . and volcanic-hosted tourmaline-bearing strata-bound sulphide deposits," and has tabulated 14 deposits from four continents that he considers to be of this character. He concludes that the common presence of tourmaline nearby as well as within these deposits makes it a prime candidate for use of both itself and boron "as an exploration guide for strata-bound mineral deposits in metamorphic terrains" (p. B82).

Slack and Plimer (1983) have classified tourmalines of stratiform sulfide deposits as follows:

1. disseminations in massive sulfides and adjacent country rocks;
2. concentrations within discordant feeder zones or pipes;

3. fillings in syn- or late-metamorphism fractures (including joints) in wall rocks;
4. crystals in stratabound tourmaline-rich pegmatites; and
5. layers and laminations in stratiform tourmalinites.

(Each of these except category (4) is described by Slack, 1982.)

Brown and Ayuso (1984), as a result of their investigation of the "Significance of tourmaline-rich rocks in the [Precambrian] Grenville complex of St. Lawrence County, New York," have suggested that tourmaline constituents "such as low TiO_2 content, variable Na/(Na+Ca+K) ratios, and the antipathetic correlation of Na and Ca might also be useful in discriminating tourmalines associated with [particular kinds of] ore deposits."

ADDITIONAL CRITERIA FOR DIVERSE ORIGINS

The following are additions to the possible genetic criteria already mentioned.

As a result of their investigation of the chemical composition of tourmalines in diverse rocks, Vladykin et al. (1975) found a general correlation that correlates well with the statements in the introduction to this chapter.

In a study of "geochemical indicator minerals and their significance in prospecting for rare metal pegmatites," Ginzburg (1954) found that black tourmalines indicate the probable presence of Sn-Nb-Ta ores, and rose and red tourmalines indicate the probable presence of Cs-Li-Rb-Ta ores.

Korovushkin, Kuz'min, and Belov (1979), on the basis of Mössbauer studies of tourmalines of diverse geneses, concluded that tourmalines may be distinguished according to genesis on the basis of the coefficient of oxidation and the distribution by site of the contained iron. Indeed, Gorelikova Perfil'eva, and Bubeshkin (1978) found that the oxidation coefficient ($Fe^{3+}/Fe^{2+} + Fe^{3+}$) for tourmaline does appear to differ with environment of formation; that is to say, schorls from tin ore-containing pegmatites have little or no Fe^{3+} so the oxidation coefficient ranges between 0.96 and 1.00, whereas tourmalines in cassiterite-silicate deposits have oxidation coefficients that range between 0.7 and 0.94. They also concluded: (1) the oxidation coefficient tends to decrease with increased depth of formation (in the Sonechnoye deposit of Far Eastern U.S.S.R., the coefficient for tourmaline is 0.88 in the upper metasomatic zone, 0.85 in the underlying main ore-bearing formations, and 0.83 in still lower zones); and (2) recrystallization and redeposition of tourmaline are generally accompanied by increased proportions of Fe^{2+}. (This, of course, follows LeChatelier and Goldschmidt's general principle whereby the deeper a mineral is formed the smaller the radii of the included cations.)

It appears, in fact, that site population determinations may very well be of value in geothermometry and geobarometry studies, and perhaps also in investigations of the mechanics and kinetics of processes such as diffusion; for example, Al occupancy of *Y*-sites may be indicative of high pressure (Dolomanova et al., 1978) and/or disorder caused by formation at high temperatures.

MISCELLANEOUS OCCURRENCES

Tourmaline, otherwise unidentified, is included in a fulgurite found near Uhrineves, southwest of Prague, Czechoslovakia (Kratochvil, 1957). It seems likely that the tourmaline was detrital and persisted through the vitrification responsible for formation of most of the fulgurite.

"Elbaite(?)" has been identified as a constituent of Libyan desert glass now generally interpreted to represent vitrification associated with impact (i.e., with formation of an astrobleme). The presence of the tourmaline served as one of the criteria used to support the hypothesis whereby the glass is considered to have been derived from nearby Nubian Sandstone (Barnes and Underwood, 1976).

I'm sorry, Virginia, but to the present no tourmaline (or any other OH-bearing mineral) has been identified in any of the moon rocks or soils (Odette James, 1983, personal communication) or in any other nonterrestrial materials.

Appendix A

"Other Forms" (see Table 2-1) and Angles Between Relatively Common Crystal Faces

The forms in Table A-1 are listed in the order in which they are given in Goldschmidt's (1922) tabulation (which also provides references). More recently reported forms and earlier reported forms not on Goldschmidt's list are inserted in their appropriate places; the sources, indicated by superscripts, are as follows: [g]—Grill (1922), [k]—Kulaszewski (1921), [s]—Slivko (1963), and [t]—Tarnovskii (1961). Several forms are vicinal forms; in fact, Kulaszewski (1921) reported some of the forms she listed as such. They are included for the sake of making the record complete.

In Table A-2, angles between relatively commonly occurring faces are listed as reported by Dana (1892) but given as angles between indexed forms rather than between letter designators (e.g., $(0001) \wedge (10\bar{1}1)$ instead of *c*:*r*). Measured angles may differ slightly from the values given because diverse tourmalines have differenct *c* and *a* dimensions. For example, the angle between $(10\bar{1}1)$ and $(\bar{1}101)$, which is given as 47°, has been recorded as ranging from 46°48′ to 47°16′5″ (Wülfing, 1900, and Becht, 1913). Angles marked with an asterisk (*) have been repeated in several introductory textbooks, determinative tables, and so forth, presumably because they are relatively common and thus may often aid identification of tourmaline crystals. The two angles preceded by a dagger (†) are additions to Dana's list that have also been cited in some introductory texts (e.g., Hurlbut and Klein, 1977).

Table A-1
Other Forms (see also Table 2-3)

$137.2.\overline{139}.0^{k}$	$103.27.\overline{130}.0^{k*}$	$23.18.\overline{41}.0$	$1.0.\overline{1}.3$
$95.2.\overline{97}.0^{k}$	$15.4.\overline{19}.0^{k}$	$68.53.\overline{121}.0^{k\dagger}$	$\overline{7}.0.7.20$
$63.2.\overline{65}.0^{k}$	$11.3.\overline{14}.0$	$5.4.\overline{9}.0$	$2.0.\overline{2}.5$
$30.1.\overline{31}.0$	$79.22.\overline{101}.0^{k}$	$31.25.\overline{56}.0^{k}$	$\overline{4}.0.4.9$
$25.1.\overline{26}.0$	$7.2.\overline{9}.0$	$16.13.\overline{29}.0$	$4.0.\overline{4}.7$
$62.3.\overline{65}.0^{k}$	$122.35.\overline{157}.0^{k}$	$50.41.\overline{91}.0^{k}$	$\overline{5}.0.5.8$
$18.1.\overline{19}.0$	$10.3.\overline{13}.0$	$6.5.\overline{11}.0$	$2.0.\overline{2}.3$
$61.4.\overline{65}.0^{k}$	$74.23.\overline{97}.0^{k}$	$7.6.\overline{13}.0^{k}$	$5.0.\overline{5}.7$
$15.1.\overline{16}.0$	$49.16.\overline{65}.0^{k}$	$23.20.\overline{43}.0^{k}$	$3.0.\overline{3}.4$
$40.3.\overline{43}.0$	$3.1.\overline{4}.0$	$10.9.\overline{19}.0$	$11.0.\overline{11}.14^{g}$
$13.1.\overline{14}.0$	$106.37.\overline{143}.0^{k}$	$15.14.\overline{29}.0^{k}$	$\overline{4}.0.4.5$
$12.1.\overline{13}.0$	$8.3.\overline{11}.0^{k}$	$31.28.\overline{59}.0^{k}$	$10.0.\overline{10}.11$
$11.1.\overline{12}.0$	$64.25.\overline{89}.0^{k}$	$21.19.\overline{40}.0$	$13.0.\overline{13}.14$
$73.7.\overline{80}.0^{k}$	$7.3.\overline{10}.0$	$13.12.\overline{25}.0^{s}$	$20.0.\overline{20}.19$
$10.1.\overline{11}.0$	$58.25.\overline{83}.0^{k}$		$11.0.\overline{11}.10$
$29.3.\overline{32}.0^{k}$	$11.5.\overline{16}.0$	$1.1.\overline{2}.9$	$\overline{8}.0.8.7$
$35.4.\overline{39}.0^{k}$	$40.19.\overline{59}.0^{k}$	$2.2.\overline{4}.11$	$6.0.\overline{6}.5$
$26.3.\overline{29}.0^{k}$	$101.50.\overline{151}.0$	$1.1.\overline{2}.3$	$5.0.\overline{5}.4$
$106.13.\overline{119}.0^{k}$	$25.13.\overline{38}.0^{k}$	$5.5.\overline{10}.14$	$\overline{5}.0.5.4$
$65.8.\overline{73}.0^{k}$	$19.10.\overline{29}.0^{k}$	$5.5.\overline{10}.12$	$\overline{4}.0.4.3$
$8.1.\overline{9}.0$	$97.52.\overline{149}.0^{k}$	$1.1.\overline{2}.2$	$10.0.\overline{10}.7$
$121.16.\overline{137}.0^{k}$	$20.11.\overline{31}.0^{k}$	$2.2.\overline{4}.3$	$3.0.\overline{3}.2$
$150.21.\overline{171}.0^{k}$	$76.43.\overline{119}.0^{k}$	$2.2.\overline{4}.1$	$5.0.\overline{5}.3$
$7.1.\overline{8}.0$	$5.3.\overline{8}.0$		$13.0.\overline{13}.7$
$68.10.\overline{78}.0^{k}$	$91.55.\overline{146}.0^{k}$	$1.0.\overline{1}.22$	$7.0.\overline{7}.4$
$67.10.\overline{77}.0^{k}$	$8.5.\overline{13}.0$	$1.0.\overline{1}.17$	$9.0.\overline{9}.5$
$6.1.\overline{7}.0$	$76.49.\overline{125}.0^{k}$	$1.0.\overline{1}.12^{g}$	$2.0.\overline{2}.1$
$109.19.\overline{128}.0^{k}$	$3.2.\overline{5}.0$	$1.0.\overline{1}.10$	$7.0.\overline{7}.3$
$11.2.\overline{13}.0$	$83.56.\overline{139}.0^{k}$	$1.0.\overline{1}.8^{s}$	$5.0.\overline{5}.2$
$41.8.\overline{49}.0^{k}$	$16.11.\overline{27}.0^{k}$	$1.0.\overline{1}.6^{s}$	$\overline{5}.0.5.2$
$5.1.\overline{6}.0^{s}$	$10.7.\overline{17}.0$	$2.0.\overline{2}.11$	$\overline{11}.0.11.4$
$29.6.\overline{35}.0^{k}$	$17.12.\overline{29}.0^{k}$	$1.0.\overline{1}.5$	$14.0.\overline{14}.5$
$9.2.\overline{11}.0$	$53.38.\overline{91}.0^{k}$	$2.0.\overline{2}.9^{s}$	$3.0.\overline{3}.1$
$101.23.\overline{124}.0^{k}$	$11.8.\overline{19}.0$	$1.0.\overline{1}.4$	$\overline{3}.0.3.1$
$91.22.\overline{113}.0^{k}$	$78.57.\overline{135}.0^{k}$	$\overline{1}.0.1.4$	$13.0.\overline{13}.4$
$4.1.\overline{5}.0$	$21.16.\overline{37}.0^{k}$	$2.0.\overline{2}.7$	$\overline{7}.0.7.2$

Table A-1 *(cont.)*

Other Forms (see also Table 2-3)

$\overline{15}.0.15.4$	$7.1.\overline{8}.6$	$4.3.\overline{7}.5$	$\overline{11}.\overline{4}.15.14$
$\overline{19}.0.19.5$	$6.1.\overline{7}.5$	$2.1.\overline{3}.2$	$\overline{3}.\overline{2}.5.2$
$17.0.\overline{17}.4$	$5.1.\overline{6}.4$[g]	$\overline{8}.\overline{3}.11.7$	$\overline{5}.\overline{4}.9.2$
$\overline{22}.0.22.5$	$4.1.\overline{5}.3$	$\overline{10}.\overline{3}.13.8$	$\overline{4}.\overline{1}.5.4$
$9.0.\overline{9}.2$	$\overline{4}.\overline{1}.5.3$	$\overline{14}.\overline{3}.17.10$	$\overline{49}.\overline{1}.50.32$
$\overline{9}.0.9.2$	$11.3.\overline{14}.8$	$\overline{10}.\overline{2}.12.7$	$\overline{8}.\overline{3}.11.4$
$\overline{5}.0.5.1$	$7.2.\overline{9}.5$	$\overline{6}.\overline{1}.7.4$	$7.2.\overline{9}.2$
$\overline{16}.0.16.3$	$10.3.\overline{13}.7$	$\overline{8}.\overline{1}.9.5$	$15.4.\overline{19}.2$
$11.0.\overline{11}.2$	$3.1.\overline{4}.2$	$\overline{26}.\overline{1}.27.14$	$\overline{6}.\overline{2}.8.3$
$\overline{17}.0.17.3$	$14.5.\overline{19}.9$	$\overline{29}.\overline{1}.30.14$	$\overline{5}.\overline{2}.7.12$
$6.0.\overline{6}.1$	$8.3.\overline{11}.5$	$\overline{11}.\overline{1}.12.5$	$3.2.\overline{5}.4$
$\overline{20}.0.20.3$	$17.7.\overline{24}.10$	$\overline{7}.\overline{1}.8.3$	$11.2.\overline{13}.12$
$7.0.\overline{7}.1$	$7.3.\overline{10}.4$	$\overline{5}.\overline{1}.6.2$	$\overline{2}.\overline{1}.3.5$
$\overline{7}.0.7.1$	$11.5.\overline{16}.6$	$\overline{13}.\overline{3}.16.5$	$\overline{7}.\overline{2}.9.4$
$\overline{8}.0.8.1$	$\overline{2}.\overline{1}.3.1$	$\overline{8}.\overline{2}.10.3$	$\overline{10}.\overline{9}.19.5$
$\overline{17}.0.17.2$	$13.7.\overline{20}.6$	$8.2.\overline{10}.3$	$4.2.\overline{6}.5$
$\overline{19}.0.19.2$	$9.5.\overline{14}.4$	$\overline{11}.\overline{3}.14.4$	$12.6.\overline{18}.5$
$10.0.\overline{10}.1$	$7.4.\overline{11}.3$	$\overline{7}.\overline{3}.10.2$	$\overline{10}.\overline{3}.13.5$
$\overline{10}.0.10.1$	$5.3.\overline{8}.2$	$4.2.6.1$	$5.4.\overline{9}.7$
$\overline{11}.0.11.1$	$7.5.\overline{12}.2$	$\overline{17}.\overline{9}.26.4$	$\overline{7}.\overline{1}.8.7$
$12.0.\overline{12}.1$	$18.13.\overline{31}.5$	$\overline{5}.\overline{3}.8.1$	$\overline{12}.\overline{4}.16.7$
$13.0.\overline{13}.1$	$15.11.\overline{26}.4$	$\overline{11}.\overline{7}.18.2$	$\overline{14}.\overline{3}.17.7$
$20.0.\overline{20}.1$	$4.3.\overline{7}.1$	$\overline{6}.\overline{4}.10.1$	$14.1.\overline{15}.7$
$\overline{27}.0.27.1$	$9.7.\overline{16}.2$	$\overline{7}.\overline{5}.12.1$	$20.7.\overline{27}.7$
$\overline{32}.0.32.1$	$5.4.\overline{9}.1$	$\overline{8}.\overline{6}.14.1$	$\overline{3}.\overline{2}.5.8$
$33.0.\overline{33}.1$	$6.5.\overline{11}.1$	$11.2.\overline{13}.3$	$7.2.\overline{9}.8$
	$7.6.\overline{13}.1$	$5.2.\overline{7}.1$	$\overline{7}.\overline{2}.9.9$
$1.3.\overline{4}.3$	$15.13.\overline{28}.2$	$9.1.\overline{10}.2$	$9.2.\overline{11}.10$
$4.3.\overline{7}.10$	$13.12.\overline{25}.1$	$5.1.\overline{6}.1$	$9.4.\overline{13}.11$
$2.1.\overline{3}.4$	$15.14.\overline{29}.1$	$6.2.\overline{8}.1$	$13.2.\overline{15}.14$
$16.1.\overline{17}.15$	$18.17.\overline{35}.1$	$\overline{4}.\overline{3}.7.2$	$18.2.\overline{20}.19$
$12.1.\overline{13}.11$	$20.19.\overline{39}.1$	$\overline{6}.\overline{3}.9.2$	
$10.1.\overline{11}.9$	$45.42.\overline{87}.1$[t]	$\overline{6}.\overline{1}.7.10$	

*Order of forms in original report suggests this was meant to be $103.47.\overline{150}.0$.

†Reported as $68.33.\overline{121}.0$; order indicates $68.53.\overline{121}.0$.

Table A-2
Angles Between Commonly Occurring Faces

Faces	Angle	Faces	Angle
$(11\bar{2}0) \wedge (13\cdot 1\cdot\overline{14}\cdot 0)$	= 26°48′	$(05\bar{5}4) \wedge (\bar{5}054)$	= 56°04′
$(11\bar{2}0) \wedge (1\cdot 10\cdot\overline{11}\cdot 0)$	= 26°02′	*$(02\bar{2}1) \wedge (\bar{2}021)$	= 77°00′
$(11\bar{2}0) \wedge (71\bar{8}0)$	= 23°25′	$(31\bar{4}2) \wedge (\bar{3}4\bar{1}2)$	= 58°51′
$(11\bar{2}0) \wedge (41\bar{5}0)$	= 19°06½′	$(05\bar{5}1) \wedge (\bar{5}051)$	= 107°44½′
$(11\bar{2}0) \wedge (21\bar{3}0)$	= 10°53½′	$(31\bar{4}2) \wedge (4\bar{3}\bar{1}2)$	= 18°51′
$(0001) \wedge (10\bar{1}4)$	= 7°22′	$(21\bar{3}1) \wedge (\bar{2}3\bar{1}1)$	= 63°48′
$(0001) \wedge (10\bar{1}2)$	= 14°29½′	$(21\bar{3}1) \wedge (3\bar{2}\bar{1}1)$	= 30°38½′
*$(0001) \wedge (10\bar{1}1)$	= 27°20′	$(0001) \wedge (32\bar{5}1)$	= 66°04′
$(0001) \wedge (70\bar{7}4)$	= 42°08′	*$(32\bar{5}1) \wedge (\bar{3}5\bar{2}1)$	= 66°01′
$(0001) \wedge (50\bar{5}2)$	= 52°16′	*$(32\bar{5}1) \wedge (5\bar{3}\bar{2}1)$	= 42°36′
$(0001) \wedge (40\bar{4}1)$	= 64°11½′	$(12\bar{3}2) \wedge (\bar{1}3\bar{2}2)$	= 21°18′
$(0001) \wedge (10\cdot 0\cdot\overline{10}\cdot 1)$	= 79°03′	$(12\bar{3}2) \wedge (3\bar{1}\bar{2}2)$	= 43°22½′
$(0001) \wedge (01\bar{1}2)$	= 14°29½′	$(24\bar{6}1) \wedge (\bar{2}6\bar{4}1)$	= 35°48½′
$(0001) \wedge (01\bar{1}1)$	= 27°20′	$(24\bar{6}1) \wedge (6\bar{2}\bar{4}1)$	= 75°53′
$(0001) \wedge (05\bar{5}4)$	= 32°52′	$(23\bar{5}2) \wedge (\bar{2}5\bar{3}2)$	= 34°35½′
*$(0001) \wedge (02\bar{2}1)$	= 45°57′	$(23\bar{5}2) \wedge (5\bar{2}\bar{3}2)$	= 52°57′
$(0001) \wedge (07\bar{7}2)$	= 61°04′	$(45\bar{9}2) \wedge (\bar{4}9\bar{5}2)$	= 46°50′
$(0001) \wedge (0\cdot 15\cdot\overline{15}\cdot 4)$	= 62°43′	$(45\bar{9}2) \wedge (9\bar{4}\bar{5}2)$	= 59°35′
$(0001) \wedge (09\bar{9}2)$	= 66°44′	$(13\bar{4}1) \wedge (\bar{1}4\bar{3}1)$	= 24°26′
$(0001) \wedge (05\bar{5}1)$	= 68°51′	$(13\bar{4}1) \wedge (4\bar{1}\bar{3}1)$	= 78°50′
$(0001) \wedge (0\cdot 11\cdot\overline{11}\cdot 1)$	= 80°01½′	$(11\bar{2}0) \wedge (32\bar{5}1)$	= 24°46′
$(10\bar{1}4) \wedge (\bar{1}104)$	= 12°45′	$(11\bar{2}0) \wedge (21\bar{3}1)$	= 37°34′
$(10\bar{1}2) \wedge (\bar{1}102)$	= 25°02′	$(11\bar{2}0) \wedge (31\bar{4}2)$	= 49°05′
*$(10\bar{1}1) \wedge (\bar{1}101)$	= 47°	$(11\bar{2}0) \wedge (10\bar{1}1)$	= 66°34′
†$(10\bar{1}1) \wedge (10\bar{1}0)$	= 62°40′	$(11\bar{2}0) \wedge (13\bar{4}1)$	= 32°09′
$(70\bar{7}4) \wedge (\bar{7}704)$	= 71°02′	$(11\bar{2}0) \wedge (17\bar{8}3)$	= 43°19′
$(50\bar{5}2) \wedge (\bar{5}502)$	= 86°27½′	$(11\bar{2}0) \wedge (02\bar{2}1)$	= 51°30′
$(40\bar{4}1) \wedge (\bar{4}401)$	= 102°27′	$(10\bar{1}1) \wedge (02\bar{2}1)$	= 38°30′
$(10\cdot 0\cdot\overline{10}\cdot 1) \wedge (\overline{10}\cdot 10\cdot 0\cdot 1)$	= 116°29′	$(10\bar{1}0) \wedge (32\bar{5}1)$	= 33°00′
$(01\bar{1}2) \wedge (\bar{1}012)$	= 25°02′	$(10\bar{1}0) \wedge (02\bar{2}1)$	= 68°56′
$(01\bar{1}1) \wedge (\bar{1}011)$	= 46°52′	†$(01\bar{1}0) \wedge (02\bar{2}1)$	= 44°03′

Appendix B

Ideal Midpoint Analyses for Tourmalines of the Common Solid-Solution Series

The following table of ideal analyses for midpoint compositions in the liddicoatite-elbaite-schorl-calcium analog of schorl and uvite-dravite-schorl-calcium analog of schorl series—expressed in weight percentages of oxides—will facilitate appropriate naming of a tourmaline belonging to either of these series from partial or complete chemical analyses.

Table B-1
Ideal Midpoint Analyses

	Elbaite$_{50}$-*Liddicoatite*$_{50}$	*Elbaite*$_{50}$-*Schorl*$_{50}$	*Schorl*$_{50}$-*Ca-Schorl*$_{50}$	*Liddicoatite*$_{50}$-*Ca-Schorl*$_{50}$	*Elb*$_{25}$*Sch*$_{25}$-*Lid*$_{25}$*CaS*$_{25}$
SiO_2	38.345	36.355	33.51	35.50	35.93
B_2O_3	11.105	10.535	9.71	10.28	10.41
Al_2O_3	39.315	34.93	26.10	30.48	32.70
Fe_2O_3	—	—	—	—	—
FeO	—	10.23	23.30	13.07	11.65
MgO	—	—	—	—	—
CaO	2.97	—	2.55	5.52	2.76
Na_2O	1.655	3.125	1.47	—	1.56
Li_2O	2.78	1.195	—	1.59	1.39
H_2O	3.83	3.63	3.36	3.56	3.60
Total	100.00	100.00	100.00	100.00	100.00

	Schorl$_{50}$-*Ca-Schorl*$_{50}$	*Schorl*$_{50}$-*Dravite*$_{50}$	*Dravite*$_{50}$-*Uvite*$_{50}$	*Ca-Schorl*$_{50}$-*Uvite*$_{50}$	*Sch*$_{25}$*Dra*$_{25}$-*CaS*$_{25}$*Uvi*$_{25}$
SiO_2	33.51	35.91	37.33	34.93	35.42
B_2O_3	9.71	10.41	10.81	10.12	10.26
Al_2O_3	26.10	30.48	29.05	24.67	27.56
Fe_2O_3	—	—	—	—	—
FeO	23.30	10.23	—	13.07	11.65
MgO	—	6.30	14.59	8.28	7.30
CaO	2.55	—	2.88	5.43	2.72
Na_2O	1.47	3.08	1.61	—	1.54
Li_2O	—	—	—	—	—
H_2O	3.36	3.59	3.73	3.50	3.55
Total	100.00	100.00	100.00	100.00	100.00

Appendix C

Localities That Have Yielded Noteworthy Tourmaline Specimens and/or Gem Materials

The following locality list, like those for most relatively common minerals, is strictly historical—that is, tourmaline no longer occurs at some of the localities and tourmaline will undoubtedly be found at other ("new") localities in the future. Also, because of both space restrictions and the widespread occurrence of tourmaline, especially schorl, the list is not comprehensive. It is, instead, an abridgment of the catalog of tourmaline specimens, including gemstones and carvings, in the collections at U.S. National Museum of Natural History plus a few additional localities represented by noteworthy specimens in other collections seen by the compiler—for example, in the collections of the Harvard University Mineralogical Museum and the Royal Ontario Museum.

The localities are listed alphabetically by country under each species heading. Individual entries are in the order of greater and greater specificity—for example: Country, State, County, and nearest municipality. Localities included in parentheses or brackets, unless noted otherwise, have not been located by the compiler on maps or in gazetteers. Localitites in *italic type* have yielded fine quality gem material.

Specimens for some of the localities reported on this list were identifed solely on the basis of macroscopic examination. Consequently, some of the species-to-locality correlations may be incorrect. For example: some of the localities listed for schorl probably represent occurrences of black-appearing elbaite-liddicoatite or dravite-uvite specimens; some of the localities under elbaite may be for liddicoatites; some of those under dravite may be for uvites; . . . etc. This does not mean, however, that individual localities recorded for more than one species of tourmaline necessarily represent such misidentifications. Many localities have indeed yielded more than one species—for example, both elbaite and schorl.

BUERGERITE

Mexico: San Luis Potosi—Mexquitic.

CHROMDRAVITE

U.S.S.R.: Karelia—Onezhkii Depression.

DRAVITE

Australia: South Australia—Painter Mt.; Western Australia—Yinnietharra.
Austria: Carinthia—Dobrowa and (Pravalia).
Brazil: Bahia—Brumado; Minas Gerais—Boa Vista Ranch near Galileia and Diamantina Region.
Burma: Kaya—Loi-Kaw.
Canada:
Ontario: Frontenac Co.—Olden & Oso Twps., Sharbot Lake; Haliburton Co.—Cardiff Twp., Pusey Lake; Monmouth Twp., Wilberforce; Hastings Co.—Faraday, Huntington, and Madoc twps.; Lanark Co. —Front of Leeds and Lansdowne Twp., Gananoque; Leeds Co.; Lennox and Addington Co.—Sheffield Twp., Enterprise; Peterborough Co.—Galway Twp.; Renfrew Co.—Bagot, Lyndoch, and Ross twps.; Victoria Co.—Laxton Twp.; Nipissing Dist.—Mattawan Twp.
Quebec: Argenteuil Co.—e.g., Grenville Twp. and Calumet; Hull Co.—Hull Twp., Pink Lake; Terrebonne Co.—St. Jerome.
Ceylon—see Sri Lanka.
Czechoslovakia: Cechy (formerly Bohemia)—Pisek; Moravia—(Cyrilov).
Finland: Kuopion—Kaavi and Outokumpu.
France: Isère—Bourg Doisans, Le Grave.
India: Keonjhar District—Orissa.
Ireland: Dublin Co.—(Dalkey).
Italy: Piemonte—Valli di Lanzo, near Torino.
Kenya: Narok District—*Osarara.*
Madagascar.
Mexico: Sonora—Santa Cruz.
New Zealand: South Island—Golden Bay.
Norway: Aust-Agder—Arendal; Buskerud—(Snarum); Telemark—Bamle, Bjordamen; Vestfold—Stavern (formerly Fredriksvarn).
Sri Lanka (formerly *Ceylon*): near Ratnapura.
Switzerland: Graubünden (formerly Grisons)—(Ems); Valais—Simplon Tunnel.
U.S.S.R.: Sverdlovsk Oblast—Murzinka and Sysert'.
United States:
Alabama: Cleburne Co.—west of Chulifinnee.

California: Imperial Co.—Cargo Muchacho Mts.; Mariposa Co.—Silver Knob; Nevada Co.—Colfax; Plumas Co.—Thomson Peak; Riverside Co.—Crestmore.

Connecticut: Fairfield Co.—Bethel (Cod Fish Hill); Litchfield Co.—Thomaston dam; Middlesex Co.—Haddam and Middlesex; New Haven Co.—Cheshire.

Maine: Hancock Co.—Blue Hill; Oxford Co.—Newry and Rumford.

Maryland: Baltimore Co.—Cockeysville and Texas; Cecil Co.; Howard Co.—Marriotsville; Montgomery Co.—Etchison.

New Hampshire: Grafton Co.—Orford and Warren.

New Jersey: Hunterdon Co.—Lambertville; Sussex Co.—Franklin and Hamburg.

New York: Bronx—Kings Bridge; Essex Co.—Crown Point; *St. Lawrence Co.*—Canton, DeKalb, Edwards, Fowler, Gouverneur, Macomb, Norwood, Pierrepont, and Richville.

North Carolina: Cherokee Co.—Murphy.

Pennsylvania: Bucks Co.—New Hope; Chester Co.—e.g., London Grove, New Garden, Newlin, Parkersburg, and Unionville; Delaware Co.—Aston.

Virginia: Fairfax Co.—Centreville.

Yugoslavia: ?Dravograd (Jobronia).

ELBAITE

Afghanistan: *Laghman*—Nuristan (e.g., Pech and Nangarhar, northeast of Jalalabad).

Australia: South Australia—*Kangaroo Island;* Western Australia—Ravensthorpe, *Spargoville* and Wodgina.

Austria: Moravia—Rozena; Salzburg—Rauris, Hoher Tauer.

Belgian Congo—See Zaire.

Brazil: Bahia—Bom Jesus dos Meiras; *Minas Gerais*—Several localities (e.g., the Aracuai-Itinga and Aracuai-Salinas districts and the area around Governador Valadares) are shown on maps published by, for example, Proctor (1984).

Bulgaria: Vitosha Mts.

Burma: Sagaing District—Ava, Mogok and Nampai.

Canada:

Manitoba: Bernic Lake (east of Lac du Bonnet).

Ontario: Kenora Dist., Patricia Portion—Lilypad Lakes; Thunder Bay Dist.—Beavis Lake.

Quebec: Gatineau Co.—Wakefield Twp., Wakefield. Papineau Co.—Villeneauve.

Ceylon—see Sri Lanka.

Czechoslovakia: Moravia—Dobra Voda, near Velke' Mezirici; Jihlava (near Jeclev); Rozna.
Germany (East-D.D.R.): Saxony—Karl-Marxstad (formerly Chemnitz)—Penig.
India: West Bengal—(Harari-bagh—Pahira).
Ireland: Donegal Co.—(Dunfanaghy); Sligo Co.—(Ox Mt.).
Italy: Piemonte—Traversella; Toscany—*Elba,* Grotto d'Oggi near San Piero.
Japan: Kyushu Island, Fukuoka Pref.—Hakata Bay near Fukuoka.
Kenya: *Voi.*
Madagascar (Malagasy Republic): e.g., *Anjanabonoina, Tananarive-Antsirabe* and Mt. Bity Region.
Manchuria: *no locality given.*
Mexico: Baja California—*El Alamo,* Ensenada (La Verde mine near Ojos Negros), (La Huerta).
Mozambique: *Alto Ligonha*—Nacala (Pedro Baiessa).
Namibia (formerly Southwest Africa): e.g., *Karibib, Klein Spitzkopie,* Omaruru, *Usakos.*
Nepal: Kosi—Dhankuta.
Norway: Nordland—Askardet on Holandsfjord.
Pakistan (India?): Kashmir—Gilgit; Stak Nala.
Rwanda: Birenga.
Southern Rhodesia—see Zimbabwe.
South West Africa—see Namibia.
Sri Lanka.
Sweden: Salermanland—Utö; Varuträsk; Norrbotten—(Kluntarna).
Switzerland: Grisons Canton—Saflit, between Gibenhorn and Grauhorn areas; Ticino—Campolongo, near St. Gotthardt; Valais—Binnenthal, Langenbach.
Tanzania (formerly Tanganyika): *no locality given.*
U.S.S.R.: Kola Peninsula—(Khibiny Tundra); Transbaikalia—Chita Oblast—Nertchinsk; Ural Mts.—Sverdlovsk Oblast—e.g., Alabaschka, Alapayeusk, *Lipovka,* Murzinka, Pervoural'sk (formerly *Schiantansk*), Sarapulka and Sverdlovsk [formerly (Y)Ekaterinburg].
United Kingdom: England, Devonshire—(Meldon).
United States:
Alaska: Circle.
California: *San Diego Co.*—Several localities (e.g., Mesa Grande, Pala, and Ramona) are shown on maps published by, for example, Foord (1977); *Riverside Co.*—Hemet, (Sage), and Temecula.
Connecticut: Fairfield Co.—Bethel; Litchfield Co.—Thomaston Dam; Middlesex Co.—*Haddam Neck*—Gillette quarry and Portland-Strickland quarry.
Georgia: Dawson Co.—along Etowah River.

Maine: *Androscoggin Co.*—Auburn, Mt. Apatite and Mt. Rubellite; Cumberland Co.—Portland; *Oxford Co.*—Black Mt. near Andover and Rumford, Hebron, Mt. Mica near Paris and Norway, Noyes Mt. near Greenwood and Plumbago Mt. near Newry; Sagadahoc Co.—Topsham.

Massachusetts: Hampden Co.—Chester; Hampshire Co.—Chesterfield and Goshen.

Montana: Lewis and Clark Co.—Helena (Davis Gulch); Silver Bow Co.—Homestake area.

Nevada: Pershing Co.—Lovelock; (Rozena).

New Hampshire: Grafton Co.—North Groton.

New Mexico: Catron Co.—Elk Mts. (Upper Picos).

New York: New York City.

North Carolina: Alexander Co.—Hiddenite; Burke Co.—Silver Creek; *Macon Co.*—(Corundum Hill Mine); Mitchell Co.—Bakersville.

Pennsylvania: Northampton Co.—Bushkill.

South Dakota: Pennington Co.—Keystone.

Vermont: Windsor Co.—Norwich.

Zaire (formerly Belgian Congo): Katanga province—Musonoi.

Zimbabwe (formerly Southern Rhodesia): Mashonaland North—Karoi; Midlands—Somabula Forest.

FERRIDRAVITE

Bolivia: Cochabamba—Villa Tunari.

LIDDICOATITE

Madagascar: (Anjanabonoena), *Antsirabe* and (Tsilaizina).

SCHORL

Afghanistan: Laghman—Nuristan (e.g., Nangarhar and Kunar).

Australia: New South Wales—Ardlethan and (Stanley River); Northern Territory—Harts Range; South Australia—Flinders Range and (Umberatana).

Austria: Tyrol—(Pfitscher Joch) and along the Zillerthal.

Boliva: Cochabamba Dept.—Ayopayo—Icoya (formerly Kami); La Paz Dept.—Larecaja Province—(Millipaya) on Cotacajes (= Sorata) River; Potosoi Dept.—(Llallagua).

Borneo: along Koetei (= Mahakam) River.

Brazil: Minas Gerias—e.g., Bom Jesus de Lapa and Mendes Pimental.

Burma: Sagaing District—Mogok.

Canada:

British Columbia: Kootenay Dist.—Cranbrook.

Manitoba: Shatford Lake (east of Lac du Bonnet).

Northwest Territories—District of Franklin—Baffin Island—(Niantitic) on Cumberland Sound.

Ontario: Lanark Co.—Bathurst Twp.; Lennox and Addington Co.—Sheffield Twp., Enterprise; Peterborough Co.—Methuen Twp., Nephton; Renfrew Co.—Bagot Twp., Calabogie; Lyndoch Twp., Quadeville; Raglan & Lyndoch twps., (Bruceton); Simcoe Co.—Matchedash Twp.; Kenora Dist., Patricia Portion—Setting Net Lake.

Quebec: Papineau Co.—Wakefield Twp., St. Pierre de Wakefield.

Yukon Territory: Emerald Lake.

Ceylon—see Sri Lanka.

Chile: Atacama; (Lucero Mine near Cupiapo).

Czechoslovakia: Cechy—Nejdek, Pisek and Zd'ar; Moravia—Velkè Mezírící (Cyrilov and Delni Bery).

Finland: (Evajarvi—Viitaniemi and Kaatiala).

France: Rhône—(Mesveres) near Beaujeu.

Germany (East-D.D.R.): Saxony—Karl-Marxstad (formerly Chemnitz)—e.g., Eibenstock, Fichtelgebirg (Epprechstein), Penig and Vogtland.

Germany (West-B.D.): Bavaria—Aschaffenburg (Gailbach and Galgenbuckel), Bodenmais and Hörlberg (Lam); Lower Saxony—Harz Mts.—Andreasberg.

Greenland: Godthaab Dist.—Ameralikfjord; Julianshaab Dist.

India (Pakistan?): Kashmir—(Korund).

Italy: Toscany—Elba, Grotto d'Oggi near San Piero.

Japan:

Honshu Island: Fukushima Pref.—Ishikawa; Gifu Pref.—Mino; Ibaraki Pref.—Yokogawa; Yamanashi Pref.—Kimposan and (Koma Gori).

Kyushu Island: Oita Pref.—(Obira).

Madagascar (Malagasy Republic): e.g., Tananarive—Antsirabe and (Tangafend).

Mexico: Baja California—(La Huerta); Guanajuata; Sonora—Bacanora, (Milopilas and Santa Cruz).

Namibia (formerly Southwest Africa): Damarland—Usakos (Ameib); Windbock—between Tsumeb and Windhoek.

Norway: Aust-Agder—Arendal, Sondeled and Tvedestrand; Buskerud—Geithus (formerly Modum-this may be the often cited Snarum locality-see dravite) and Kongsberg; Rogaland—Egersund; Telemark—Bamle (Odegaard), Brevik (Laaven), Kragero (Rinde), Larvik (Klastad) and Tordal (Hoydalen).

Peru: Lima (near).

Poland: Silesia—Stiregau (= Strzegom?).

Southern Rhodesia—see Zimbabwe.
South West Africa—see Namibia.
Sri Lanka.
Sweden: Vastmanland—(Karingbricka) and Norberg.
Switzerland: Ticino—St. Gotthardt; Valais—e.g., Binnenthal, (Gehrenthal, Leisald, Piz Cavradi, Selkingertal and Valtellina).
Tanzania (formerly Tanganyika): Morogoro Prov.—Morogoro.
U.S.S.R.: Caucasas—Azerbajdzhanskaya District; Ural Mts.—Sverdlovsk Oblast—e.g., Alabaschka, Kamensk, Murzinka, Pervoural'sk (formerly Shiantanka), Sarapulka, Ufalei Mts. and Zavodouspenskoye.
United Kingdom:
England: Cornwall—(Huel Unity) and St. Just; Devonshire—Bovey Tracey, Exeter (Chudleigh) and Tavistock.
Scotland: Aberdeen—Aberdeen; Ayr—Carrick Fell.
United States:
Arizona: Maricopa Co.; Yavapai Co.
California: Plumas Co.—(Thomson Peak); Riverside Co.—Crestmore, Nuevo and Sage; San Bernardino Co.—Wrightwood and Sheep Creek; San Diego Co.—e.g., Coahuila, Mesa Grande and Pala.
Colorado: Park Co.—Wilkerson Pass (Abe Springs).
Connecticut: Middlesex Co.—Haddam and Portland; Monroe Co.—(Lane's Mine); New Haven Co.—Newton.
District of Columbia: Washington—Rock Creek Park.
Georgia: Clayton Co.—Jonesboro; Dawson Co.—along Etowah River.
Idaho: Valley Co.—Ola.
Maine: Androscoggin Co.—Auburn-Mt. Apatite and (Buckfield-Bennett Mine); Cumberland Co.—Freeport, Mechanics Falls, (Minot), and Portland; Oxford Co.—e.g., Greenwood (Harvard Mine), Newry (Plumbago Mt.), Paris (Mt. Mica) and Rumford (Black Mt.).
Maryland: Baltimore Co.—Baltimore and Towson; Cecil Co.; Howard Co.—(Brighton Dam Road); Montgomery Co.—Kensington.
Massachusetts: Franklin Co.—Northfield and Warwick; Hampden Co.—Chester; Hampshire Co.—Chesterfield and Pelham; Worcester Co.—Fitchburg (Rollstone Hill), Lancaster and Sturbridge.
Michigan: Marquette Co.—Champion.
Montana: Jefferson Co.—Hay Creek and Whitehall.
Nevada: Nye Co.—Luning.
New Hampshire: Carroll Co.—Conway; Cheshire Co.—Keene; Grafton Co.—Grafton, North Groton and Oxford; Strafford Co.—Barrington; Sullivan Co.—Acworth, Alstead, Newport and Springfield.
New Jersey: Hunterdon Co.—Lambertville; Sussex Co.—Ogdensburg; Warren Co.—Oxford.
New York: Essex Co.—Crown Point and Port Henry; Jefferson Co.—

Thousand Islands area; St. Lawrence Co.—Little Hammond, Macomb, Point Comfort, Oxbow and Pierrepont; Saratoga Co.—(Day) and Overlook.

North Carolina: Alexander Co.—Stony Point; Ashe Co.; Cleveland Co.—Kings Mt. and Lawndale; Iredell Co.—Statesville; Lincoln Co.; Macon Co.; McDowell Co.—Little Switzerland; Mitchell Co.—Bakersville and Spruce Pine; Yancey Co.—Burnsville.

Pennsylvania: Chester Co.; Delaware Co.; Montgomery Co.—Bustleton.

Rhode Island: Kent Co.—Warwick.

South Dakota: Black Hills region.

Texas: Hudspeth Co.—Sierra Blanca.

Utah: Beaver Co.

Vermont: Windham Co.—Brattleboro.

Virginia: Amelia Co.—Amelia Court House; Culpeper Co.—Mitchells; Floyd Co.—Willis; Mecklenburg Co.—Kerr Lake.

Zimbabwe (formerly Southern Rhodesia): Mashonaland North—Karoi and Miami.

UVITE

Brazil: *no locality given.*

Burma: Sagaing District—Mogok.

Ceylon—see Sri Lanka.

Kenya: *Magadi.*

Sri Lanka (formerly *Ceylon*): *Uva area.*

Tanzania (formerly Tanganyika): Kilimanjaro or Arusha—Moshi (60 mi. SE of); Tanga Prov.—Gerevi Hills.

United States:

Connecticut: Middlesex Co.—East Hampton.

Maine: Oxford Co.

New Jersey: Sussex Co.—Franklin, Hamburg, Ogdensburg and Newtown.

New York: *St. Lawrence Co.*—Balmat, DeKalb, Fowler, Gouverneur, Macomb, Pierrepont; Warren Co.—Horicon (Brant Lake).

Bibliography

This bibliography is a list of nearly 1000 general and cited references, most of which were published before 1981. As mentioned in the preface, a more complete bibliography—approximately 2500 entries—is available on 3″ × 5″ cards at the Department of Mineral Sciences of the U.S. National Museum of Natural History.

Although attempts were made to obtain all of the cited articles, some were not found. Each of those cited on the basis of references in other articles but not read by the writer is followed by *ns;* a few of them are incomplete references. Each of those cited but the contents of which are known to the writer only through abstracts in either *Chemical Abstracts* or *Mineralogical Abstracts* is followed by a *c* or *m*, respectively.

Titles of, for example, Japanese and Russian language articles are translated into English. Titles of, for example, English, German, Swedish, and Romance language articles are given as published except for the titles that are translated in the abstract volumes or other indirect sources from which they were taken. The language of the articles, if different from the cited title, is given if known. Periodicals are, except as modified by the editor, cited as given in *Chemical Abstracts Service Source Indices.*

Entries preceded by an asterisk (*) contain chemical analyses that were considered in the study of solid-solution series reported in Chapter 4.

*Abraham, K., Mielke, H., and Povondra, P., 1972, On the enrichment of tourmaline in metamorphic sediments of the Arzberg Series, W-Germany (NE-Bavaria), *Neues Jahrbuch für Mineralogie, Monatshefte* **5:**209-219. *c*

Ackermann, W., 1915, Beobachtung über Pyroelektrizität in ihrer Abhängigkeit von der Temperatur, *Annalen der Physik* (Leipzig) **46:**197-230.

Adams, J. E., and Matelski, R. P., 1955, Distribution of heavy minerals and soil development in Scott silt loam, *Soil Science* **79:**59-69.

Adrian, H., 1962, Von Beryllen und Turmalinen, *Mitteilungen der Naturforschenden Gesellschaft* (Bern) **19:**83-85.

Æpinus, F. U. T., 1756, Mémoire concernant quelque nouvelles expériences électriques remarquables, *Histoire de l'Académie Royale des Sciences et Belles Lettres, Berlin* **12:**101-121.

Æpinus, F. U. T., 1762, *Recueil de différents mémoires sur la tourmaline,* St. Petersbourg. *ns*

Afonina, G. G., and Vladykin, L. A., 1980, Determination of the composition of tourmaline based on data from powder X-ray diffraction patterns, *Rentgenografiya Mineral'nogo Syr'ya Kristallokhimiya Mineralov,* Moscow, 1979, pp. 99-112. (Russian) *c*

Afonina, G. G., Makagon, V. M., Glebov, M. P., Vladykin, N. V., and Petrova, M. G., 1974, X-ray diffraction study of tourmalines with different chemical compositions, *Akademiya Nauk SSSR, Sibirskoe Otdelenie. Institut Geokhimii, Ezhegodnik,* pp. 336-339. (Russian)

Afonina, G. G., Makagon, V. M., Bogdanova, L. A., Vladykin, N. V., and Glebov, M. P., 1979, Determination of the composition of tourmalines based on data from powder x-ray diffraction patterns, *Rentgenografiya Mineral'nogo Syr'ya i Kristallokhimiya Mineralov,* Moscow, pp. 99-112. *c*

Afonina, G. G., Makagon, V. M., Bogdanova, L. A., and Vladykin, N. V., 1980, Parameters of unit cells of tourmalines with different compositions, *Zapiski Vsesoyuznogo Mineralogicheskogo Obshchestvo.* (Leningrad) **109:**105-112. (Russian)

Agafonoff, V., 1896, Absorption des rayons ultra-violets par les cristaux et polychroisme dans la partie ultra violette du spectre, *Bibliothéque Universelle Archives des Sciences Physiques et Naturelles* **2:**349-364.

*Agafonova, T. N., 1947, La composition chimique et la couleur des tourmalines de la montagne Borstchovotchny, *Comptes Rendus (Doklady) de l'Académie des Sciences de l'URSS* **55:**849-852.

Agafonova, T. N., 1949, Crystallographic study of tourmalines, *Doklady Akademiia Nauk SSSR (Leningrad) Comptes Rendus* **65:**207-209. (Russian)

*Agrell, S. O., 1941, Dravite-bearing rocks from Dinas Head, Cornwall, *Mineralogical Magazine* **26:**81-93.

Aitkens, I., 1932, Tourmaline, *United States Bureau of Mines, Information Circular 6539* (1931).

Alexanian, C., Morel, P., and LeBouffant, L., 1966, Sur les spectres d'absorption infra-rouge des minéraux naturels, *Bulletin de la Société Française de Ceramique* **71:**3-38.

Allais, G., and Curien, H., 1959, Mesure de la teneur en bore-10 des minéraux, *Geochimica et Cosmochimica Acta* **17:**108-112.

Allen, P., Sutton, J., and Watson, J. V., 1974, Torridonian tourmaline-quartz pebbles and the Precambrian crust northwest of Britain, *Geological Society of London Journal* **130:**85-91.

Allen, P., Parker, A., Fitch, F. J., and Miller, J. A., 1972, Wealden detrital tourmaline: implications for northwestern Europe, *Geological Society of London Journal* **128:**273-294.

Alling, H. L., 1944, Grain analyses of minerals of sand size in ball mills, *Journal of Sedimentary Petrology* **14:**103-114.

Althaus, E., 1979, Wassermelonen und Mohrenköpfe, *Lapis* **4:**8-11.
Alty, S. W., 1933, Some properties of authigenic tourmaline from Lower Devonian sediments, *American Mineralogist* **18:**351-355.
Alvarez, M. A., and Coy-Yll, R., 1978, Raman spectra of tourmaline, *Spectrochimica Acta* **34A:**899-908.
Alvarez, M. A., Tornero, J., Vara, J. M., Coy-Yll, R., 1975, Moessbauer spectroscopy of tourmaline varieties, *Anales de Quimica (Madrid)* **71:**498-502. (Spanish)
Anderson, B. W., 1980, *Gem Testing,* 9th ed., Butterworth Scientific, London.
Aoki, M., Konno, H., and Abe, H., 1975, *Magnesium tourmaline (dravite) from Yakeishidake district, Iwate Prefecture,* Professor T. Takeuchi Memorial Volume, pp. 43-51. (Japanese)
Arfvedson, A., 1818, Undersökning af nagra vid Uto jernmalmsbrott förekommande fossilier och af ett deri funet eldfast alkali, *Abhandlungen i Fysik, Kemi och Mineralogi* **6:**145. *ns* (Also published in German in *Schweiger's Journal* **22:**111. *ns*)
Arons, A. B., and Cole, R. H., 1950, Design and use of piezoelectric gages for measurement of large transient pressures, *Review of Scientific Instruments* **21:**31-38.
Arons, A. B., Cole, R. H., Kennedy, W. D., and Wilson, E. B., Jr., 1947, Design and use of tourmaline gages for piezoelectric measurement of explosion phenomena, *Physical Review* **72:**176-177.
Ayuso, R. A., and Brown, C. E., 1984, Manganese-rich red tourmaline from the Fowler talc belt, New York, *Canadian Mineralogist* **22:**327-331.

Babu, S. K., 1970, Mineralogy of achroite (colorless tourmaline) from a pegmatite near Ajmer, *Current Science* (India) **39:**154-156. *c*
*Badalov, S. T., 1951*a,* New variety of tourmaline, *Vsesoiuznoe Mineralogicheskoe Obshchestvo. Uzbekistanskoe Otdelenie, Tashkend. Zapiski.* **2:**84-87. (Russian)
*Badalov, S. T., 1951*b,* Vanadium containing turmaline and garnet, *Zapiski Vsesoyuznogo Mineralogicheskogo Obshchestvo* (Leningrad) **80:**212-213. (Russian)
Balzer, R., and Sigrist, A., 1972, Diskriminierung schwerer Ionen mit Trackdetektoren, *Helvetica Physica Acta* **45:**921-922.
Bank, H., 1965, Lightbrechung, doppelbrechung, dispersion, dichte und gitterkonstanten bei turmalinen, *Zeitschrift der Deutschen Gesellschaft für Edelsteinkunde* (Deutsche Gemmologische Gesellschaft) **14:**1-32.
Bank, H., 1974*a,* Rote Turmaline mit Hoher Licht- und Doppelbrechung aus Kenya, *Zeitschrift der Deutschen Gemmologischen Gesellschaft* **23:**89-92.
Bank, H., 1974*b,* Durchsichtiger grüner Grossular-Unterschiebiengen bezw. Imitationen, *Zeitschrift der Deutschen Gemmologischen Gesellschaft* **23:**195-198.
Bank, H., 1975, Hellbraune und gelbe sowie grüne schleifwürdige Turmaline niedriger Licht- und Mittlerer Doppelbrechung aus Kenya, *Zeitschrift der Deutschen Gemmologischen Gesellschaft* **24:**166.
Bank, H., and Berdesinski, W., 1967, Chromhaltiger Turmaline, *Zeitschrift der Deutschen Gemmologischen Gesellschaft Edelsteinkunde* **61:**30-32.
Bank, H., and Berdesinski, W., 1975, Über Phänomene bei der Refraktometerablesung gewisser Turmaline (Vorlängfiza Mitteilung), *Zeitschrift der Deutschen Gemmologischen Gesellschaft* **24:**20-22. *m*
Bantly, A. W., 1964, Tourmaline whiskers, *Rocks and Minerals* **39:**138-140.
Barbosa, A. L. deM., 1974, The "consanguineous" origin of a tourmaline-bearing gold deposit: Passagem de Mariana (Brazil), *Economic Geology* **69:**416-422.

Bariand, P., and Poullen, J. F., 1979, Nicht nur Lapis!, *Lapis* **4:**20-24.

Barnes, V. E., and Underwood, J. R., 1976, New investigations of the strewn field of Libyan desert glass and its petrography, *Earth and Planetary Science Letters* **30:**117-122.

Barnes, W. H., 1950, An electron-microscopic examination of synthetic tourmaline crystals, *American Mineralogist* **35:**407-411.

Barnes, W. H., and Wendling, A. V., 1934, Space group of tourmaline, *Royal Society of Canada Transactions* **27:**169-175.

*Barsanov, G. P., and Plyusnina, I. I., 1971, Infrared spectroscopic study of aluminum → silicon isomorphic replacements in some silicates, *Vestnik Moskovskogo Universiteta Geologiya* **26:**64-71. (Russian)

Barsanov, G. P., and Sheveleva, V. A., 1952, Materials for study of luminescence of minerals, *Akademiya Nauk SSSR, Mineralogicheskii Muzei. Trudy* **4:**3-35. (Russian)

*Barsanov, G. P., and Yakovleva, M. E., 1964, Tourmalines of dravite composition, *Akademiya Nauk SSSR, Mineralogicheskii Muzei. Moscow* **15:**39-80. (Russian)

*Barsanov, G. P., and Yakovleva, M. E., 1965, Tourmalines of schorl composition, *Akademiya Nauk SSSR, Mineralogicheskii Muzei. Moscow* **16:**3-44. (Russian)

*Barsanov, G. P., and Yakovleva, M. E., 1966, Elbaite and certain rare varieties of tourmaline, *Akademiya Nauk SSSR, Mineralogicheskii Muzei. Moscow* **17:**3-25. (Russian)

Bartlett, J. M., 1923, Report of boron in fertilizers, *Journal of the Association of Official Agricultural Chemists* **6:**381-384.

Barton, R., Jr., 1968, *Refinement of the crystal structure of buergerite and the absolute orientation of tourmaline,* unpublished dissertation, Johns Hopkins University, Maryland.

Barton, R., Jr., 1969, Refinement of the crystal structure of buergerite and the absolute orientation of tourmalines, *Acta Crystallographica* **B25:**1524-1533.

Barton, R., Jr., and Donnay, G., 1968, *Absolute orientation of tourmaline by anomalous dispersion of X rays* (abst.), Geological Society of America, Special paper 101, pp. 12-13.

Basset, H., 1953, A vanadiferous variety of tourmaline from Tanganyika, *Geological Survey Tanganyika Records* **3:**93-96.

Bastin, E. S., 1911, Geology of the pegmatites and associated rocks of Maine, including feldspar, quartz, mica, and gem deposits, *U.S. Geological Survey Bulletin 445.*

Bauer, M., 1890, Beiträge zur Mineralogie, *Neues Jahrbuch für Mineralogie, Geologie und Palaeontologie* **1:**10-48.

Bauer, M. H., 1896, *Precious Stones* (1904 translation by L. J. Spencer), C. E. Tuttle, Rutland, Vermont.

Baumhauer, H., 1876, Die Ätzfiguren am Lithioglimmer Turmalin, Topas und Kieselzinkerz, *Neues Jahrbuch für Mineralogie, Geologie und Palaeontologie* **3:**1-8.

Bayramgil, O., 1945, Erzlagerstätte von Isik dag (Turkey), *Schweizerische Mineralogische und Petrographische Mitteilungen* **25:**67-75.

Bazhenov, A. I., and Poluektova, T. I., 1969, Changes in the accessory mineral assemblages at the contact of the Turochak granites and diorites of the Ulmen inrusion (Altai Mountains), *Izvestiya Tomskogo Politekhnicheskogo Instituta* **166:**10-21. (Russian) *c*

*Becht, K., 1913, *Beiträge zur Kenntnis der Magnesia-Turmalin (Inaugural-Dissertation),* Rössler & Herbert, Heidelberg, Germany.

Beesley, C. R., 1975, Dunton mine tourmaline: an analysis, *Gems & Gemology* **15:**19-24.

Beising, R., 1972, Die Minerals der neiderrheinischen Braunkohle und ihr Verhalten bei der Verbrennung in Kraftwerken, *Neues Jahrbuch für Mineralogie, Abhandlungen* **117:**96-115.

Belov, A. F., Korneev, E. V., Khimich, T. A., Belov, V. F., Korovushkin, V. V., Zheludev, I. S., and Kolesnikov, I. M., 1975, Electron-nuclear interactions and nonequivalent positions of iron ions in tourmalines, *Zhurnal Fizicheskoi Khimii* **49:**1683-1688.

Belov, N. V., and Belova, E. N., 1949, Crystal structure of tourmaline, *Doklady Akademiya Nauk SSSR* **69:**185-188. (Russian)

Belov, N. V., and Belova, E. N., 1950, Structure of tourmaline, *Doklady Akademiya Nauk SSSR* **75:**807-810. (Russian)

Belov, V. F., Khimich, T. A., Shipko, M. N., Voskresenskaya, I. E., and Okulov, E. N., 1973, Gamma-resonance investigations of ferruginous tourmalines, *Soviet* Physics-Crystallography **18:**19-20.

Benesch, F., 1981, *Apokalypse: Die Verwandlung der Erde, Eine Okkulte Mineralogie,* Verlag Urachhaus Johannes M. Mayer GmbH, Stuttgart.

Berger, F., Hoberg, H., and Schneider, F., 1980, Neue Untersuchungen und Entwicklungstendenzen in der Oxidflotation, *Erzmetall* **33:**314-319.

Berger, J., 1978, Kostbare Klippen, *Lapis* **4:**28-30.

Bergmann, T. O., 1766, Commentarius de indole Electrica Turmalini, *Philosophical Transactions of the Royal Society of London* **56:**236-243.

Bershov, L. V., Martirosyan, V. O., Marfunin, A. S., Plantonov, A. N., and Tarashchan, A. N., 1969, Color centers in lithium tourmaline (elbaite), *Soviet Physics-Crystallography* **13:**629-630.

Berthois, L., 1939, Remarques sur l'origine de la tourmaline dans les roches sédimentaires, *Comptes Rendus Hebdomadaires des Séances de l'Académie des Sciences (Paris)* **208:**207-209.

Bessonova, V. Ya., 1970, Lithology of Upper Precambrian dolomite-terrigenous stratum in the Orsha syncline, *Litologiya Geokhimiya i Poleznye Iskopaemye Belorussii i Pribaltiki,* pp. 17-31. (Russian) *c*

Beus, A. A., 1950, Aplitoid pegmatite zones, *Trudy Mineralogicheskogo Muzei. Akademiya Nauk SSSR,* no. 2, pp. 64-71. (Russian) *c*

Bezborodov, M. A., Zaporozhtseva, A. S., and Moiseeva, G. G., 1940, The solubilities of quartz-sand minerals in the glass melt, *Soveshchaniya po Eksperimental'noi Mineralogii i Petrografii, Trudy Tret'ego, Akademiya Nauk SSSR (Moscow)* **1940:**195-203. *c*

Bhandari, S. S., and Varma, J., 1975, Mössbauer studies of natural tourmaline, *Nuclear Physics and Solid State Physics Symposium, Proceedings* **18C:**552.

Bhaskara-Rao, A., and de Assis, A. D., 1968, Chatoyant and pseudomorphosed tourmalines in northeastern Brazil, *Journal de Mineralogia (Brazil)* **6:**31-36. *m*

Birch, F., 1950, A simple technique for the study of the elasticity of crystals, *American Mineralogist* **35:**644-650.

Bischoff, G., 1885, *Elements of Chemical Physical Geology* (vol. 2: Translation of "Lehrbuch der chemischen und physikalischen Geologie"), Cavendish Society, London.

*Black, P. M., 1971, Tourmalines from Cuvier Island, New Zealand, *Mineralogical Magazine* **38:**374-376.

Blakemore, R., 1975, Magnetotactic bacteria, *Science* **190:**377-379.

Blatt, H., and Sutherland, B., 1969, Intrastratal solution and non-opaque heavy minerals in shales, *Journal of Sedimentary Petrology* **39:**591-600.

Bleekrode, L., 1903, On several investigations with liquid air, *Annalen der Physik* (Leipzig) **12:**218-. *ns*

*Blokhina, N. A., 1964, Greisen of the May-Khurinskoye deposit, *Trudy Instituta Geologii, Akademiya Nauk Tadzhikskoi SSR* **8:**89-117. (Russian)

Blum, J. R., 1843, *Die Pseudomorphosen des Mineralreichs,* E. Schweizerbart'sche, Stuttgart.

Boeke, H. E., 1916, Eine Anwendung Mehrdimensionaler Geometrie auf chemische-mineralogische Fragen, die Zusammensetzung der Turmalins, *Neues Jahrbuch für Mineralogie* **II:**109-148.

Bonney, T. G., 1877, On the microscopic structure of luxillianite, *Mineralogical Magazine* **1:**215-221.

*Boriskov, F. F., 1969, Tourmaline asbestos from quartz veins of the southern Urals, *Trudy Sverdlovskogo Gornogo Instituta* **57:**99-102. (Russian)

Born, I., 1772, *Index fossilium quae collegit . . .* (vol. 1), W. Gerle, Prague (1778 ed. (Wein) read).

Born, I., 1790, *Catalogue methodique et raisonné de la Collection des Fossiles de Mlle. Eléonore de Raab* (Tome 1), Vienne.

Bourgeois, J., Ertand, A., and Jacquesson, J., 1953, Étude de quelques bétons spéciaux de protection, *Journal de physique et le radium (Paris)* **14:**317-322.

*Bouška, V., Povondra, P., and Lisý, E., 1973, Uvit z Hnúště, *Acta Universitatis Carolinae, Geologica (Prague),* **3:**163-170.

*Bozhenko, G. M., and Lisa, N. Ya., 1949, Tourmaline from pegmatitic formations of western Volhynia, *Mineralogicheskii Sbornik L'vovskogo Geologicheskogo Obshchehestvo* **3:**181-188. (Russian)

Bradley, J. E. S., and Bradley, O., 1953, The colouring of pink and green tourmaline, *Mineralogical Magazine* **30:**26-38.

Bragg, W. L., 1937, *Atomic Structure of Minerals,* Cornell University Press, Ithaca, New York.

Braitsch, O., 1959, Über den Mineralbestand der Wasserlöslichen Rückstände von Salzen der Strassfurtserie im sudlichen Leinetal, *Freiberger Forschungschefte* (Reihe) **A123:**160-165.

Brammall, A., 1921, Reconstitution processes in shales, slates, and phyllites, *Mineralogical Magazine* **19:**211-224.

Brammall, A., and Harwood, A. F., 1925, Tourmalinization in the Dartmoor granite, *Mineralogical Magazine* **20:**319-330.

Brammall, A., and Leech, J. G. C., 1937, A note on the hydrolysis of rock-forming minerals, *Nature* **139:**754-755.

Branche, G., and Ropert, M.-E., 1956, Sur une association tourmaline-glaucophane, *Académie des Sciences (Paris) Comptes Rendus* **243:**387-389.

Bravais, A., 1851, Études cristallographiques, *Journal de l'École Polytechnique (Journal)* **20:**101-276.

Bray, P. J., Edwards, J. O., O'Keefe, J. G., Ross, V. F., and Tatsuzaki, I., 1961, Nuclear magnetic resonance studies of B^{11} in crystalline borates, *Journal of Chemical Physics* **35:**435-442.

Breithaupt, J. F. A., 1847, *Vollständiges Handbuch der Mineralogie,* vol. 3, Arnoldische Buchhandlung, Dresden and Leipzig. *ns*

Breskovska, V., and Eskenazi, G., 1959-1960, Tourmaline from Bulgarian deposits, *Godishnik ha Sofiiskiya Universitet, Biologo-Geologo-Geografski Fakultet* **54:** 15-46. *c*

*Bridge, P. J., Daniels, J. L., and Pyrce, M. W., 1977, The dravite crystal bonanza of Yinnietharra, Western Australia, *Mineralogical Record* **8:**109-110.

Bridgman, P. W., 1949, Linear compressions to 30,000 kg/sq. cm., including relatively incompressible substances, *American Academy of Arts and Sciences Proceedings* **77:**187-234.

Brisson, M. J., 1787, *Pesanteur Spécifique des Corps,* Ouvrage utile à l'Histoire Naturelle, à la Physique, aux Arts & au Commerce (Paris). *ns*

Brown, C. E., and Ayuso, R. A., 1984, Significance of tourmaline-rich rocks in the Grenville complex in St. Lawrence County, New York, *U.S. Geological Survey Bulletin 1626c* (in press).

Bruce, E. L., 1916, Magnesian tourmaline from Renfrew, Ontario, *Nature* **97:**374.

*Bruce, E. L., 1917, Magnesian tourmaline from Renfrew, Ontario, *Mineralogical Magazine* **18:**133-135.

Brückmann, F. E., 1727 and 1730, *Magnalia Dei in locis subterraneis* (parts I and II), Braunschweig und Wolffenbüttel, Germany (1773 printing read).

Brückmann, V. F. B., 1773, *Abhandlung von Edelsteinen,* Mansenhaus, Braunschweig, Germany.

Buerger, M. J., and Parrish, W., 1937, Unit cell and space-group of tourmaline: Inspective equi-inclination treatment of trigonal crystals, *American Mineralogist* **22:**1139-1150.

Buerger, M. J., Burnham, C. W., and Peacor, D. R., 1962, Assessment of the several structures proposed for tourmaline, *Acta Crystallographica* **15:**583-590.

Burgelya, N. K., 1961, Tourmaline from the Bakal iron deposit, *Ocherki po Metallogenii Osadochnykh Porod,* pp. 139-144. (Russian)

Burns, R. G., 1972, Mixed valencies and site occupancies of iron in silicate minerals from Moessbauer spectroscopy, *Canadian Journal of Spectroscopy* **17:**51-59.

Burns, R. G., and Simon, H. F., 1973, Cation disorder in tourmalines (abst.), *Geological Society of America Annual Meeting (Dallas),* pp. 563-564.

Butler, J. R., 1953, The geochemistry and mineralogy of rock weatherings (1) The Lizard area, Cornwall, *Geochimica et Cosmochimica Acta* **4:**157-178.

Cady, W. G., 1921, The piezoelectric resonator (abst.), *The Physical Review* **17:**531.

Cady, W. G., 1946, *Piezoelectricity,* McGraw-Hill, New York (reprinted by Dover, N.Y., 1964).

Cameron, E. N., Jahns, R. H., McNair, A. H., and Page, L. R., 1949, Internal structure of granitic pegmatites, *Economic Geology,* monograph 2.

Carobbi, G., and Pieruccini, R., 1947, Spectrographic analysis of tourmalines from the Island of Elba, with correlation of color and composition, *American Mineralogist* **32:**121-130.

Casasopra, S., 1939, Petrographic study of the Leventina granite gneiss (Valle Riviera and Valle Leventina, Canton Ticino), *Schweizerische Mineralogische und Petrographische Mitteilungen* **19:**449-710. (Italian) *c*

Cassedanne, J. P., and Lowell, J., 1982, The Virgem da Lapa pegmatites, *Mineralogical Record* **13:**19-28.

Cazeau, C. J., 1962, Value of heavy mineral investigations in the coastal plain of South Carolina, *South Carolina State Development Board, Division of Geology, Geological Notes #6,* pp. 93-94.

Čech, F., 1958, Crystallographic study of the colored tourmalines of the lithium pegmatite from Ctidružice near Moravské Budějovice, *Rozpravy Ceskoslovenske Akademie Věd, Rada Matematickych a Přirodnich Věd* **68:**33-43. *c*

Čech, F., Litomiský, J., and Novotný, J., 1965, Contribution to the chemis of tourmaline, *Sbornik Geologických Věd: Technologie, Geochemie. Prague* **5:**45-74. (Czech, English summary)

Čech, V., Fediukova, E., Kotrba, Z., and Taborsky, Z., 1975, Occurrence of barium-pharmacosiderite in tourmalinite from southern Bohemia, *Casopsis pro Mineralogii a Geologii* **20:**423-426. (Czech, English summary) *m*

*Chatard, T. M., 1890, Minerals and rocks from Maryland, *United States Geological Survey Bulletin 64,* pp. 41-42.

*Chaudhry, M. N., and Howie, R. A., 1976, Lithium tourmalines from the Meldon aplite, Devonshire, England, *Mineralogical Magazine* **40:**747-751.

Chauris, L., 1965, Les minéralisations pneumatolytiques du massif Armoricain, *Memories Bureau de Recherches Géologiques et Minières (France), no. 31.*

*Chebotarev, G. M., and Chebotareva, G. P., 1971, Dravite variety of tourmaline from Muruntan (Western Uzbekistan), *Zapiski Uzbekistanskogo Otdeleniya Vsesoiuznoe Mineralogicheskoe Obshchestvo* **24:**112-115. (Russian)

Chester, A. H., 1896, *A Dictionary of the Names of Minerals Including their History and Etymology,* Wiley, New York.

Chipanin, I. V., and Koshukhovskaya, A. N., 1967, Methods for the flotation separation of beryl, tourmaline, and garnet, *Nauchnye Trudy, Irkutskii Gosudarstvennyi: Nauchno-Issledovatel'skii Institut Redkikh i Tsvetnykh Metallov* **16:**186-190. *c*

Cho, I. W., 1974, Crystalline and optical properties of tourmaline, *Chijil Kwa Chiri* **14:**22-27.

Chorlton, L. B., and Martin, R. F., 1978, The effect of boron on the granite solidus, *Canadian Mineralogist* **16:**239-244.

Chou, Y. T., and Sha, G. T., 1971, Dislocation energies in anisotropic trigonal crystals, *Physica Status Solidus* (A) **6:**505-513.

Clark, A. H., 1970, Early beryllium-bearing veins, South Crofty mine, Cornwall, *Institution of Mining and Metallurgy Transactions* (Sec. B: Applied Earth Sciences) **79:**173-175. *m*

Clarke, F. W., 1899, Constitution of Tourmaline, *American Journal of Science,* 4th ser., **8:**111-121.

Clifford, T. N., 1958, A note on kyanite in the Moine Series of southern Ross-shire, and a review of related rocks in the Northern Highlands of Scotland, *Geological Magazine* **95:**333-346.

Cocco, G., 1952, Tormalina filiforme dei graniti elbani e di Alzo, *Peridico di Mineralogia* (Rome) **21:**231-234. (Italian)

Codarcea, V., 1977, Percentage distribution of heavy minerals in loess profiles at Paks and Mohacs, *Foeldtani Koezlony* **101:**138-143. *ns*

*Coelho, I. S., 1948, Turmalina fibrosa da "Mina do Cruzeiro," Santa Maria do Suassui, Minas Gerais, *Mineração e Metalurgia* **13:**49-53.

Coetzee, C. B., 1966, An ancient volcanic vent on Boschplaat 396 in Bloemfontein district, Orange Free State, *Geological Society of South Africa Transactions* **69:**127-137. *m*

Conklin, N. M., and Slack, J. F., 1983, Trace element analyses of tourmalines from Appalachian-Caledonian massive sulfide deposits, *U.S. Geological Survey, Open File Report 83-890.*

Constantinescu, E., 1976, Tourmaline from the Cioaca Inalta zone (SW of Banat), *Revue Roumaine de Géologie, Géophysique et Géographie, Série de Geologie* **20:**147-153. (French) *c*

Cook, R. C., 1940, Absolute pressure calibrations of microphones, *Bureau of Standards Journal of Research* **25:**489-505.

Cooper, A. F., 1971, Piemontite schists from Haast River, New Zealand, *Mineralogical Magazine* **38:**64-71.

Corin, F., 1940-1941, Observations nouvelles sur les inclusions à halos pleochroiques, *Bulletin de la Société Belge de Géologie, de Paléontologie, et d'Hydrologie* **50:**48-61.

*Cossa, A., and Arzruni, A., 1883, Ein Chromturmalin aus den Chromeisenlagers des Urals, *Zeitschrift für Kristallographie und Mineralogie von P. Groth* **7:**1-16.

Crampton, C. B., 1958, Heavy minerals in the Permian magnesian limestone of Yorkshire, *Yorkshire Geological Society Proceedings* **31:**383-390. *c*

Cronstedt, A., 1758, *Mineralogie; eller Mineral-Rikets Upstallning,* Stockholm (originally published anonymously). *ns*

Crowningshield, R., 1974, Commercial implications of Gamma Radiation on Gem Materials, *Zeitschrift der Deutschen Gemmologischen Gesellschaft* **23:**95-101.

Crowley, J. A., 1984, Nature's rainbow mineral: tourmaline, *Merchandiser,* January 1984, pp. 10-12, 14, and 124.

Cuppels, N. P., and White, A. M., 1973, Fluvial monazite deposits in the drainage basins of the Enoree, Tygee, and Pacolet rivers, South Carolina, *South Carolina Division of Geology, Mineral Resources Series, 1.*

Curie, J., 1889, Recherches sur le pouvoir inducteur spécifique et la conductibilité des corps cristallisés, *Annales de Chimie et de Physique* (Paris) **17:**385-434.

Curie, J., and Curie, P., 1880, Développment par compression de l'electricité polaire dans les cristaux hémièdres à faces inclinées, *Bulletin de la Société Mineralogie de France* **3:**90.

D'Achiardi, G., 1893 and 1896, Le Tormaline del Granito Elbano (Parts I and II), *Atti della Societa Toscana Scienze Naturali Pisa, Memories* **13:**95p.; **15:**74p.

D'Achiardi, G., 1897, Osservazioni sulle tormaline, *Annali delle Università Toscáne* **22:**3-17.

Dakshinamurti, C., Satyanarayana, K. V. S., and Singh, B., 1955, Trace-element status of Delhi soils, *National Academy of Sciences (India) Proceedings* sect. A, **24:**566-572. *c*

Dambly, M., Pollak, H., Quartier, R., and Bruyneel, W., 1976, IR-irradiation enhanced effects in tourmaline, *Journal de Physique. Colloque (Paris)* **6:**807-810.

Damon, P. E., and Kulp, J. L., 1958, Excess helium and argon in beryl and other minerals, *American Mineralogist* **43**:433-459.
Dana, E. S., 1892, *The System of Mineralogy of James Dwight Dana 1837-1868,* 6th ed., Wiley, New York.
Dana, J. D., 1837, 1844, 1850, 1854, and 1868, *The System of Mineralogy,* 1st, 2nd, 3rd, 4th, and 5th eds., New York.
d'Andrada, B. J., 1800, Des caractères et des propriétés de plusieurs nouveaux minéraux de Suède et de Norwège, avec quelques observations chimiques faites sur ces substances, *Journal de Physique, de Chimie, d'Histoire Naturelle et des Arts* **51**:243-244.
Darragh, P. J., Gaskin, A. J., and Taylor, A. M., 1967, Crystal growth: gemstones, *Australia. Commonwealth Scientific Industrial Research Organization, Division Applied Mineralogy, Annual Report, 1966-1967* **8**:AM8.
Dashdavaa, S., 1970, Tourmaline from Mongolian quartz pegmatites, *Mineralogicheskii Sbornik* (L'vov) **24**:355-358. (Russian) *c*
DeBoodt, A. B., 1609, *Gemmarum et Lapiduum Historia,* Hanover Typis Wechelianis, 2nd and 3rd eds., Leiden-Joannis Maire, 1636 and 1647; first published in Prague; there are later Latin editions and also translations; 1647 reprint seen.
DeBuffon, G. L. L., 1801, *Histoire Naturelle des Minéraux* (Tome II), Deterville, Paris.
de Camargo, W. G. R., and Souza, I. M., 1970, Nŏvo Hábito da Turmalina, *Academia Brasileira de Ciêncais Anais* (Rio de Janeiro) **42**:219-222.
deCarvalho, G. S., 1955, New observations on the sedimentology of the Plio-Quaternary deposits of the mouth of the Mondego, Portugal, *Publicaciones del Museu e Laboratorio Mineralogico e Geologico . . . Univ. Coimbra, Mem. e not.,* no. 39, pp. 13-24. *c*
deCoster, M., Pollak, H., and Amelinckx, S., 1963, Moessbauer absorption in iron silicates, *Physica Status Solidi,* **3**:283-288.
*Deer, W. A., Howie, R. A., and Zussman, J., 1962, *Rock-Forming Minerals,* vol. 1, Wiley, New York.
deLaet, J., 1647, *De Gemmis et Lapidibus . . . Theophrasti Liber de Lapidibus,* Lugduni Batavorum (also Leiden: J. Maire). (Incorporated in 3rd ed. of deBoodt).
Delamétherie, J. C., (also cited as Lametherie, J. C. de), 1797, *Théorie de la Terre,* 2nd ed., 5 vol., Chez Maradan, Paris. *ns*
Delesse, A. E. O. J., 1859, Recherches sur les pseudomorphoses, *Annales des Mines (Paris)* **16**:317-392.
Demay, A., 1938, Granites, microgranites and rhyolites of east end of the Guéret massif: Tourmaline in some rhyolites, *Compte rendu Sommaire des Séances de la Société Géologique de France* **206**:1905-1907. (French) *c*
de Noya, C., 1759, *Lettre du Duc de Noya Carafa sur la tourmaline,* Paris. *ns* It appears that at least the substantive part of this "lettre" was by Michel Adamson—see Home, 1976.
Deprat, J., 1909, The tourmaline-bearing granulite of the vicinity of Erula (Sardinia), *Bulletin de la Société Géologique de France* **7**:440-443. (French) *c*
Derby, O. A., 1898, On the accessory elements of the itacolumite, and the secondary enlargement of tourmaline, *American Journal of Science* **5**:187-192.
Déverin, L., 1934, Sur la turmaline authigène dans les roches sédimentaires, *Schweizerische Mineralogische und Petrographische Mitteilungen* **14**:528-529.

Devore, G. W., 1966, Elastic strain energy and mineral recrystallization: a commentary on rock deformation, *University of Wyoming Contributions to Geology* **5**:19-42.

Diehl, W., 1970, Beiträg zur statischen Eichung von piezoelektrischen Druckaufnehmen mit Quarz und Turmalin, *Messtechnik (Braunschweig)* **78**:4-9.

Dietrich, R. V., 1974, Migmatites—A Résumé, *Journal of Geological Education* **22**:144-156.

Dietrich, R. V., 1982, Untitled letter to the Editor, *Geotimes*, June 1982, p. 14.

Dietrich, V., de Quervain, F., and Nissen, H. V., 1966, Turmalinasbest aus alpinen Mineral-klüften, *Schweizerische Mineralogische und Petrographische Mitteilungen* **46**:695-697.

Dillon, S., 1983, Gem News, *Gems and Gemology* **19**:59-62.

Dmitriev, S. D., 1955, The mechanism of tourmaline-muscovite intergrowth, *Kristallografiya* **4**:110-115. (Russian)

Dobrovol'skaya, N. V., 1972, Magnetic susceptibility of some dia- and paramagnetic minerals and their response to NIKH ultrasound, *Elektrofizicheskie Metody, Obrabotki Redkometal'nogo Syr'ya*, pp. 37-41. (Russian)

Dobrovol'skaya, N. V., and Kuz'min, V. I., 1975, Magnetic properties of tourmaline, *Konstitutsiia i Svoistva Mineralov* **9**:124-130. (Russian)

*Dodatko, A. D., 1969, Tourmaline from the weathering crust of ultrabasic rocks of the Dneiper Basin, *Mineralogicheskiy Sbornik* (L'vov) **23**:192-195. (Russian) *c*

Doelter, C., 1909, Über den Einfluss der Radiumstrahlen auf die Mineralfarben, *Tschermaks Mineralogische und Petrographische Mitteilungen* (Wein) **28**:171-178.

Doelter, C., 1915, *Die Farben der Mineralien insbesonders der Edelsteine*, Friedrich Vieweg und Sohn, Braunschweig.

*Doelter, C., 1917, *Handbuch der Mineral-chemie*, vol. 2, Theodor Steinkopff, Dresden and Leipzig, pp. 736-791.

Dolgov, Yu. A., and Shugurova, N. A., 1968, Composition of gases from individual inclusions in various minerals, *Academiia Nauk SSSR Mineralogicheskaya Termometriya i Barometriya* **1**:290-298. (Russian)

Döll, E., 1886, Ueber einen Riesenpegmatit bei Pisek. Pyrit nach Turmalin, eine neue Pseudomorphose, *Verhandlungen der K. K. Geologischen Reichsanstalt (Wein)*, pp. 351-356.

Dolomanova, E. I., Loseva, T. I., and Tsepin, A. I., 1974, Chemical composition of solid residues in vacuoles of cassiterite, tourmaline and quartz from tin ore deposits, *Problemy Endogennogo Rudoobrozovaniia*, pp. 138-149. (Russian)

Dolomanova, E. I., Ziborova, T. A., Loseva, T. I., Martynova, A. F., and Panova, M. A., 1978, Tourmaline from the tin ore deposits of Transbaikalia and its typomorphic significance, *Trudy Mineralogicheskogo Muzeya. Academiya Nauk SSSR* **26**:40-69. (Russian) *c*

Domeyko, I., 1860, *Elementos de Mineralojia* . . . 2nd ed., Santiago (Chile). *ns*

*Donnay, G., 1963, Tourmaline, *Carnegie Institute of Washington Year Book* **62**:166-169.

Donnay, G., 1969, Crystalline heterogeneity, evidence from electron-probe study of Brazilian tourmaline, *Carnegie Institute of Washington Year Book* **67**:219-220.

Donnay, G., 1977, Structural mechanism of pyroelectricity in tourmaline, *Acta Crystallographica* **A33**:927-932.

Donnay, G., and Barton, R., Jr., 1967, Absolute orientation of the tourmaline crystal structure, *Carnegie Institute of Washington Year Book* **65:**299.
*Donnay, G., and Barton, R., Jr., 1972, Refinement of the crystal structure of elbaite and the mechanism of tourmaline solid solution, *Tschermaks Mineralogische und Petrographische Mittelungen* **18:**273-286.
Donnay, G., and Buerger, M. J., 1950, The determination of the crystal structure of tourmaline, *Acta Crystallographica* **3:**379-388.
*Donnay, G., Ingamells, C. O., and Mason, B., 1966, Buergerite, a new species of tourmaline, *American Mineralogist* **51:**198-199.
Donnay, G., Senftle, F. E., Thorpe, A., and White, S., 1967, Magnetic properties of tourmaline, *Carnegie Institute of Washington Year Book* **65:**295-299.
Donnay, G., Wyart, J., and Sabatier, G., 1959, Structural mechanism of thermal and compositional transformations in silicates, *Zeitschrift für Kristallographie* **112:**161-168.
Donova, K. V., 1972, *Treasures of the U.S.S.R. Diamond Fund,* Isobrazitelbnoe Iskysstvo (Moscow). (Russian)
Dorofeev, A. V., 1974, Mineralogy, distribution, and origin of boron mineralization in skarn deposits, as illustrated by one of the Siberian regions, *Voprosy Rudonosnosti Yakutii,* pp. 156-167. (Russian) *c*
Dromasko, S. G., Lukashev, K. I., and Matveeva, L. I., 1966, Heavy minerals in arenaceous silty fraction of the Quaternary cover of the Belorussian Polese, *Doklady Akademiya Nauk Belorusskoy SSR* **10:**472-476. (Russian) *c*
Dropsy, U., 1943, A granulometric study of some sands from Mauritania, *Bulletin de la Société Française de Minéralogie* **66:**251-263. *c*
Drozhdin, S. N., Novik, V. K., Koptsik, V. A., and Kobyakov, I. B., 1974, Pyroelectric properties of tourmaline and cancrinite crystals in a broad temperature range, *Fizika Tverdogo Tela* (Leningrad) **16:**3266-3269. (Russian) *c*
Drozhdin, S. N., Novik, V. K., Gavrilova, N. D., Koptsik, V. A., and Popova, T. V., 1975, Behavior of polar crystals at low temperatures, *Izvestiya Akademiya Nauk SSSR Seriya Fizicheskaya* **39:**990-994.
Dryden, L., and Dryden, C., 1946, Comparative rates of weathering of some common heavy minerals, *Journal of Sedimentary Petrology* **16:**91-96.
*Dubinina, V. N., and Kornilovich, I. A., 1968, Behavior of tourmaline in the supergene zone of polymetallic deposits (Eastern Transbaikal), *Zapiski Vsesoyuznogo Mineralogicheskogo Obschchestvo* **97:**301-308. (Russian) *c*
deBois-Reymond, R., and Schaefer, D. C., 1908, *Ludwig Darnstaedter's Handbuch zur Geschichte der Naturwissenschaften und der Technik,* Julius Springer, Berlin.
Dunn, P. J., 1974, Gem spodumene and achroite tourmaline from Afghanistan, *Journal of Gemmology* **14:**170-174.
Dunn, P. J., 1975*a,* Elbaite from Newry, Maine, *Mineralogical Record* **6:**22-25.
Dunn, P. J., 1975*b,* Notes on inclusions in tanzanite and tourmalinated quartz, *Journal of Gemmology* **14:**335-338.
Dunn, P. J., 1977*a,* Uvite, a newly classified gem tourmaline, *Journal of Gemmology* **15:**300-307.
Dunn, P. J., 1977*b,* Chromium in dravite, *Mineralogical Magazine* **41:**408-410.
Dunn, P. J., 1978, Blue-green gem dravite, *Journal of Gemmology* **16:**92-93.
*Dunn, P. J., Appleman, D. E., and Nelen, J. E., 1977, Liddicoatite, a new calcium end-member of the tourmaline group, *American Mineralogist* **62:**1121-1124.

Dunn, P. J., Arem, J. E., and Saul, J., 1975, Red dravite from Kenya, *Journal of Gemmology* **14:**386-387.

*Dunn, P. J., Nelen, J. A., and Appleman, D. E., 1978, Liddicoatite, a new gem tourmaline species from Madagascar, *Journal of Gemmology* **16:**172-176.

*Dunn, P. J., Appleman, D., Nelen, J. A., and Norberg, J., 1977, Uvite, a new (old) common member of the tourmaline group and its implications for collectors, *Mineralogical Record* **8:**100-108.

*DuParc, L., and Sigg, H., 1914, Sur un gisement de tourmalines dans une serpentine de l'Oural, *Bulletin de la Société Française de mineralogie* **37:**14-19. *ns*

*DuParc, L., Wunder, M., and Sabot, R., 1910, Les minéraux des pegmatites des environs d'Antisirabe à Madagaskar, *Mémoires de la Société de Physique et d'Histoire Naturelle de Genéve* **36:**281-410. *ns*

*DuRietz, T., 1935, Peridotites, serpentines, and soapstones of northern Sweden, with special references to some occurrences in northern Jämtland, *Geologiska Föreningen i Stockholm Förhandlingar* **57:**133-260.

DuRietz, T., 1955, Geology and ores of the Kristineberg deposit, Vesterbotten, Sweden, *Sveriges Geologiska Undersökning Årsbok 45* (no. 524) 1951, ser. C. Avhandlingar och Uppsatser, no. 5, 90p.

*Dzhamaletdinov, N. K., 1973, New occurrence of dravitic tourmaline mineralization in Western Uzbekistan, *Uzbekskiy Geologicheskiy Zhurnal* **17**(4):50-54. (Russian)

Edwards, A. B., 1936, The occurrence of quartz-tourmaline nodules in the granite of Clear Creek, near Everton, *Royal Society of Victoria Proceedings* (Australia) **49:**11-16.

*El-Hinnawi, E. E., and Hofmann, R., 1966, Optische und chemische untersuchungen am neun Turmalinen (Elbaiten), *Neues Jahrbuch für Mineralogie, Monatshefte* **3:**80-89.

Eliseeva, O. P., 1958, Accessary minerals and accessary elements of the Samgarsk intrusive quartz porphyries from southern Kuruminsk crest, *Uzbekskiy Khimicheskiy Zhurnal* **2:**29-45. *c*

Emel'yanova, E. N., and Zigareva, T. A., 1960, Production of tourmaline under hydrothermal conditions, *Kristallografiya* **5:**955-957.

*Engelmann, T., 1878, Über den Dolomit des Binnenthales und seine Mineralien verglichen mit dem des Campolongo, *Zeitschrift für Kristallographie* **2.** *ns*

English, G. L., 1939, *Descriptive list of the new minerals, 1892-1938, containing all new mineral names not mentioned in Dana's System of Mineralogy,* 6th ed., 1892, McGraw-Hill, New York.

Eppler, W. F., 1958, Notes on asterism in spinel and chatoyancy in chrysoberyl, quartz, tourmaline, zircon and scapolite, *Journal of Gemmology* **6:**251.

Epprecht, W., 1953, Die Gitterkonstanten der Turmaline, *Schweizerische Mineralogische und Petrographische Mitteilungen* **33:**481-505.

Ercker, L., 1672, *Aula Subterranea . . . ,* J. D. Zunners, Frankfurt. *ns*

Erd, R. C., 1980, The minerals of boron, in *Supplement to theoretical chemistry,* vol. 5, *Boron,* Longman, New York, pp. 13-58.

Erofeyev, M. (also cited as Jerofejew, etc.), 1871, Crystallographic and crystallo-optical studies of tourmalines, *Zapiski Imperatovskago St.-Peterburgskago Mineralogicheskago Obshchestvo* **6:**81-342. (Russian)

Esikov, A. D., and Esikova, G. S., 1974, Measurement of the isotopic composition of boron with an MI-1305 mass spectrometer, *Iadernaia Geologiia,* pp. 219-229. (Russian)

Ethier, V. G., and Campbell, F. A., 1977, Tourmaline concentrations in Proterozoic sediments of the southern Cordillera of Canada and their economic significance, *Canadian Journal of Earth Science* **14:**2348-2363.

Eugster, H. P., 1980, Geochemistry of evaporitic lacustrine deposits, *Earth and Planetary Sciences Annual Review* **8:**35-63.

Fabel, G. W., and Henisch, H. K., 1971, Technique for measurement of pyroelectric coefficients, *Solid State Electronics* **14:**1281-. *ns*

Fairbanks, E. E., 1973, Tourmaline, *Earth Science* **26:**198-201.

Faye, G. H., Manning, P. G., and Gosselin, J. R., 1974, The optical absorption spectra of tourmaline; importance of charge-transfer processes, *Canadian Mineralogist* **12:**370-380.

Faye, G. H., Manning, P. G., and Nickel, E. H., 1968, The polarized optical absorption spectra of tourmaline, cordierite, chloritoid and vivianite—Ferrous-ferric electronic interaction as a source of pleochroism, *American Mineralogist* **53:**1174-1201.

Fellinger, R., 1919, Über die Dielektrizitätskonstante einiger natürlicher und synthetischer Edelsteine, *Annalen der Physik* (Leipzig), ser. 4, **60:**181-195. *ns*

Ferguson, H. G., and Bateman, A. M., 1912, Geologic features of tin deposits, *Economic Geology* **7:**209-262.

Fersman, A. E., 1915, An example of regular grouping of tourmaline and feldspar at Murinska, Ural, *Mémoires de la Société Ouralienne d'Amateurs des Sciences Naturelles* (Ekaterinebourg) **35:**19-22. *ns*

Fieremans, C., 1955, Preliminary geological study of the diamantiferous conglomerates of Mesozoic age of Kasai, Belgian Congo, *Mémoires de l'Institut Geologique de l'Université de Louvain,* no. 19, pp. 225-293. *c*

Fischer, K., 1979, Knallerbsen und Katzenaugen, *Lapis* **4:**43-44.

Fitch, F. J., and Miller, J. A., 1972, $^{40}Ar/^{39}Ar$ dating of detrital tourmaline and tourmalinized rock schists and micas, *Journal of the Geological Society of London* **128:**291-294.

Fleischer, M., 1983, *Glossary of Mineral Species,* Mineralogical Record, Tucson (Arizona).

Fletcher, C. J. N., 1977, The geology, mineralization, and alteration of Ilkwang mine, Republic of Korea. A Cu-W-bearing tourmaline breccia pipe, *Economic Geology* **72:**753-768.

Foit, F. F., Jr., and Rosenberg, P. E., 1975, Aluminobuergerite, $Na_{1-x}Al_3Al_6B_3Si_6O_{27}O_{3-x}(OH)_{1+x}$, a new end-member of the tourmaline group (abst.), *American Geophysical Union Transactions* (EOS) **56:**461.

Foit, F. F., Jr., and Rosenberg, P. E., 1977, Coupled substitutions in the tourmaline group, *Contributions to Mineralogy and Petrology* **62:**109-127.

Foit, F. F., Jr., and Rosenberg, P. E., 1979, The structure of vanadium-bearing tourmaline and its implications regarding tourmaline solid solutions, *American Mineralogist* **64:**788-798.

Foord, E. E., 1976, *Mineralogy and Petrogenesis of Layered Pegmatite-Aplite Dikes in the Mesa Grande District, San Diego County, California,* unpublished Ph.D. thesis, Stanford University, Stanford, California.

Foord, E. E., 1977, The Himalaya dike system, Mesa Grande district, San Diego County, California, *Mineralogical Record* **8:**475-478.

Foord, E. E., and Mills, B. A., 1978, Biaxiality in 'isometric' and 'dimetric' crystals, *American Mineralogist* **63:**316-325.

Foord, E. E., Heyl, A. V., and Conklin, N. M., 1981, Chromium minerals at the Lime Pit, State Line chromite district, Pennsylvania and Maryland, *Mineralogical Record* **12:**149-156.

Fortier, S., and Donnay, G., 1975, Schorl refinement showing composition dependence of the tourmaline structure, *Canadian Mineralogist* **13:**173-177.

Fouqué, F., and Michel-Lévy, A., 1882, *Synthèse des Minéraux et des Roches,* G. Masson, Paris.

Foxworth, R. D., Priddy, R. R., Johnson, W. B., and Moore, W. S., 1962, Heavy minerals of sand from Recent beaches of the Gulf Coast of Mississippi and associated islands, *Mississippi State Geological Survey Bulletin 93.*

Frankel, J. J., 1950, Flattened tourmaline crystals in muscovite from Miami, Southern Rhodesia, *South African Journal of Science* **47:**109-111.

Fraser, H. J., 1930, Paragenesis of the Newry pegmatite, Maine, *American Mineralogist* **15:**349-364.

Frey, M., 1969, Die Metamorphose des Keupers von Tafeljura bis zum Lukmanier-Gebeit (Veränderungen tonig-mergeliger Gesteine vom Bereich der diagenese bis zur Staurolith-Zone), *Beiträge zur Geologischen Karte der Schweiz,* new ser. 137.

Freyberg, B. von, 1934, Die Bodenschätze des Staates Minas Geraes (Brasilien), p. 68. *ns*

Frondel, C., 1935, Vectorial chemical alteration of crystals, *American Mineralogist* **20:**852-862.

Frondel, C., 1936, Oriented inclusions of tourmaline in muscovite, *American Mineralogist* **21:**777-799.

Frondel, C., 1948, Tourmaline pressure gages, *American Mineralogist* **33:**1-17.

Frondel, C., 1962, The System of Mineralogy of James Dwight Dana . . . (7th ed.), vol. III: *Silica Minerals,* Wiley, New York.

Frondel, C., and Collette, R. L., 1957, Synthesis of tourmaline by reaction of mineral grains with NaCl-H_3BO_3 solution, and its implications in rock metamorphism, *American Mineralogist* **42:**754-758.

*Frondel, C., Biedl, A., and Ito, J., 1966, New type of ferric iron tourmaline, *American Mineralogist* **51:**1501-1505.

Frondel, C., Hurlbut, C. S., Jr., and Collette, R. C., 1947, Synthesis of tourmaline, *American Mineralogist* **32:**680-681.

Fryer, C., ed., 1982, Gem Trade Lab Notes, *Gems and Gemology* **18:**107.

Fuge, R., and Power, G. M., 1969, Chlorine in tourmalines from SW (southwest) England, *Mineralogical Magazine* **37:**293-294.

Fuh, T. M., 1965, A hydrothermal breakdown of tourmaline at high temperatures, *Science Reports of the National Taiwan University, First Series Acta Geologica Taiwanica,* pp. 21-29.

Furbish, W. J., 1968, Tourmaline of acicular and filiform morphology from pyrophyllite deposits of North Carolina, *Rocks and Minerals* **43:**584-586.

Gaines, R. V., and Thadeu, D., 1971, The minerals of Panasqueira, Portugal, *Mineralogical Record* **2:**73-78.

*Gallitelli, P., 1937, Analisi chimica del granito de Montorfano e del granito e della tormalina di Alzo, *Accademia nazionale dei Lincei, Rome* **26:**103-106.

Gallitelli, P., 1941, Terre rare granito de Alzo: osservazioni spettrografiche, *Accademia d'Italia, Rome. Classe di scienze fisiche, matematiche e naturali, Rendiconti* ser. 7, **2:**87-92.

Gapon, H. E., and Gapeeva, M. M., 1969, Minerals of pyroclastic genesis in the Dogaldynsky suite of the Bodiabo Basin, *Akademiia Nauk SSSR, Sibirskogo Otdelenie, Institut Geokhimii, Ezhegodnik (Irkutsk)* **1970:**263-265. (Russian)

Gaugain, J.-M., 1856, Note sur les propriétés Électriques de la tourmaline, *Comptes Rendus Hebdomadaires des Séances de l'Académie des Sciences (Paris)* **42:**1264-. *ns*

Gaugain, J.-M., 1859, Mémoire sur l'électricité des tourmalines, *Annales de Chimie et de Physique* **57:**5-11. *ns*

Gault, H. R., 1939, The heavy minerals of the Mansfield sandstone of Indiana, *Indiana Academy of Science Proceedings* (1938) **48:**129-136.

Gautier, D. L., 1979, Preliminary report of authigenic euhedral tourmaline crystals in a productive gas reservoir of the Tiger Ridge field, north-central Montana, *Journal of Sedimentary Petrology* **49:**911-916.

Gavrilova, N. D., 1965, Temperature dependence of pyroelectric coefficients as studied by a static method, *Kristallografiya* **10:**346-. *ns*

Gebert, W., and Zemann, J., 1965, Messung des Ultrarot-Pleochroismus von Mineralen, II. Der Pleochroismus der OH-Streckfrequenz in Turmalin, *Neues Jahrbuch für Mineralogie Monatshefte* **8:**232-235.

Geletsyan, G. G., 1967, Authigenic tourmaline from Paleozoic carbonate sequences of the Siberian platform, *Geologiya i Geofizika* **7:**155-159. (Russian) *c*

Genkin, A. D., 1954, Corrosion and replacement of quartz by sulfides, *Zapiski Vsesoyuznogo Mineralogicheskogo Obshchestvo* **83:**398-399. *c*

Gerhard, K. A., 1777, Mémoire sur les Principes de la Tourmaline, *Nouveaux Mémoires de L'Académie Royale des Sciences et Belles-Lettres à Berlin,* pp. 14-24.

Gerling, E. K., Shukolyukov, Yu. A., and Matveeva, I. I., 1962, Age determination by the Rb/Sr method of beryl and other minerals containing inclusions, *Geokhimiya,* no. 1, pp. 67-72. *c*

Gesner, K., 1565, *De Omne Rerum Fossilium,* Tiguri, Zurich.

Getmanskaya, T. I., Zabolotnaya, N. P., Novikova, M. I., and Panteleyev, A. I., 1970, Beryllium mineralization in a polymetallic ore district in Siberia (English translation), *International Geology Review* **14:**829-836.

Giampaolo, S., 1963, Minor constituents of tourmalines of the Strona Valley (Novara), *Atti della Societa dei Naturalisti e Matematici di Modena* **94:**25-32. (Italian) *c*

Giesecke, C. L., 1832, *A Descriptive Catalogue of a New Collection of Minerals in the Museum of the Royal Dublin Society,* Royal Dublin Society, Dublin. *ns*

*Gill, A. C., 1889, Note on some minerals from the chrome pits of Montgomery County, Maryland, *Johns Hopkins University Circular 75,* pp. 100-101. *ns*

Gillot, J. E., 1955, Metamorphism of the Manx slates, *Geological Magazine* **92:**141-153. *c*

Ginzburg, A. I., 1954, The geochemical indicator minerals and their significance for the prospecting of rare metal ores in pegmatites, *Doklady Akademii Nauk SSSR* **98:**233-235. (Russian) *c*

Ginzburg, A. I., Kruglova, N. A., and Moleva, V. A., 1951, Magnesiotriplite, a new mineral of the triplite group, *Doklady Akademii Nauk SSSR* **77:**97-100. (Russian) *c*

Gladkii, V. V., and Zheludev, I. S., 1965, Methods and results of an investigation of the pyroelectric properties of some single crystals, *Kristallografiya* **10:**63-67.

Gladkikh, Yu. F., and Pol'kin, S. I., 1963, Effect of salts of multivalent metals on floatability of tantalite-columbite, tourmaline and garnet, *Izvestiya Vysshikh Uchebnykh Zavedenii, Tsvetnaya Metallurgiya* **6:**33-37. (Russian)

Glebov, M. P., Petrova, M. G., Shiryaeva, V. A., and Grigor'eva, V. A., 1975, Tourmaline as an indicator mineral of the ore specialization of pegmatites, *Trudy Mineralogicheskogo Muzeya Akademiya Nauk SSSR* **24:**40-47. (Russian)

Gmelin, C. G., 1827, Chemische Untersuchungen über den Turmalin, (2nd ser.), *Annalen der Physik und Chemie* **9:**172-176.

Gnativ, G. M., 1959, Mineralogy of hybrid quartz diorites from the vicinity of the village Golycheva in eastern Volhynia, *Mineralogicheskii Sbornik, L'vovskii Gosudarstvennyi Universitet, USSR* **13:**316-327. *c*

Gokhale, K. V. G. K., and Bagchi, T. C., 1959, Authigenic tourmaline in the Banganapally State (Kurnool system), India, *Journal of Sedimentary Petrology* **29:**468-469.

Göksu, E., 1953, Geological, genetic, and mineralogical observations on the bauxite deposit of Akseki (Vil. Antalya) and comparison with other Turkish and European bauxites, *Bulletin of the Geological Society of Turkey* **4:**79-140. (Turkish, German summary) *c*

Goldschimdt, V., 1891, *Index der Krystallformen der Mineralien* (vol. 3), J. Springer, Berlin, pp. 243-248.

Goldschmidt, V., 1922, *Atlas der Kristallformen,* Carl Winters Universitäts Buchhandlung, Heidelberg (Germany), **IX:**16-36 and Tfl. 7-29.

Goldschmidt, V. M., and Peters, C., 1932, Zur Geochemie des Bors. II. *Nachrichten von der Gesellshaft der Wissenschaften (Göttingen), Mathematisch-Physikalische Klasse* **28:**528-545.

Gomm, P. S., 1973, Electrical conduction studies in transition metal complexes, *Pentacol* (Chemical Journal of the University of Wales) **11:**22-26.

Goni, J. C., and Guillemin, C., 1964, Étude de la paragenèse de quelques mineraux des pegmatites d'Alto Ligonha (Mozambique), *Bulletin de la Société Française de Minéralogie et de Cristallographie* **87:**553-556.

*Gorelikova, N. V., Perfil'eva, Yu. D., and Bubeskin, A. M., 1976, Mössbauer data on distribution of Fe ions in tourmaline (translation), *International Geology Review* **20:**982-990.

Goroshnikov, B. I., 1956, Tourmaline from the Saksagansk zone of overthrust folding, *Geologichniy Zhurnal* (Kiev) **16**(3):83-86. *c*

Gould, J. L., Kirschvink, J. L., and Deffeyes, K. S., 1978, Bees have magnetic remanence, *Science* **201:**1026-1028.

Govorov, I. N., 1971, Thermodynamic calculations and the interpretation of tourmaline systems, *Trudy Soveshchaniya po Eksperimental'noi i Tekhnicheskoi Mineralogii i Petrografii, Eksperimental'noe Modelirovanie Prirodnykh,* Protsessov, 8th, Novosibirsk, 1968. (Russian)

Graham, E. R., 1957, The weathering of some boron-bearing materials, *Soil Science Society of America Proceedings* **21**:505-508.

Grant, R. W., 1964, *Metamorphism of the Triassic sediments near Lambertville, New Jersey,* unpublished manuscript.

Graves, S. B., 1967, A tourmaline bonanza at old Mt. Mica, Maine location, *Lapidary Journal* **21**:10, 12, 13, 16, and 17.

Graziani, G., Gübelin, C. G., and Lucchesi, S., 1982, Tourmaline chatoyancy, *Journal of Gemmology* **18**:181-193.

Greene, J. C., and Burke, J. G., 1978, The science of minerals in the Age of Jefferson, *American Philosophical Society Transactions* **68**(4):113p.

Greg, R. P., 1855, On two doubtful British species, *Philosophical Magazine* **10**:118-119.

Griffith, W. P., 1970, Raman studies on rock-forming minerals. Part I. Orthosilicates and cyclosilicates, *Journal of the Chemical Society* (A) (London), pp. 1372-1377.

Grill, E., 1922, Nuove forme cristalline della tourmalina elbana, Atti della *Societa di Scienze Naturali Toscana (Pisa) Memorie* **34**:243-248.

*Groddeck, Z., 1887, Ueber Turmalin enthaltende Kupfererze vom Tamaya in Chile nebst einer Uebersicht des geologischen Vorkommens der Bormineralien, *Deutsche Geologische Gesellschaft* (Berlin) **39**. *ns*

Gross, G., 1972, Zerstörungsfreie, statistisch-morphologische Untersuchung eines Quarzfundes mit festen Einschlüssen, *Schweizerische Mineralogische und Petrographische Mitteilungen* **52**:523-535.

Gross, G., 1973, The effect of solid inclusions on the R-L twinning of Alpine quartz, *Aufschluss* (Heidelberg) **24**:336-341. *ns*

Grum-Grzhimailo, S. V., 1948, Rose color of tourmaline, *Doklady Akademii Nauk SSSR* **60**:1377-1380. (Russian) *c*

Grum-Grzhimailo, S. V., 1956, The color of tourmalines; their examination in polarized ultraviolet light, *Trudy Instituta Kristallografii, Akademiya Nauk SSSR* **12**:79-84. (Russian)

Grum-Grzhimailo, S. V., and Klimusheva, G. V., 1960, Temperature dependence of the broad absorption bands in the spectra of crystals with different structures which have been colored by isomorphic impurities, *Optika i Spektroskopiia* **8**:342-351.

*Gübelin, E., 1939, The minerals of the dolomite of Campalungo (Tessin), *Schweizerische Mineralogische und Petrographische Mitteilungen* **19**:325-442. *ns*

Gübelin, E., 1979, Einschlüsse im Turmalin, *Lapis* **4**:38-39.

Gübelin, E. J., 1956, The emerald from Habachtal, *Gems & Gemology* **8**:295-309.

Gübelin, E. J., 1976, Helvite and tourmaline accompanied by grunerite in quartz, *Journal of Gemmology* **15**:111-113. *m*

Haidinger, W., 1845, Über den Pleochroismus der Krystalle, *Abhandlungen der k. böhm., Gesellschaft der Wissenschaften (Prag),* Bd. 3, p. 7.

Hall, A. L., 1918, A remarkable occurrence of chromium tourmaline and rutile in the Barberton district, *Geological Society of South Africa Transactions* **20**:51-52. *c*

Hamburger, G. E., and Buerger, M. J., 1948, The structure of tourmaline, *American Mineralogist* **33**:532-540.

Hamid, S. A., 1980, Tourmaline as a pyroelectric infrared radiation detector, *Zeitschrift für Kristallographie* **151**:67-75.

Hamlin, A. C., 1873, *The tourmaline,* J. R. Osgood, Boston.

Hamlin, A. C., 1895, *The history of Mount Mica of Maine, U.S.A. and its wonderful deposits of matchless tourmalines,* A. C. Hamlin, Bangor, Maine.

*Hänni, H. A., Frank, E., and Bosshart, G., 1981, Golden yellow tourmaline of gem quality from Kenya, *Journal of Gemmology* **17**:437-442.

*Harada, Z., 1939, Beitrage zur Kenntnis . . . der japanischen Bormineralien, *Journal of the Faculty of Science, Hokkaido University,* 4th ser., **4**:487-500.

Hatch, F. H., Wells, A. K., and Wells, M. K., 1949, *The Petrology of the Igneous Rocks* (10th ed.), Thomas Murby & Co., London.

Hatoya, T., 1948, Boric acid from refined tourmaline (Feb. 27, 1948), *Japanese patent 174,835. c*

Hausmann, J. F. L., 1813, *Handbuch der Mineralogie* (3 vols.), Göttingen. *ns*

Haüy, R. J., 1785, Memorie sur les propriétés électriques plusieurs minéraux, *Mémoires de l'Académie Royale des Sciences,* 206. *ns*

Haüy, R. J., 1801, *Traité de Minérologie* (4 vols.), Chez Louis, Paris.

Haüy, R. J., 1822, *Traité de Cristallographie* (2 vols.), Bachelier et Huzard, Paris.

Heald, M. T., 1955, Stylolites in sandstones, *Journal of Geology* **63**:101-114.

Hearst, J. R., Irani, G. B., and Geesaman, L. B., 1965, Piezoelectric response of Z-cut tourmaline to shocks of up to 21 Kilobars, *Journal of Applied Physics* **36**:3440-3444.

Heinrich, E. W., 1963, Notes on western mineral occurrences, *American Mineralogist* **48**:1172-1174.

Helme, B. G. M., and King, P. J., 1977, Microwave acoustic relaxation absorption in iron tourmaline, *Journal de Physique* (Paris) **38**:1535-1540.

Henry, D. J., and Guidotti, C. V., 1985, Tourmaline as a petrographic indicator mineral: an example from the staurolite grade matapelites of NW Maine, *American Mineralogist,* vol. 70, in press.

Hermann, R., 1845, Untersuchungen russischer Mineralien: Ueber die Zusammensetzung der Turmaline, so wie über die Atom-Gewichte von Bor und Kiesel, *Journal für Praktische Chemie* (Leipzig) **35**:232-247.

Hermon, E., Simkin, D. J., and Donnay, G., 1973, The distribution of Fe^{2+} and Fe^{3+} in iron-bearing tourmalines: a Mössbauer study, *Tschermaks Mineralogische und Petrographische Mitteilungen* **19**:124-132.

Hess, F. L., 1943, The rare alkalies in New England, *U.S. Bureau of Mines Information Circular 7232.*

Heyl, A. V., Foord, E. E., and Heyl, M. L., 1977, Chromian dravite, a new variety of tourmaline from the Line Pit (Lows Mine) in Pennsylvania and Maryland, *Friends of Mineralogy,* Pennsylvania Chapter Newsletter, fall issue, pp. 3-5.

Hibben, J. H., 1939, *The Raman Effect and its Chemical Applications,* Reinhold, New York.

Hill, J., 1771, *Fossils arranged according to their obvious characters. . . .* R. Baldwin (etc.), London. *ns*

Hiller, J.-E., and Keller, P., 1965, Untersuchungen an den Lösern der Kalisalzlagerstätte Buggingen, *Kali und Steinsalz* **4**:190-203.

*Hintze, C., 1897, *Handbuch der Mineralogie* (vol. II), Verlag von Veit, Leipzig, pp. 310-367.

Hitchen, C. S., 1935, The pegmatites of Fitchburg, Massachusetts, *American Mineralogist* **20:**1-24.

Hoel, A., and Schetelig, J., 1916, Nepheline-bearing pegmatitic dykes in Leiland, *Festskrift til Professor Amund Helland,* Det Mallingske Bogtrykkeri, Kristiania (Norway), pp. 110-131.

Hofmann, H., 1970, Einfluss von Piezowerkstoff, Form und Aufbau auf das Signalverhalten Piezoelektrischer Druckwandlersonden, *Symposium—Explosive Cladding Papers Presented to the 1st International Symposium,* pp. 177-192.

*Holgate, N., 1977, Tourmaline from amphibolized gabbro at Hamter Hill, Radnorshire, *Mineralogical Magazine* **41:**124-127.

Holmquist, P. J., 1920, Die Härte von Mischkristallen, *Geologiska Föreningen i Stockholm Förhandlingar* **2:**393.

Home, R. W., 1976, Aepinus, the tourmaline crystal, and the theory of electricity and magnetism, *Isis* **67:**21-30.

Honess, A. P., 1927, *The nature, origin and interpretation of the etch figures on crystals,* Wiley, New York.

Hoover, D. B., 1983, The GEM Diamond Master and the thermal properties of gems, *Gems & Gemology* **29:**77-86.

Horai, K., 1971, Thermal conductivity of rock forming minerals, *Journal of Geophysical Research* **76:**1278-1308.

Horn, E. E., and Schulz, H., 1968, Bestimmung des Turmalin-Chemismus auf röntgenographis-chem Wege, *Neues Jahrbuch für Mineralogie, Abhandlungen* **108:**20-35.

Houser, B. B., 1979, The ratio of zircon to tourmaline: an indication of the relative age of alluvial deposits in SW Virginia (abst.), *Geological Society of America Abstract volume* **11**(4):183.

Hubert, J. F., 1962, A zircon-tourmaline-rutile maturity index and the interdependence of the composition of heavy mineral assemblages with the gross composition and texture of sandstones, *Journal of Sedimentary Petrology* **32:**440-450.

Hunt, G. R., Salisbury, J. W., and Lenhoff, C. J., 1973, Visible and near infrared spectra of minerals and rocks: VI. Additional silicates, *Modern Geology* **4:**85-106.

Hunt, T. S., 1886, *Mineral Physiology and Physiography,* S. E. Cassino, Boston. *ns*

Huntington, H. B., 1958, The elastic constants of crystals, *Solid State Physics* **7:**213-353.

Hurlbut, C. S., Jr., and Klein, C., 1977, *Manual of Mineralogy* (19th ed.), Wiley, New York.

Hutcheon, I., Gunter, A. E., and Lecheminant, A. N., 1977, Serendibite from Penrhyn Group Marble, Mellville Peninsula, District of Franklin, *Canadian Mineralogist* **15:**108-112.

Hutton, C. O., 1939, The significance of tourmaline in the Otago schists, *Royal Society of New Zealand Transactions* **68:**599-602.

Iijima, S., Cowley, J. M., and Donnay, G., 1973, High resolution electron microscopy of tourmaline crystals, *Tschermaks Mineralogische und Petrographische Mitteilungen* **20:**216-224.

Il'inskii, G. A., and Shinkarev, N. F., 1964, Types of sodalite rocks in the Alai Ridge, *Voprosy Magmatizma i Metamorfizma* (Leningrad) **2:**216-221. (Russian)

Irving, E. M., 1937, Tourmalinization in the vicinity of the Cajalco tin mine, near Corona, California (abst.), *Geological Society of America Proceedings 1936,* pp. 300-301.

Ito, T., and Sadanaga, R., 1951, A Fourier analysis of the structure of tourmaline, *Acta Crystallographica* **4:**385-390.

Ivanov, O. P., 1970, Crystallization of a natural silicate melt rich in boron, *Uchenye Zapiski Tsentral'nyi Nauchno-Issledovatel'skii Institut Oloviannoi Promyshlennosti,* no. 1, pp. 40-49. (Russian) *c*

Ivanov, V. V., and Rozbianskaya, A. A., 1961, Geochemistry of indium in cassiterite-silicate-sulfide ores (English translation), *Geochemistry* (USSR), pp. 71-83. (Russian) *m*

Ivanova, T. N., 1981, Microhardness of minerals of the tourmaline group, *Diagnostika i Diagnosticheskie Svoistva Mineralov Proceedings,* pp. 237-239. (Russian)

*Iyengar, K. Y. S., 1937, Fibrous tourmalines from the Mysore State, *Current Science* (India) **5:**534-535.

Ja, Y. H., 1972, G = 4.3 isotropic EPR line in tourmaline, *Journal of Chemical Physics* (Amsterdam) **57:**3020-3022.

Jacobson, M. I., and Tilander, N. G., 1982, A lithium-bearing pegmatite in the Clear Creek District, Clear Creek County, Colorado, *Rocks and Minerals* **57:**241-244.

Jahns, R. H., 1953, The genesis of pegmatites. I: Occurrence and origin of large crystals, *American Mineralogist* **38:**563-596.

Jahns, R. H., and Wright, L. A., 1951, Gem and lithium-bearing pegmatites of the Pala district, San Diego County, California, *California Department of Natural Resources, Special Report 7-A.*

Jain, A. K., 1972, Heavy minerals in Precambrian quartzite of the Lesser Himalaya, Garhural, India, *Journal of Sedimentary Petrology* **42:**941-960.

Jain, V. K., and Mitra, S., 1977, Thermoluminescence studies on some silicate minerals, *Thermochimica Acta* **18:**241-244.

Jakob, J., 1937, Analysen dreier Tessiner Turmaline, *Schweizerische Mineralogische und Petrographische Mitteilungen* **17:**146-148. *c*

*Jakob, J., 1938, Der Turmalin von Karharia stream, Kodarma, Britisch Indien, *Schweizerische Mineralogische und Petrographische Mitteilungen* **17:**605-606.

Jameson, R., 1804, *A System of Mineralogy,* Edinburgh. *ns*

Jan, M. Q., Kempe, D. R. C., and Symes, R. F., 1972, Chromian tourmaline from Swat, West Pakistan, *Mineralogical Magazine* **38:**756-759.

*Jannasch, P., and Kalb, G. W., 1889, Ueber die Zusammensetzung des Turmalines, *Berichte der Deutsche Chemischen Gesellschaft, Berlin* (Heidelberg) **22:**216-221.

Jansen, W., 1933, Röntgenographische Untersuchungen über die Kristallorientierung in Sphärolithen, *Zeitschrift für Kristallographie* **85:**239-270.

Jaroš, Z., 1936, Příspěvek k Øtázce původu pseudomorfos v dolnoborských pegmatitech, *Veda Prirodni, Praha* **17:**160-164.

*Jedwab, J., 1962, Tourmaline zincifère dans une pegmatite de Muika (Congo), *Société Belge de Geologie, de Paleontologie et d'Hydrologie Bulletin* **71:**132-135.

Jérémine, E., and Sandréa, A., 1957, Contribution a l'étude lithologique de l'Ile d'Ouessant, *Bulletin du Service de la Carte Géologique de la France* **55**(252):1-7.

Johanssen, A., 1931-1939, *A Descriptive Petrography of the Igneous Rocks* (vol. II), University of Chicago Press, Chicago, Illinois.

Johnsen, A., 1907, Untersuchungen über Kristallzwillinge und denen Zusammenhang mit anderen Erscheinungen, *Neues Jahrbuch für Mineralogie . . . ,* Bd. 23, p. 237. *ns*

Johnston, J. H., and Duncan, J. F., 1973, Manganese iron site distribution studies in tourmaline by anomalous X-ray scattering methods, *Journal of Applied Crystallography* **8:**469-472.

Joint Committee on Powder Diffraction Standards, 1974-1980, Powder Diffraction File, Swarthmore, Pennsylvania.

*Kalb, G. W., 1890, Über die chemische Zusammensetzung und Constitution des Turmalins, (Inaug. dissertation, Göttingen), *Neues Jahrbuch für Mineralogie, Geologie und Palaeontologie* **2:**199-203.

Kalenchuk, G. E., 1964, Flame-photometric determination of K, Na, Li, Rb, and Cs in rocks and minerals, *Khimicheskii Analiz Mineralov i Ikh Khimicheskii Sostav, Akademiia Nauk SSSR, Institut Geologii Rudnykh Mestorozdenii, Petrografii, Mineralogii i Geokhimii,* pp. 16-32. *c*

Kálmán, S., 1941, Az ékkövek színének nemesítése, *Természettudomanyi Kozlony* **73:**445-449.

Kamiko, M., 1944, Preparation of alumina and boric acid from tourmaline, *Yogyo Kyokai Shi* **52:**131-136. (Japanese)

Kaneko, T., Yasuhara, N., and Toshio, I., 1943, Test making of ampul glass of low boric acid content with tourmaline, *Journal of the Japanese Ceramic Association* **51:**666-667. (Japanese)

Kang, H. U., and Pak, B., 1971, Manufacture of borax from tourmaline 1. X-ray analysis of the reaction of tourmaline with calcite, *Hwahak Kwa Hwahak Kongop* **5:**227-231. (Korean)

Karishima, Y. et al., 1946, Boric acid from tourmaline, *Japanese Patent 173,829,* October 9, 1946. (Japanese) *c*

Karnojitzky, A., 1890, Kristallographisch-optische studienam Turmalin, *Zeitschrift für Kristallographie (Kristallgeometrie, Kristallphysik, Kristallchemie)* **20:**78-80.

Karnojitzky, A., 1894, Uber Trichroismus beim Turmalin, *Zeitschrift für Kristallographie (Kristallgeometric, Kristallphysik, Kristallchemie)* **22:**77-78.

Karpova, O. V., 1965, Tourmaline from an area of basic rocks in the western slopes of the southern Urals, *Trudy Mineralogicheskogo Muzeya Akademiya Nauk SSSR* **16:**101-113. (Russian) *c*

Karskii, B. E., Zorin, B. I., Fortunatov, S. P., and Abromov, A. V., 1977, Tourmaline from Mama mica-containing pegmatites, *Izvestiya Vysshikh Uchebnykh Zavedenii. Seriya Geologiya i Razvedka* **20:**172-174. (Russian)

Karyakina, N. F., Novik, V. K., and Gavrilova, N. D., 1971, Measurement of pyroelectric coefficient in the 40-600 range, *Pribory i Tekhnika Eksperimenta* **197:**227-230 (Russian) *c*

Katalymov, M. V., 1941, The use of tourmaline as a B fertilizer, *Zhurnal Khimicheskoi Promyshlennosti* **18:**15-17. (Russian)

Kauffman, A. J., Jr., and Dilling, E. D., 1950, D.t.a. curves of certain minerals, *Economic Geology* **45:**222.

Kazanskii, Yu. P., 1961, Stability of relict minerals in the profile of a kaolin weathering crust, *Trudy Sibirskogo Nauchno-Issledovatel'skogo Instituta, Geologii, Geofiziki i Mineral'nogo Syr'ya* **14:**80-94. *c*

Keferstein, C., 1849, *Mineralogia polyglotta,* Eduard Anton, Halle (Germany).

Kenngott, A., 1892, Über die Formel der Turmaline, *Neues Jahrbuch für Mineralogie, Geologie und Palaeontologie* **2:**4-57.

Keys, D. A., 1921, A piezoelectric method of measuring explosion pressures, *Philosophical Magazine* (London, Edinburgh, and Dublin) **42:**473-488.

Keys, D. A., 1923, The adiabatic and isothermal piezo-electric constants of tourmaline, *Philosophical Magazine* **46:**999-1001.

Khan, D. H., 1960, Examination of the sedimentary nature of some soil profiles on limestones by using the tourmaline:zircon ratio as an index, *Pakistan Journal of Biological and Agricultural Sciences* **3:**8-13.

Khasnova, A. Kh., 1960, Formation of red granites during the postmagmatic stage of the Gissar pluton, *Trudy Tadshikskii Gosudarstvennyi Universitet* **28:**89-99. *c*

Khlopin, V. G., and Abidov, Sh. A., 1941, Radioactivity and helium content of beryllium, boron and lithium minerals of the U.S.S.R., *Comptes Rendus de l'Academie des Sciences de l'URSS* **32:**637-640.

*Khorvat, V. A., 1964, Quartz-tourmaline rocks in the Balyksui ore field and some of their characteristics, *Sredneaziatskii Nauchno-Issledovatel'skii Institut Geologii i Mineral'nogo Syr'ya,* pp. 253-257. *c*

Kim, Y. S., 1972, A study on the reactivity of component minerals of heavy sands, *Kumsok Hakhoe Chi* **10:**106-113. (Korean) *c*

Kirschvink, J. L., and Lowenstam, H. A., 1979, Mineralization and magnetization of chiton teeth: paleomagnetic, sedimento-logic, and biologic implications of organic magnetite, *Earth and Planetary Science Letters* **44:**193-204.

Kirwan, R., 1794, *Elements of Mineralogy,* (vol. I, 2nd ed.), J. Nichols/P. Elmsly, London, pp. 265, 271, and 286-288. *ns*

*Kitahara, J., 1966, On dravite (Mg-tourmaline) from the Hirose mine, Tattori Prefecture, *Japanese Association of Mineralogists, Petrologists and Economic Geologists Journal* **56:**228-233.

Klemens, P. G., 1973, Radiation damage in solids and phonon scattering, *U.S. National Technology Information Service, A D Report, No. 7755 1618 GA. ns*

Knapp, W. J., 1943, Thermal conductivity of non-metallic single crystals, *Journal of the American Ceramic Society* **26:**48-55.

Koch, P. P., 1906, Beobachtungen über Elektricitätserrung an Krystallen durch nichthomogene und homogene Deformation, *Annalen der Physik* **19:**567-586.

Koch, S., 1957, Hydrothermal tourmaline from Nagybörzsöny, *Acta Universitatis Szegediensis, Acta Mineralogica Petrographica* **10:**47-50. *c*

Koga, I., 1933, Elements of Piezoelectric Oscillating Crystal Plate, *Institute of Electrical Engineers of Japan. ns*

Koivula, J. I., 1982, Tourmaline as an inclusion in Zambian emeralds, *Gems & Gemology* **18:**225-227.

Koja, S., and Zheku, V., 1968, Përeaktimi i konstantes dielektrike të mineraleve, *Bulletin: Shkencave të Natyrore, Universiteti Shteteror i Tiranes* **22:**3-18.

Konovalenka, S. I., and Bozin, A. B., 1976, Distribution of rare alkaline elements in tourmalines of the schorlite-elbaite-tsilaisite series of pegmatite origin, *Trudy Sibirskogo Nauchno-Issledovatel'skogo Instituta Geologii, Geofiziki i Mineral'nogo Syr'ra* **179:**87-90. (Russian) *c*

Koomans, C. M., 1938, A tourmaline-zoisite rock from Loh-oclo, Java, *Leidse Geologische Mededelingen* **10:**104-109. *c*

*Kornetova, V. A., 1975, Classification of minerals of the tourmaline group, *Zapiski Vsesoyuznogo Mineralogicheskogo Obshchestvo* **104:**332-336. (Russian)

Korovushkin, V. V., Kuz'min, V. I., and Belov, V. F., 1979, Mössbauer studies of structural features in tourmaline of various geneses, *Physics and Chemistry of Minerals* **4:**209-220.

*Korzhinskii, A. F., 1958, Thermoöptical studies of epidote group minerals and some tourmalines, *Trudy Soveshchaniya po Eksperimental'noi i Tekhnicheskoi Mineralogii i Petrografii, 5th* (Leningrad), 1956, pp. 97-113. (Russian)

Kosoĭ, L. A., 1939, Tourmaline in the Karelian pegmatites, *Uchenye Zapiski Leningrad Gosudarstvennyĭ Universitet* **1939**(7):54-59. *c*

Kovyzhenko, N. A., 1974, Synthetic tourmalines obtained in high-concentration chloride media, *Sbornik Nauchnogo Studencheskogo Obshchestvo; Geologicheskii Fakul'tet Moskovskii Gosudarstvennyi* **10:**113-119. (Russian)

*Kramer, H., and Allen, R. D., 1954, Analyses and indexes of refraction of tourmaline from fault gouge near Barstow, San Bernardino County, California, *American Mineralogist* **39:**1020-1022.

Kranz, R., 1967, Geochemical significance of organic compounds in inclusions of uranium-containing minerals, *Naturwissenschaften* **54:**469. *c*

Kratochvil, F., 1957, The fulgurite from Křenice near Uhřineves, southwest of Prague, *František Slavik Memorial Volume, Ceskoslovenska Akademie Ved 1957,* pp. 197-202. *c*

Krokström, T., 1946, Feldspathization and boudinage in a quartzite boulder from the Västervik area, *Bulletin of the Geological Institution of the University of Uppsala* **31:**389-400. *c*

Krotova, N. A., and Karasev, V. V., 1953, Investigation of electron emission upon cleavage of solids in vacuum, *Doklady Akademii Nauk SSSR* **92:**607-610. (Russian)

Kruglyakova, G. I., 1954, On the magnetic properties of minerals, in *Aspects of Theoretical Mineralogy in the U.S.S.R.* (English translation), M. H. Battey and S. I. Tomkeieff, eds., 1964, Macmillan, New York, pp. 435-450.

Krynine, P. D., 1946, The tourmaline group in sediments, *Journal of Geology* **54:**65-87.

Kulazewski, C., 1920, Studien über chemische Wirkungen an Kristallen: I. Ätz-und Lösungserscheinungen an Turmalin, *Berichte Akademie der Wissenschaften, (Leipzig)* **72:**48-55.

Kulaszewski, C., 1921, Über die Kristallstruktur des Turmalines, *Abhandlungen die Mathematisch-Physische Klasse dei Sächsische Akademie der Wissenschaften (Leipzig)* **38:**81-117.

*Kulikov, I. V., Petrova, M. G., Royzenman, F. M., and Glebov, M. P., 1976, Rare-metal rubellite-quartz-feldspar pegmatite veins and their importance in exploration (English translation), *International Geological Review* **18:**205-208.

Kundt, A., 1883, Ueber eine einfache Methode zur Untersuchung der Thermo-; Actino- und Piëzoëlektricität der Krystalle, *Weidmann Annalen der Physik und Chemie* **20:**592-601.

Kunitz, W., 1926, Chemische Untersuchungen an der Turmalin-und glimmergruppe, *Tschermaks Mineralogische und Petrographische Mitteilungen* **37:**258-259.

*Kunitz, W., 1929, Die Mischungsreihen in der Turmalingruppe und die genetischen Beziehungen zwischen Turmalinen und Glimmern, *Chemie der Erde* **4:**208-251.

Kunitz, W., 1936, Die Rolle des Titans und Zirkoniums in den gesteinsbildenden Silikaten, *Neues Jahrbuch für Mineralogie, Geologie und Palaeontologie Abteilungen A*, **70:**385-466.

Kunz, G. F., 1902, The composition of tourmaline, *Engineering and Mining Journal* **73:**482.

Kunz, G. F., 1913, *The Curious Lore of Precious Stones*, J. B. Lippincott, Philadelphia.

Kupková, D., 1966, Effect of the structure of silicates on the spectral emission of the main components, *Collection of Czechoslovak Chemical Communications* **31:**4047-4056.

Kuroda, P. K., and Sandell, E. B., 1953, Chlorine in igneous rocks (some aspects of the geochemistry of chlorine), *Geological Society of America Bulletin* **64:**879-896

Kurtz, S. K., and Perry, T. T., 1968, A powder technique for the evaluation of nonlinear optical materials, *Journal of Applied Physics* **39:**3798-3813.

Kurylenko, C., 1949, Contribution a l'étude de la rubellite, *Bulletin de la Société Français de Minéralogie et de Cristallographie* **72:**319-321.

Kurylenko, C., 1950, Analyse thermique de quelques tourmalines, *Bulletin de la Société Française de Minéralogie et de Cristallographie* **73:**49-54.

Kurylenko, C., 1951, Transformation de la dravite de Doubrova (Moravie) de 375°C à 1350°C, *Comptes Rendus de l'Academie des Sciences (Paris)* **232:**2109-2111.

Kurylenko, C., 1953, Analyses des tourmalines noires à l'aide de la microbalance de Chevenard, *Comptes Rendus Academie des Sciences (Paris)* **239:**391-393.

Kurylenko, K. I., 1957, Densité des tourmalines et leur transformation au cours du chauffage jusquà 1350°C (title of French summary), *Mineralogicheskii Sbornik, L'vovskii Gosudarstvennyi Universitet, L'vov, USSR* **11:**69-80. (Russian, French summary)

*Kuz'min, V. I., Dobrovol'skaya, H. V., and Solntseva, L. C., 1979, *Tourmaline and its use in prospecting-evaluation work*, Nedra, Moscow. (Russian)

*Labuntsov, A. N., 1930, Geology and mineralogic studies in western Pamirs and Badakhshan Province in 1928, *Trudy Tadzhiksko-Pamirskaia Ekspeditsiia* **4**. *ns*

LaCroix, A., 1893, *Minéralogie de la France et de ses Colonies* (vol. I), Librairie Polytechnique, Paris.

LaCroix, A., 1910, *Minéralogie de la France et de ses Colonies* (vol. IV), Librairie Polytechnique, Paris.

LaCroix, A., 1914, A propos de la tourmaline des serpentines, *Bulletin de la Société Française de Minéralogie* **37:**75-76.

LaCroix, A., 1918, Sur l'identité de l'iochroite et de la tourmaline, *Bulletin de la Société Française de Minéralogie* **41:**130-131.

*LaCroix, A., 1922, *Minéralogie de Madagascar* (vol. I), Librairie Maritime et Coloniale, Paris.

Ladurner, J., 1951, A recrystallized aplite mylonite (helsinkite type) from the Stuba Alps, Tyrol, *Neues Jahrbuch für Mineralogie Monatshefte 1951*, pp. 241-247. *c*

Lal, N., Parchad, R., and Nagpaul, K. K., 1977, Fission track etching and annealing of tourmaline, *Nuclear Track Detection* **1:**145-148. *m*

Lallemont, A., 1978, New tourmaline discovery in Brazil, *Mineralogical Record* **9:**298.

Landergren, S., 1945, Contribution to the geochemistry of boron. II. The distribution of boron in some Swedish sediments, rocks, and iron ores. The boron cycle in the upper lithosphere, *Arkiv für Kemi, Mineralogi och Geologi* **19A:**1-31.

Landgrebe, G., 1841, *Über die Pseudomorphosen in Mineralreiche und verwandte Erscheinungen,* Verlag-J. J. Bohné, Cassel.

Landolt-Börnstein numerical data and funcational relationships in science and technology, 1962-, new ser., Springer-Verlag, New York.

Lang, S. B., 1969, Thermal expansion coefficients and the primary and secondary pyroelectric coefficients of animal bone, *Nature* **224:**798-799.

Lang, S. B., 1974, *Sourcebook of Pyroelectricity,* Gorton and Breack, London.

Lareida, S., 1978, Man mietet sich einen Esel . . . *Lapis* **4:**14-19.

*Lebedev, A., 1937, On lead-bearing turmaline from the Maly Khingan, *Comptes Rendus de l'Académie des Sciences de l'URSS* **14:**127-128.

Lebedev, V. I., 1945, Chemical formulae of tourmaline, axinite, and dumontierite, *Comptes Rendus de l'Académie des Sciences de l'URSS* **47:**635-639.

Leckebusch, R., 1978, Chemical composition and colour of tourmaline from Darře Pěch (Nuristan, Afghanistan), *Neues Jahrbuch für Mineralogie, Abhandlungen* **133:**53-70.

Leela, M., 1954, Crystal Magnetism of Tourmaline, *Journal of Mysore University (India)* **14b:**155-165.

Leidy, J., 1871, Remarks on the minerals of Mount Mica, Maine, *Proceedings of the Academy of Natural Sciences of Philadelphia,* pp. 245-247. *ns*

Lemery, L., 1717, Diverses observations de physique générale, *Histoire de l'Académie Royale des Sciences (Paris)* **7**. *ns*

Leoni, L., and Troysi, M., 1975, Ricerche sulla microdurezza dei silicati II. Le tourmaline, *Atti della Societa Toscana di Scienze Naturali (Pisa) Memorie,* ser. A, **82:**177-184. (Italian)

Lévy, A., 1838, *Description d'une Collection de Mineraux formée par M. Henri Heuland . . .* (3 vols. and 83 plates), F. Richter and Haas, London. *ns*

Lewis, M. F., and Patterson, E., 1972, Assessment of tourmaline as an acoustic-surface-wave-delay medium, *Applied Physics Letters* **20:**275-276.

Lewis, M. F., and Patterson, E., 1973, Microwave ultrasonic attenuation in topaz, beryl, and tourmaline, *Journal of Applied Physics* **44:**10-13.

l'Hermina, C. (also cited as Lermina, C.), 1799, (Title not given), *Journal de l'École Polytechnique* **1:**439. *ns*

Liddicoat, R. T., Jr., 1975, *Handbook of Gem Identification* (10th ed.), Gemological Institute of America, Santa Monica, California.

Liese, H. C., 1975, Selected terrestrial minerals and their infrared absorption spectral data (4000-300 cm^{-1}), in *Infrared and Raman Spectroscopy of Lunar and Terrestrial Materials,* C. Carr, Jr., ed., Academic Press, New York, pp. 197-227.

Linck, G., 1899, Die Pegmatite des oberen Veltlin, *Jenaische Zeitschrift für Medizin und Naturwissenschaften* **33:**345-360.

Lind, S. C., and Bardwell, D. C., 1923, The coloring and thermophosphorescence produced in transparent minerals and gems by radium radiation, *American Mineralogist* **8:**171-181.

Linnaeus, C., 1768, *Systema Naturae* (12th ed.), L. Salvii, Holmiae. (Edition number not preserved on copy used.)

Lister, C. J., 1978, Some tourmalinized rocks from Cornwall and Devon, *Ussher Society Proceedings* **4:**211-216.

Lister, C. J., 1979, Quartz-cored tourmaline from Cape Cornwall and other localities, *Ussher Society Proceedings* **4:**402-418.

Lodochnikov, V. N., 1933, Serpentine and serpentinites and the petrological problems connected with them, *Problemy Sovetskoi Geologii,* new ser. **2:**119-144. (Russian)

Loewenstein, W., 1956, Boron in tetrahedra of borates and borosilicates, *American Mineralogist* **41:**349-351.

*Lokka, L., 1943, Beiträge sur Kenntnis des Chemismus der Finnischen Mineralé . . . , *Bulletin de la Commission Géologique de Finlande* **129:**72.

López de Azcona, J. M., 1947, Is there lead of radioactive origin in tourmaline? in *Report of the Committee on the Measurement of Geologic Time, 1943-1949,* J. P. Marble, chairman, National Research Council, Division of Geology and Geography (Annual Report) 1943-1946 Exhibit C, pp. 60-62.

Lum, T. Y., 1972, Geochemistry of the chromium tourmaline from Kaavi, Finland, *Malaysia Geological Survey Annual Report* (1970) pp. 157-168.

Lyakhovich, V. V., 1963, Cesium in granitic rocks, *Doklady Akademii Nauk SSSR* **153:**1424-1427. (Russian) *c*

Lyakhovich, V. V., 1965, Characteristics of tin and boron contents in granitic rock, *Geokhimiya* **1965:**25-31. (Russian) *c*

MacCarthy, G. R., 1928, Tourmaline-bearing quartz from Amelia, Virginia, *American Mineralogist* **13:**531.

Machatschki, F., 1929, Die Formeleinheit des Turmalins, *Zeitschrift für Kristallographie* **70:**211-233; **71:**45-46; **76:**475-476.

Machatschki, F., 1941, Notiz über die Entwässerung des Turmalins, *Zentralblatt für Mineralogie, Geologie und Palaeontologie (Stuttgart), Abteilung A.* **A1941:**135-137.

MacRae, N. D., and Kullerud, G., 1971, Experimental investigations of some sulfide-silicate reactions in the hydrothermal temperature range, *Carnegie Institution of Washington Year Book* **70:**292-295.

McCrillis, D. A., 1975, Gem tourmaline rediscovered at Newry, *Mineralogical Record* **6:**14-21.

McCurry, P., 1971*a,* Pseudomorphic quartz-tourmaline relation from northern Nigeria, *American Mineralogist* **56:**1474-1476.

McCurry, P., 1971*b,* Relation between optical properties and occurrence of some black tourmalines from northern Nigeria, *Mineralogical Magazine* **38:**369-373.

Madelung, A., 1883, Beobachtungen mit Breithaupts Polarisationsmikroskop, *Zeitschrift für Kristallographie* **7:**73-76.

Mader, D., 1978, Turmalinauthigenese im Buntsandstein von Oberbettingen (Westeifel), *Neues Jahrbuch für Mineralogie Monatshefte* **5:**233-240.

Mafveyev, K. K., 1946, Observations on polychromatic tourmalines, *Voprosy Geokhimii, Mineralogii i Petrografii* (Moscow-Leningrad), pp. 82-88. (Russian)

Makagon, V. M., 1972, Tourmaline chemical composition change during the formation of Siberian rare-metal and muscovite pegmatites, *Ezhegodnik, Institut Geokhimii, Sibirskoe Otdelenie, Akademiya Nauk SSSR* (Published 1973), pp. 154-158. (Russian)

Makagon, V. M., Afonnia, G. G., and Bogdanova, L. A., 1977, X-ray diffraction characteristics of tourmalines from pegmatites of diverse mineralizations, *Geokhimicheski Metod' Poishov Metod' Analiza,* pp. 114-118. (Russian)

Makarov, V. N., and Kondrat'eva, D. M., 1965, Alterations of tourmaline in weathering profile of the Yakovlev iron-ore deposits in the Kursk Magnetic Anomaly area, *Akademiya Nauk Ukrainskoi RSR (Kiev) Dopovidi* **1:**84-87. (Russian)

*Malireddi, R., and Gordienko, V. V., 1968, Pseudomorphs of muscovite after tourmaline in mica-bearing pegmatites of northern Karelia, *Mineralogiya i Geokhimiya* (Leningrad) **3:**27-36. (Russian)

Mallet, F. R., 1866, On the gypsum of Lower Spiti, with a list of minerals collected in the Himalayas, 1864, *Geological Survey of India Memoir* **5:**154-172.

Manning, P. G., 1969*a,* An optical absorption study of the origin of color and pleochroism in pink and brown tourmalines, *Canadian Mineralogist* **9:**678-690.

Manning, P. G., 1969*b,* Optical absorption spectra of chromium-bearing tourmaline, black tourmaline, and buergerite, *Canadian Mineralogist* **10:**57-70.

Manning, P. G., 1973, Effect of second-nearest neighbor interaction on manganese (3+) absorption in pink and black tourmalines, *Canadian Mineralogist* **11:**971-977.

Marfunin, A. S., Mkrtchyan, A. R., Nadzharyan, G. N., Nyussik, Ya. M., and Platonov, A. N., 1970, Optical and Mössbauer spectra of iron in tourmalines, *Izvestiya Akademiya Nauk SSSR, Seriya Geologicheskaya* **2:**146-150. (Russian)

Martin, A. J. P., 1931, A new method for detecting pyroelectricity, *Mineralogical Magazine* **22:**519-523.

Martin, J. P., and Witten, L., 1961, Temperature dependence of microwave phonon attenuation in quartz and tourmaline (abst.), *Bulletin of the American Physical Society* **11:**447. *ns*

Martirosyan, R. A., 1962, The Kedabelk secondary quartzites containing colloform tourmaline, *Uchenye Zapiski Baku. Azerbaidzhanskii Gosudarstvennyi Universitet. Seriia Geologo-Geograficheskikh* **1:**25-35. (Russian)

Mason, B., Donnay, G., and Hardie, L. A., 1964, Ferric tourmaline from Mexico, *Science* **144:**71-73.

Mason, W. P., 1950, *Piezoelectric Crystals and their Application to Ultrasonics,* Van Nostrand, New York.

*Mateos, J. P., 1944, El color en la tourmaline, *Instituto Geológico y Minero de España. Notas y comunicaciones* **13:**217-287.

Mathesius, J., 1564, *Sarepta oder Bergpostill: Sampt der Jochimsthalischen Kurtschen Chroniken,* Nürnberg, 2 vols., unnumbered. Reprinted by Návroni Technické Muzeum Praha (Prague), 1975.

Matias, V. V., and Karmanova, I. G., 1963, Tantalum and niobium contents in tourmalines from the granitic pegmatites, *Mineral'noe Syr'e (Vsesoiuznyi Nauchnoissledovatel'skii Institut Mineral'nogo Syr'ya)* **8:**78-81. (Russian)

Matousova, O., 1967, Isotopic composition of lithium in some Czechoslovak minerals, *Sbornik Vysoke Skoly Chemicko-Technologicke v Praze, Mineralogie* **9:**15-32. (Czech) *c*

Maurice, M. E., 1930, Demonstration of electric lines of force and a new method of measuring the electric moment of tourmaline, *Cambridge Philosophical Society Proceedings* **26:**491-. *ns*

Mawe, J., 1818, On the tourmaline and apatite of Devonshire, *Quarterly Journal of Science, Literature and Art (London)* **4:**369-372.

Mawson, D., and Dallwitz, W. B., 1945, Soda-rich leucogranite cupolas of Umberatania, *Royal Society of South Australia Transactions* **69:**22-49.

Maxwell, J. C., 1873, *A Treatise on Electricity and Magnetism,* Oxford Press, Clarendon, England.

Meissner, A., and Bechmann, R., 1928, Untersuchung und Theorie der Pyroelektrizitat, *Zeitschrift für Technische Physik* **9:**175-186.

Melent'ev, G. B., Sharybkin, A. M., and Mart'yanov, N. N., 1971, Pegmatites of Central Asia as a source of tourmaline crystals, *Redkie Elementy* **6:**128-134. (Russian) *c*

Melent'ev, G. B., Novikova, Yu. I., Kataeva, Z. T., Bykova, A. V., and Cherepivskaya, G. A., 1967, Characteristics of changes in chemical composition of tourmalines during formation of rare-metal granitic pegmatites, *Nauchnykh Konferentsiya Molodykh Sotrudnyk, Institut Mineralogii, Geokhimii i Kristallodkhimii Redkikh Elementov Akademiya Nauk SSSR* 5th, pp. 41-43. (Russian) *c*

Mellor, J. W., 1925, *A comprehensive treatise on inorganic and theoretical chemistry* (vol. 6), Longmans, Green, London.

Melon, J., 1930, Sur deux mineraux du Congo belge: 1-Tourmaline non pyroélectrique à facies special; 2-Chrysoberyl incolor et non maclé, *Académie Royale de Belgique* **16:**996-1000.

Merritt, E., 1895, Ueber den Dichroismus von Kalkspath, Quartz und Turmalin für ultrarothe Strahlen, *Annalen der Physik* (Leipzig) **55:**49-64.

Messina, C., 1940, I mineral: di boro de granito di Baveno, *Atti della Societa Italiana di Scienze Naturali e del Museo Civico di Storia Naturale (Milano)* **79:**31-48.

Metz, R., 1964, Wie gross werden Kristalle?, *Der Aufschluss* **15:**319-324.

Mian, I., 1970, Chromium-bearing minerals of Northwest Frontier Province, *Peshawar University Geological Bulletin* **5:**131-134.

Michel-Lévy, A., 1889, Propriétés optiques des aureoles polychroiques, *Comptes Rendus de l'Académie des Sciences* (Paris) **109:**973-976.

Michel-Lévy, M. C., 1949, Synthèse de la tourmaline et de la jérémé iéwité, *Comptes Rendus de l'Académie des Sciences* (Paris) **228:**1814-1816.

Michel-Lévy, M. C., 1953, Reproduction artificielle de minéraux qui apparaissent dans le métamorphisme de contact du granite, *Bulletin de la Société Française de Minéralogie et de Cristallographie* **76:**237-293.

Michel-Lévy, M. C., and Kurylenko, M. C., 1952, Un cas d'altération de la tourmaline dans une rocke, *Bulletin de la Société de Minéralogie et de Cristallographie* **75:**446.

Miethe, A., 1906, Über die Farbung von Edelsteinen durch Radium, *Annalen der Physik* **4**(19):633-638.

Miko, O., and Hovorka, D., 1978, Kremito-turmalinické horniny veporidného kryštalinka Nızkych Tatier, Západné Karpaty, Šeria Mineralógia, Petrografia, Geochémia Metalogenéza (Geol. Úst. D. Stúra, Bratislava) **5:**7-22.

Miller, W. H., 1839, *A Treatise on Crystallography,* J & J. J. Deighton, Cambridge, England.

Milone-Tamburino, S., and Stella, A., 1952-1954, Radioactivity of crystalline rocks of the Siculo-Calabrian system, *Atti dell'Accademia Gioenia di Scienze Naturali in Catania* **9:**139-150. *c*

Milton, C., 1971, Authigenic minerals of the Green River Formation, *University of Wyoming Contributions to Geology* **10:**57-63.

Mirgabitov, R. M., 1970, Accessory minerals in granitic rocks of the Akchashava intrusion, *Nauchnye Trudy, Tashkentskii Gosudarstvennyi Universitet* **372:**71-74. (Russian) *c*

Mitchell, R. K., 1967, Refraction anomalies in tourmaline, *Journal of Gemmology* **10:**194.

Mitchell, R. K., 1976, R. I. anomalies in tourmaline and a strange zircon, *Journal of Gemmology* **15:**17-18.

Mitchell, R. S., 1964, Matted filiform tourmaline from Amelia County, Virginia, Rocks and Minerals **39:**236-237.

Moenke, H., 1962, Nachweis von BO_3 und BO_4 Gruppen in dem häufigsten natürlich gebilden silikaten, *Silikat Technik* **13:**287-288.

Moenke, H. H. W., 1974, Silica, the three-dimensional silicates, borosilicates and beryllium silicates, in *The Infrared Spectra of Minerals,* V. C. Farmer, ed., Mineralogical Society, London, England, pp. 365-382.

Mohs, F., 1824, *Grundriss der Mineralogie,* vol. II, Dresden, Germany. (1825 English translation by W. Haidinger, Edinburgh, used.)

Moine, B., 1979, La recherche d'anciennes séries évaporitiques dans les ensembles métamorphiques méthodes et résultats, *Sciences de la Terre* (Nancy) **23:**(2)-85-(2)-94.

Mons, W., and Paulitsch, P., 1972, Preferred orientation of tourmaline prism in rocks, *Geological Survey of India Records* **99:**121-126.

Moon, R. J., 1948, Inorganic crystals for the detection of high-energy particles and quanta, *Physical Review* **73:**1210.

Mücke, A., 1975, Epitaxie und Diataxie, *Aufschluss* **26:**272-275.

Mügge, O., 1903, Die regelmässigen Verwachsungen von mineralen verschiedener Art, *Neues Jahrbuch für Mineralogie, Geologie und Paleontologie Beilage* **16:**335-475. *ns*

Muir, R. O., 1963, Petrography and provenance of the Millstone Grit of central Scotland, *Edinburgh Geological Society Transactions* **19:**439-485. *c*

*Mukherjee, S., 1968, Investigations on the chrome-tourmaline from Nausahi, Keonjhar district, Orissa, *Quarterly Journal of the Geological, Mining and Metallurgical Society of India* **40:**119-121.

Müller, H., 1912, Kristallographische Untersuchungen am Turmalin aus Brasilien, *Verhandlungen der Physikalisch-Medicinische Gesellschaft zu Würzburg* N. F., Bd. 42.

Nagai, S., 1944, Leaching of boric acid from various boron minerals, I. Tourmaline, *Journal of the Japanese Ceramic Association* **52:**87-90. (Japanese)

Naifonov, T. B., Pol'kin, S. I., and Shafeev, R. Sh., 1963, The condition of double electrically charged layer of tantalite and some accompanying minerals during flotation, *Izvestiia Vysshikh Uchebnykh Zavedenii. Tsvetnaia Metallurgiia* **6:**40-46. (Russian)

Nassau, K., 1975, Gamma ray irradiation induced changes in the color of tourmalines, *American Mineralogist* **60:**710-713.

Nassau, K., 1984, *Gemstone enhancement,* Butterworth Scientific, London.

Nassau, K., and Schonhorn, H., 1978, The contact angle of water on gems, *Gems & Gemology,* winter 1977-1978, pp. 354-360.

Neiva, A. M. R., 1974, Geochemistry of tourmaline (schorlite) from granites, aplites and pegmatites from northern Portugal, *Geochimica Cosmochimica Acta* **38:** 1307-1317.

Němec, D., 1951, Minor elements of tourmalines muscovites, and lepidolites from West Moravian pegmatites, *Académie Tchèque des Sciences, Bulletin International, Classe des Sciences, Mathematiques, Naturelles, et de la Medecine,* **52:**425-436 (published 1953).

Němec, D., 1954, Jednosměrny polární růst krystalu turmalinu (The unipolar growth of tourmaline crystals), *Spisy Vydávané Přírodovědeckou Fakulton Massarykovy University,* no. 359. (French summary)

Němec, D., 1955, Die Berechungsindizes zonarer Turmaline von Dobrà Voda in Westmähren, *Neues Jahrbuch für Mineralogie, Monatshefte* **2:**25-32.

Němec, D., 1969, Fluorine in tourmalines, *Contributions to Mineralogy and Petrology* **20:**235-243.

Němec, D., 1973, Tin in tourmalines, *Neues Jahrbuch für Mineralogie, Monatshefte* **2:**58-63.

Němec, D., 1975, Genesis of tourmaline spots in leucocratic granites, *Neues Jahrbuch für Mineralogie, Monatshefte* **7:**308-317.

Němec, D., 1981, Ein Pegmatit mit Li-Mineralisierung von Dolní Bory in Westmähren (ČSSR), *Chemie der Erde* **40:**146-177.

Newberry, E., and Lupton, H., 1918, Radio-activity and the coloration of minerals, *Manchester Literary and Philosophical Society Memoirs* **62:**1-16.

Newcomet, W., 1941, Some radium reactions on minerals, *Rocks and Minerals* **16:**126.

Newhouse, W., and Holden, E., 1925, Graphic intergrowths of quartz and black tourmaline from Maine, *American Mineralogist* **10:**42-43.

Newnham, R. E., and Yoon, H. S., 1973, Elastic anisotropy in minerals, *Mineralogical Magazine* **39:**78-84.

Niggli, P., 1926, *Lehrbuch der Mineralogie* (Band II) (2nd ed.), Gebrüder Borntraeger, Berlin.

Nishikawa, Y., 1958, Gallium content of rocks and minerals of Japan, *Nippon Kagaku Zasshi* **79:**351-354. (Japanese) *c*

Nishio, S., Imai, H., Inuzuka, S., and Okada, Y., 1953, Temperatures of mineral formation in some deposits in Japan, as measured by the decrepitation method, *Mining Geology* (Tokyo) **3:**21-29. (Japanese)

Nisi, H., 1929, Raman effect in some crystals, *Imperial Academy (Japan) Proceedings* **5:**407-410.

Nisi, H., 1932, Further studies on the Raman effect in crystals (Phosphates and silicates). Luminescence of zircon, *Proceedings of the Physico-Mathematical Society of Japan* **14:**214-229.

Nordenskiöld, N. A. E., 1863, *Beskrifning öfver de i Finland funna Mineralier* (2nd ed.), Helsingfors. *ns*

*Novak, F., and Zak, L., 1970, Dravite asbestos from Chvaletice, *Acta Universitatis Carolinae, Geologica* **1:**27-44.

Novozhilov, A. I., Voskresenskaya, I. E., and Samoilovich, M. I., 1969, E.P.R. Study of tourmalines, *Kristallografiya* **14:**507-509.

Nuber, B., and Schmetzer, K., 1979, Die Gitterposition des Cr^{3+} im Turmalin. Strukturverfeinerung eines Cr-reichen Mg-Al-Turmalins, *Neues Jahrbuch für Mineralogie, Abhandlungen* **137:**184-197.

Nuber, B., and Schmetzer, K., 1981, Strukturverfeinerung von Liddicoatit, *Neues Jahrbuch für Mineralogie, Monatshefte* **5:**215-219.

Nündel, M., 1973, Fund eines ungewöhnlichen Turmalinkristalls auf Ceylon, *Aufschluss* **24:**259-260.

Nye, J. F., 1954, *Physical properties of crystals,* Oxford (Clarendon Press), London.

Oba, N., and Ishikawa, H., 1959, Orbicular rocks in the Takukuma granite, Osumi Peninsula, *Ganseki Kôbutsu Kôshô Gakkaishi* **43:**15-25. *c*

Okuda, N., Tomita, M., and Tanaka, K., 1973, Steel welding fluxes, Japan. Koka: *Japanese patent 74,134 545* (Cl.12 B104), December 25, 1974, Appl. 73 49,517, January 5, 1973. *c*

*Okulov, E. N., 1973*a,* Tourmaline as an indicator of alkalies in the enclosing medium, *Zapiski Uzbekistanskogo Otdeleniya Vsesoyuznogo Mineralogicheskogo Obshchestva* no. 26, pp. 136-137. (Russian) *c*

*Okulov, E. N., 1973*b*, Zinc-containing tourmaline in a pegmatite deposit in Central Asia, *Uzbekskiy Geologichesky Zhurnal* **17:**86-88. (Russia)

Oliver, D. W., and Slack, G. A., 1966, Ultrasonic attenuation in insulators at room temperature, *Journal of Applied Physics* **37:**1542-1548.

Omori, K., 1961, Infrared absorption spectra of some essential minerals, *Science Reports of the Tohoku University,* ser. 3, **7:**101-130.

*Ontoev, D. O., 1956, Composition of some ore-forming tourmalines, *Akademiia Nauk SSSR Institut Geologii Rudnykh Mestorozhdenii, Petrografii, Mineralogii i Geokhimii. Trudy* **3:**340-346. (Russian)

Otroshchenko, V. D., and Zanin, M. F., 1965, Some common characteristics of accessory boron mineralization in magnesian and calcic skarns of central Asia, *Doklady Akademiya Nauk Uzbekskoi SSR* **22:**41-44. (Russian)

*Otroshchenko, V. D., Dusmatov, V. D., Khorvat, V. A., Akramov, M. B., Morozov, S. A., Otroshchenko, L. A., Khalilov, M. Kh., Kholopov, N. P., Vinogradov, O. A., Kudryavtsev, A. S., Kabanova, L. K., and Sushchinskiy, L. S., 1971, Tourmalines in Tien Shan and the Pamirs, Soviet Central Asia (English translation), *International Geological Review* **14:**1173-1181.

Ozkan, H., 1980, *Effect of thermal neutron irradiation on the elastic moduli and structure of boron-containing crystals,* IAEA-R-1800-F. *c*

*Pagliani, G., and Milani, C., 1952, La pegmatite di Candoglia (Val d'Ossola), *Atti della Societa Italiana di Scienze Naturali e del Museo Civico di Storia Naturale di Milano* **91:**190-200.

Papastamatios, J. N., 1939, New corundum rocks from Naxos (Grecian archipelago), *Compte Rendu Sommaire des Seances de la Société Geologique de France* **208:**2088-2090. *c*

Park, J., 1928, Pneumatolysis as a mineralizer, *Chemical Engineering and Mining Review* **20:**224. *c*

Parker, R. B., 1962, Blue quartz from the Wind-River Range, Wyoming, *American Mineralogist* **47:**1201-1202.

Parkhomenko, E. I., 1959, Study of piezoelectric effect of rocks and its potential uses in geology, *Sovetskaya Geologiya* **12:**101-111. (Russian)

Patek, J. M., 1934, Relative flotability of silicate minerals, *American Institute of Mining and Metallurgical Engineers, Technical Publication No. 564.*

*Pawlica, W., 1915, Polnócna wyspa krystaliczna w Tatrach-Die nördliche kristallinische Insel in der Tatra, *Polska Akademia umiejętności, Cracow* (Bulletin de l'Academy des Sciences de Cracovie), internat. ser. A, pp. 52-76.

Pehrman, G., 1962, Diadochy of silicon and phosphorous in some silicate minerals, *Meddelanden Abo Akademis Geologisk-Mineralogiska Institut* **44:**5. (German) *c*

*Peltola, E., Vuorelainen, Y., and Häkli, T. A., 1968, A chromian tourmaline from Outokumpu, Finland, *Bulletin of the Geological Society of Finland* **40:**35-38.

Penfield, S. L., 1893, On cookeite from Paris and Hebron, Maine, *American Journal of Science* **45:**393-396.

*Penfield, S. L., 1900, Chemical composition of tourmaline, *American Journal of Science,* 4th ser., **10:**19-32.

*Penfield, S. L., and Foote, H. W., 1899, On the chemical composition of tourmaline, *American Journal of Science,* 4th ser., **7:**97-125.

Perfil'eva, Yu. D., Gorelikova, N. V., Bubeshkin, A. M., 1975, Mössbauer determination of iron in tourmaline, *Vestnik Moskouskogo Universiteta Khimiya* **16:**117-118. (Russian)

Perozio, G. N., 1959, Newly formed minerals of the Mesozoic and Paleozoic rocks in the southeastern part of the West-Siberian lowland, *Trudy Sibirskogo Nauchno-Issledovatel'skogo, Institut Geologii, Geofiziki i Mineral'nogo Syr'ya* **1:**87-97. (Russian)

Ptrovskaya, N. V., and Andreeva, M. G., 1969, Klyuchevskoe deposit as an example of gold-tourmaline mineralization (eastern Transbaikal region), *Zolotorudnye Formatsii Dal'nego Vostaka,* pp. 36-60. (Russian) *c*

Petrzilka, V., 1932*a,* Tourmaline resonators for short and ultra-short waves, *Annalen der Physik* **15:**72-88. *ns*

Petrzilka, V., 1932*b*, Longitudinal and flexural vibrations of tourmaline plates, *Annalen der Physik* **15:**881-902. *ns*

Petrzilka, V., 1937, Control of transmitters by lengthwise vibrations of tourmaline plates, *Hochfrequenztechnik und Elektroakustik (Leipzig)* **50:**1-5. *ns*

Pfaffl, F., and Niggemann, M., 1967, Tracht und Habitus von Turmalinkristallen aus einigen Pegmatiten des Bayerischen Waldes, *Aufschluss* **18:**171-178.

Pichamuthu, C. S., 1944, Tourmaline schists from Holenarsipur, Hassan district, Mysore state, *Current Science* **13:**279-280. *c*

Pichavant, M., 1981*a,* An experimental study of the effect of boron on a water saturated haplogranite at 1 kbar vapor pressure, *Contributions to Mineralogy and Petrology* **76:**430-439.

Pichavant, M., 1981*b,* Application des données expérimentales aux conditions de genèse et de cristallisations des leucogranites à tourmaline, ser. 2, *Comptes Rendus de l'Académie des Sciences (Paris)* **282:**851-853.

Pieruccini, R., 1950, Distribution of boron petroleum-bearing in clays of Piacenza: spectrographic determination methods, geo-chemical discussions, *Periodico di Mineralogia* **19:**209-238. *c*

Pinet, M., 1971, Mesure des indices de réfraction d'une tourmaline verte et d'une tourmaline bleue du Bresil entre 420 et 640 mu, *Comptes Rendus du Congres National des Sociétés Savantes, Section des Sciences (Paris)* **93:**131-138.

Plimer, I. R., 1983, The association of tourmaline-bearing rocks with mineralisation at Broken Hill, NSW, *Australian Institution of Mining and Metallurgy, 1983 Conference* (Broken Hill, New South Wales), pp. 157-176.

Pliny, C. (secundi), 77, Historia Naturalis C. Plinii Secundi: (37 "books"). (See J. Bostock and H. T. Riley, 1855-1857, *The Natural History of Pliny,* 6 vols., H. G. Bohn, London.)

Plyusnina, I. I., 1961, Infra red spectra of ring silicates, III. *Zhurnal Strukturnoi Khimii* **2:**330-336. *c*

Plyusnina, I. I., and Bokiĭ, G. B., 1958, Infrared reflection spectra of ring silicates in the wave length interval 7-15μ, *Kristallografiya* **3:**752-756.

Plyusnina, I. I., and Voskresenskaya, I. E., 1974, A more exact definition of certain structure positions in tourmaline through infrared spectroscopy, *Moscow Universitete Vestnik Seriia,* **29:**38-42.

Plyusnina, I. I., Granadchikova, B. G., and Voskrensenskaya, I. E., 1969, Infra red spectroscopic study of tourmalines, *Kristallografiya* **14:**370-375, 450-455.

Pollak, H., and Bruyneel, W., 1974, Sur le rapport Fe^{++}/Fe^{+++} dans deux silicates et les transitions de Verwey, *Journal de Physique* (Société Française Physique) 35, #12 Colloque, 6 Conference Internationale sur les applications de l'effet Mossbauery (Paris), C6.571-C6.574, pp. 1-19.

Pollak, H., deCoster, M., and Amelinckx, S., 1962, *Physica Status Solidi* **2:**1653-. *ns*

Pollak, H., Dannon, J., Quatier, R., and Dauwe, C., 1979, The NGR observation of proton jumps in tourmaline (abst.), International Conference on the application of the Mössbauer Effect, *Journal de Physique* (Société Française Physique) Colloque #2, Fascicule 3, Supplement C2 p. C-480 (Conference held in Kyoto, Aug. 28-Sept. 1, 1978).

Pollard, C., Jr., and Wagner, G., 1973, Strains associated with chemical boundaries in a saltor-zoned elbaite tourmaline crystal (abst.), *Geological Society of America. Southeastern section, 22nd Annual Meeting. Abstract* **5**(5):426.

Polyak, A. M., and Devyatovskaya, L. I., 1957, Preparation of boron fertilizers from different boron ores, *Trudy Ural'skogo Nauchno-Issledovatel'skogo Khimicheskogo Instituta* **4:**50-63. (Russian) *c*

Pomârleanu, V., and Movileanu, A., 1971, Geothermometry of some pegmatites in Romania, *Society of Mining Geology, Special Issue 2 (Proceedings of Joint Symposium, IMA-IAGOD, 1970),* pp. 171-177.

Popov, V. S., 1964, Authigenic tourmaline in halogen deposits, *Litologiya, Poleznye Iskopaemneye,* pp. 158-160. (Russian) *c*

Popov, V. S., and Sadykov, T. S., 1962, Authigenic tourmaline from the Khodzha-Mumyn rock salt deposit, *Doklady Academiya Nauk SSSR* **145:**1121-1122. (Russian)

Potuzah, V., 1965, Úprava turmalinových surovin, *Sbornik Geologickych Ved, Technologie, Geochemie* **5:**97. *c*

Pough, F. H., and Rogers, T. H., 1947, Experiments in X-ray irradiation of gem stones, *American Mineralogist* **32:**31-43.

Povarennykh, A. S., 1956, On the compressibility and thermal expansion of minerals (transl.), in M. H. Battey and S. I. Tomkeieff, eds., 1964, *Aspects of Theoretical Mineralogy in the U.S.S.R.,* Macmillan, New York, pp. 421-434.

Povarennykh, A. S., 1972, *Crystal chemical classification of minerals* (vol. I), translated by J. E. S. Bradley, Plenum Press, New York.

Povarennykh, A. S., and Lebedeva, A. D., 1970, Hardness of some rare minerals determined by microtesting (part I), *Konstitutsiya i Svoistva Mineralov* **4:**121-128. (Russian)

*Povondra, P., 1981, The crystal chemistry of tourmalines of the schorl-dravite series, *Acta Universitatis Carolinae Geologica,* no. 3, pp. 223-264.

Power, G. M., 1966*a,* Secondary tourmaline from the granitic rocks of S.W. England, *Ussher Society Proceedings* **1:**256-257.

Power, G. M., 1966*b,* Strontium:calcium ratio in tourmalines from south-west England, *Nature* **211:**1072-1073.

Power, G. M., 1968, Chemical variation in tourmalines from South west England, *Mineralogical Magazine* **36:**1078-1089.

Poynting, J. H., and Thomson, J. J., 1914, Pyroelectricity and Piezoelectricity, in *A Textbook of Physics,* Charles Griffin, London. *ns*

*Prendel, R., 1892, Analyse des Turmalin von der Urulga, *Zeitschrift für Kristallographie und Mineralogie* **20:**93.

Prescott, B. E., and Nassau, K., 1978, Black elbaite from Carrego do Urucum, Minas Gerais, Brazil, *Mineralogical Magazine* **42:**357-359.

Pretorius, R., Odendaal, F., and Peisach, M., 1972, Deuteron activation analysis of geological samples, *Journal of Radioanalytical Chemistry* **12:**139-149.

Proctor, K., 1984, Gem pegmatites of Minas Gerais, Brazil: Exploration, occurrence, and aquamarine deposits, *Gems & Gemmology* **20:**78-100. [Parts II and III, dealing with Tourmaline, are "in press."]

Przibram, K., 1929, Über Piezochromie (Farbänderung durch Druck) bei natürlichen Mineralien, *Sitzungsberichte der Akademie der Wissenschaften (Wein) Mathematisch-Naturwissenschaftliche Klasse Abteilung IIa* **138:**263-269.

Pulou, R., 1947, Contribution a l'étude de l'anisotropie diélectrique des cristaux, *Annales de la Faculté des Sciences de l'Université de Toulouse pour les Sciences, Mathematiques et les Sciences Physiques* **11:**1-73.

Quensel, P., 1937, Minerals of the Varuträsk pegmatite: IV. On the occurrence of cookeite, *Geologiska Föreningens i Stockholm Förhandlingar* **59:**262-269.

Quensel, P., 1957, The paragenesis of the Varuträsk pegmatite including a review of the mineral assemblage, *Arkiv för Mineralogi och Geologi* **2**(no. 2):9-125.

*Quensel, P. and Gabrielson, O., 1939, Minerals of the Varuträsk pegmatite, XIV: The tourmaline group. *Gelogiska Föreningens i Stockholm Förhandlingar* **61:**63-90.

Radkevich, E. A., Kokorin, A. M., Korostelev, P. G., Asmanov, V. Ya., Bakulin, Yu. I., Gonevchuk, V. G., and Gonevchuk, G. A., et al., 1972, (pub. 1976), Zonality of the mineralization of the Komsomol'sk District, *Prognozirovanie Skrytogo Orudeneniya na Osnove Zonal'nosti Gidrotermal'nykh Mestorozhdenii (Doklady Vsesoyuznogo Soveshcheniya)* 2nd, Moscow, pp. 112-121. (Russian) *c*

Rajulu, B. V. G., and Nagaraja, H. R., 1969, Authigenic tourmalines from the lower Kaladgi sandstones, Jamkhandi, Mysore State, *Journal of Sedimentary Petrology* **39:**391-394.

Ramaswamy, S., and Iyengar, K. Y. S., 1937, X-ray analysis of the structure of a fibrous modification of tourmaline, *Indian Academy of Science Proceedings* **5A:**419-422. *c*

Ramiskou, T., 1969, *Solstenen, Primitiv Navigation: Norden för Kompasset,* Rhodos, Kobenhavn, Denmark.

*Rammelsberg, C. F., 1850, *Annalen der Physik und Chemie* (J. C. Poggendorff, ed.), Leipzig, **80:**449; **81:**1. *ns*

*Rammelsberg, C. F., 1890, Die chemische Natur der Turmaline, *Neues Jahrbuch für Mineralogie, Geologie, und Palaeontologie* **2:**149-162.

Ramsay, W., 1886, Om turmalinens hänförande till den romboëdrisk-tetartoëdriska formgruppen af det hexagonala systemet, *Bihang, Kongligesvenska Vetenskaps Akademien, Handlingar* **12** (Afd. II), 3-11.

Rao, D. A. A. S. N., 1949, Dielectric constants of crystals, III, *Indian Academy of Science Proceedings* **30A:**82-86.

Rao, D. A. A. S. N., 1950, Dielectric constants and elastic moduli of uniaxial crystals, *Current Science* (India), **19:**116.

Rastoin, J., 1966, Les matériaux de protection neutronique, *Bulletin de la Société Français de Ceramique* **70:**19-27.

*Rath, R., and Puchelt, H., 1957, Indigolith von Usakos, *Neues Jahrbuch für Mineralogie, Monatshefte* **42:**206-208.

*Rath, R., and Puchelt, H., 1959, Dravit von Gouverneur, *Neues Jahrbuch für Mineralogie, Monatshefte* **44:**22-24.

Reimann, G., 1907, Beiträge zur Kenntnis des Turmalins aus Brasilien, *Neues Jahrbuch für Mineralogie* **23:**91-162. *ns*

*Reiner, P., 1913, *Beiträge zur Kentniss der Turmalingruppe,* inaugural dissertation, Heidelberg. *ns*

Reinkober, O., 1911, Über Absorption und Reflexion ultraroter Strahlen durch Quarz, Turmalin und Diamant, *Annalen der Physik* (Leipzig) **34:**343-372.

Rheineck, H., 1899, Formulirung des Turmalins, *Zeitschrift für Kristallographie und Mineralogie* **31:**385-386.

Rickard, M. J., 1964, Metamorphic tourmaline overgrowths in the Oak Hill Series of southern Quebec, *Canadian Mineralogist* **8:**86-91.

Ricketts, B. D., 1978, Authigenic tourmaline from the middle Precambrian Belcher Group, Northwest Territories, Canada, *Canadian Petroleum Geology Bulletin* **26:**543-550.

Riecke, E., 1913/1914, De la pyroélectricité et de la piezoélectricité, *Bibliotheque Universelle Archives des Sciences Physiques et Naturelles* **36:**101-112, 216-238, 305-326.

Riecke, E., and Voigt, W., 1892, Die piëzoelectrischen Constanten des Quarzes und Turmalines, *Wiedemann's Annalen der Physik und Chemie* (Leipzig) **45:**523-552.

Riess, P., and Rose, G., 1843, Über die Pyroelektricität der Mineralien, *Abhandlungen der Akademie der Wissenschaften* (Berlin) **59.** *ns*

*Riggs, R. B., 1888, The analysis and composition of tourmaline, *American Journal of Science,* 3rd ser., **35:**35-51.

Rinmann, S., 1766, Tourmalinen eller Aske-Blåsare Stenen, *Kungliga Vetenskaps Academiens, Handlingar* **27:**45-. *ns*

Rinne, F., 1927, Notiz zur Veröffentlichung von P. Stamm über die Lichtabsorption und die interferenz der Röntgenstrahlung beim Turmalin, *Centralblatt für Mineralogie, Geologie und Palaeontologie,* A, pp. 217-218.

Robbins, C. R., and Yoder, H. S., Jr., 1962, Stability relations of dravite, a tourmaline, Annual Report of the Director of the Geophysical Laboratory 1961-1962 in the *Carnegie Institute of Washington Year Book* **61:**106-107.

Robbins, C. R., Yoder, H. S., Jr., and Schairer, J. F., 1959, Tourmaline, Annual Report of the Director of the Geophysical Laboratory 1958-1959 in the *Carnegie Institute of Washington Year Book* **58:**137-138.

Robert, D., 1968, Triage électrostatique en lit fluidisé de minéraux conditionnés par des tensio-actifs cationiques, *Revue de l'Industrie Minerale* **50:**363-374.

Romé de l'Isle, J. B. L., 1772, *Essai de Cristallographie ou Description des Figures Geometriques,* Didot, Paris.

Romé de l'Isle, J. B. L., 1783, *Cristallographie, ou Description des Formes Propres à Tous les Corps du Regne Mineral* (2nd ed., 4 vol.), L'Imprimerie de Monsieur, Paris. *ns*

Röntgen, W. C., 1914, Pyro- and Piezoelectric investigations, *Annalen der Physik* (Leipzig) **45:**737-. *ns*

Ropert, M. E., Geoffroy, J., and Fitte, P., 1969, Les tourmalines lithiques de Chedeville-en-Ambozac (Haute-Vienne), *Bulletin de la Société Française de Minéralogie et Cristallographie* **92:**235-236.

Rose, G., 1836, *Übersicht der Mineralien und Gebirgsarten des Ural,* Berlin. *ns*

Rose, G., 1838, *Über den Zusammenhang zwischen der Form und der elektrischen Polarität der Krystalle: Erste Abhandlung-Turmalin,* Königlichen Akademie der Wissenschaften (Berlin).

Rose, H., 1947, Spurenelemente in Gesteine des Harzes und des Sächsischen Erzgebirges, *Fortschritte der Mineralogie* **26:**108-115 (publ. 1950).

Rosenberg, P. E., and Foit, F. F., Jr., 1979, Synthesis and characterization of alkali-free tourmaline, *American Mineralogist* **64:**180-186.

Rossman, G. R., 1982, Origin of color in pegmatite minerals, in The mineralogy of pegmatites, G. E. Brown, Jr., ed., *American Mineralogist* **67:**189.

Roth, J., 1879, *Allgemeine und Chemishe Geologie,* Hertz, Berlin.

Rowley, E. B., 1942, Huge tourmaline crystals discovered, *Mineralogist* **10:**47-48, 63-64.

*Rozanov, K. I., and Lavrinenko, L. F., 1979, *Rare metal pegmatites of the Ukraine,* Akademiya Nauk SSSR (Moscow). (Russian)

Rozanov, K. I., Lavrinenko, L. F., Cherepivskaya, G. E., and Matrosova, T. I., 1979, Variations in the composition of pegmatites occurring in rocks of basic composition, *Redkometal'nye Pegmatity Ukrainy* **27:**113-120. (Russian)

Rozhkova, E. V., and Proskurovskii, L. V., 1957, Dielectric permeability determination on minerals and their dielectric separation, *Sovremennye Metody Mineralogicheskogo Issledovaniya Gornykh Porod. Rud i Mineralov,* pp. 115-138. (Russian)

Rub, A. K., 1973, Silicates. Typomorphism of topaz and tourmaline, characteristic accessary minerals of tantalum and tin ore mineralizations (as illustrated by a region in the eastern U.S.S.R.), *Tipomorphism Mineralov i Ego Prakticheskoe Znachenie,* pp. 178-185. (Russian)

Rudakova, Zh. N., 1972, Quantitative relations of some elements in the cassiterite composition, *Zapiski Vsesoyuznogo Mineralogicheskogo Obshchestvo* **101:**317-322. (Russian) *c*

*Rumantseva, E. V., 1983, Chromdravite, a new mineral, *Zapiski Vsesoyuznogo Mineralogicheskogo Obshchestvo* **112:**222-226. (Russian)

Ruskin, J., 1891, *The ethics of dust: ten lectures to little housewives on the elements of crystallization,* C. E. Merrill, New York.

Sadanaga, R., 1947, Crystal structure of tourmaline (abst.), *Journal of the Geological Society (Tokyo)* **53:**52-53. (Japanese)

Saegusa, N., Price, D. C., and Smith, G., 1979, Analysis of the Mössbauer spectra of several iron-rich tourmalines (schorls), *Journal de Physique,* Colloque C2 (suppl. au no. 3) **40:**C456-C459.

Sahama, T. G., 1945, Spurenelemente der Gesteine im Südlichen Finnisch-Lappland, *Bulletin de la Commission Géologique de Finlande* **135:**54-68.

*Sahama, T. G., Knorring, O. V., and Tornröos, R., 1979, On tourmaline, *Lithos* **12:**109-114.

*Sandréa, M. A., 1949, Sur une variété de tourmaline sodo-manganesifère dans les filons de pegmatite des environs de Roscoff, *Comptes Rendus de l'Académie des Sciences* (Paris) **228:**1142-1143.

*Sargent, G. W., 1901, Die quantitative Bestimmung der Borsäure in Turmaline, *Zeitschrift für Kristallographie und Mineralogie* **34:**205.

Sato, Y., 1969, Geological significance of zircon-garnet-tourmaline ratio of the Paleogene sandstones in northwestern Kyushu, Japan, *Chishitsu Chosajo Hokoku* **235:**1-45.

*Schaller, W. T., 1913, Beitrag zur Kenntnis der Turmalingruppe, *Zeitschrift für Kristallographie, Kristallgeometrie, Kristallphysik, Kristallchemie* **51:**321-343.

Schaller, W. T., 1914, Über "feste Lösungen" in Turmalin, *Zeitschrift für Krystallographie und Mineralogie* **53:**181.

*Scharizer, R., 1889, Ueber die chemische Constitution und über die Farbe de Turmaline von Schuttenhofen, *Zeitschrift für Krystallographie, Kristaligeometrie, Kristallphysik, Kristallchemie* **15:**337-365.

Schedtler, H., 1886, Experimentelle Untersuchungen über das elektrische Verhalten des Turmalins, *Neues Jahrbuch für Mineralogie, Geologie, Palaeontologie* **4:**519-575.

Scheidhauer, W., 1940, Gravitative Auslesevorgange bei der Sedimentation von Sanden: Korngrösen und Schwermineraluntersuchungen im Turon des Elbsandsteingebirges, *Chemie Erde* (Linck) **12:**466-507. *ns*

Scherillo, A., 1941, Fragments with boron-bearing minerals in the Ciminie volcanics, *Periodico di mineralogia* **3:**367-393. *c*

Schiffmann, C., 1972, Tourmalines. Rare multiple indices on the refractometer, *Journal of Gemmology* **13:**125-132.

Schiffmann, C., 1975, Tourmalines. Multiple indices on the refractometer; a further note, *Journal of Gemmology* **14:**324-329.

Schlossmacher, K., 1919, Beitrage zur Kenntnis der Turmalingruppe, *Neues Jahrbuch für Mineralogie, Geologie und Palaeontologie,* pp. 106-121. *ns*

Schmetzer, K., 1978, Vanadium III als Farbträger bei naturlichen Silicaten und Oxiden-ein Beitrag zur Kristallchemie des Vanadiums, Doctoral dissertation, Ruprecht-Karl-Universitat, Heidelberg. *ns*

Schmetzer, K., and Bank, H., 1979, East African tourmalines and their nomenclature, *Journal of Gemmology* **16:**310-311.

*Schmetzer, K., Nuber, B., and Abraham, K., 1979, Zur Kristallchemie Magnesiumreicher Turmaline, *Neues Jahrbuch für Mineralogie, Abhandlungen* **136:**93-112.

Schmetzer, K., Medenbach, O., Bank, H., and Krupp, H., 1977, Kristalle mit aussergewöhnlichen Einschlüssen-turmalin aus Tansania und Beryllaus Brasilien, *Zeitschrift der Deutschen Gemmologischen Gesellschaft* **26:**145-147.

Schmidt, J. G., 1707, *Curiöse Speculasiones bey schlaf-losen Nächten,* Leipzig and Chemnitz. (This item was published under the pen-name Immer Gern Speculirt.)

Schmidt, W., 1902, Dielectricitätsconstanten von Krystallen mit electrischen Wellen, *Annalen der Physik* (Leipzig) **9:**919. *ns*

Schmidt, W., 1903, Dielectricitätsconstanten von Krystallen mit electrischen Wellen, *Annalen der Physik* (Leipzig) **11:**114. *ns*

Schoep, A., 1941, Geodes en segregaties van epidoet, toermalijn, axiniet enz. in het porphyriet van Quenast, *Vlaamse Academie voor Wetenschappen Letteren en Schone Kunsten van België, Klase der Wetenshappen, Mededelingen (Brussels)* **3:**3-19.

Schrauf, A., 1861, Erklärung des Vorkommens optisch zweiaxiger Substanzen im rhomboëdrischen System-Ein Beitrag zur Krystallphysik, *Annalen der Physik* (Leipzig) **144:**221-237.

Schreyer, W., Abraham, K., and Behr, H. J., 1975, Sapphirine and associated minerals from the kornerupine rock of Waldheim, Saxony, *Neues Jahrbuch für Mineralogie, Abhandlungen* **126:**1-27.

Scorzelli, R. B., Baggio-Saitovitch, E., and Danon, J., 1976, Mössbauer spectra and electron exchange in tourmaline and staurolite, *Journal de Physique* (Paris), Colloque C6, **6:**801-805.

Seager, A., 1952, The surface structure of crystals, *Mineralogical Magazine* **30:**1-25.

*Serdyuchenko, D. P., 1956, The minerals of boron and titanium in certain sedimentary-metamorphic rocks, *Trudy, Instituta Geologicheskogo, [Dagestanskii Filial] Akademiya Nauk SSSR* **5:**53-124. (Russian)

Serdyuchenko, D. P., 1960, Boric sedimentary-metamorphic formations, *Voprosy sedimentologii, Natsional'nyi Komitet Geolgov Sovetskogo Soyuza, Doklady Sovetskikh Geolgov k 6-mu Shestomu Mezhdunarodnomu Kongressu po Sedimentologii,* Copenhagen, pp. 132-140. (Russian) *c*

Serdyuchenko, D. P., 1978, Chemical constitution of tourmaline, *Problemy Geologii Redkikh Elementov, Izdanija Nauka,* Moscow, pp. 225-251. (Russian)

Serdyuchenko, D. P., 1980, Different position of boron in the structure of tourmalines, *Doklady Akademiya Nauk SSSR* **254:**1450-1453. (Russian) *c*

Sergeev, V. N., 1963, Structural etching of oxide and silicate minerals by hydrofluoric acid, *Zapiski Vsesoyuznogo Mineralogicheskogo Obshchestvo* **92:**66-73. (Russian)

Shabynin, L. I., 1974, Chemical composition, optical properties, and parageneses of tourmalines from magnesia-skarn deposits, *Geologiya i Geofizika* **4:**34-43. (Russian) *c*

Shamos, M. H., and Levine, L. S., 1967, Piezoelectricity as a fundamental property of biological tissues, *Nature* **213:**267-269.

Shaposhnikov, G. N., 1959, The case-like form of tourmaline crystals, *Zapiski Vsesoyuznogo Mineralogicheskogo Obschestvo* **88:**336-338. (Russian) *m*

*Shcherbakov, I. B., 1961, Tourmaline from the Khoshchevato village, *Mineralogicheskii Sbornik L'vovskii Gosudarstvennyi Universitet. L'vov. USSR* **15:**338-343. (Russian) *c*

Shcherbakov, Yu. G., and Perezhogin, G. A., 1964, Gold geochemistry, *Geokhimiya* **6:**518-528. (Russian)

*Shenderova, A. G., 1955, Chromian dravite from the Krivoy Rog, *Mineralogicheskii Sbornik, L'vovskii Gosudarstvennyi Universitet, L'vov, USSR* **9:**324-326. (Russian)

Shepard, C. U., 1830, Mineralogical Journey in the northern parts of New England, *American Journal of Science* **18:**289-303.

Shima, M., 1963, Boron isotope ratio of ten Japanese boron minerals, *Rikagaku Kenkyusho Hokoku* **39:**207-210. (Japanese)

Shipley, R., 1971, *Dictionary of Gems and Gemology,* Gemological Institute of America, Los Angeles.

Shiryaeva, V. A., and Shmakin, B. M., 1969, Composition of tourmalines from the eastern Siberian muscovite pegmatites, *Zapiski Vsesoyuznogo Mineralogicheskogo Obshchestvo* **98:**166-174. (Russian)

*Shiryaeva, V. A., Novikov, V. M., and Grigoryeva, V. A., 1973, A rational analysis scheme for tourmaline by means of differential spectrometry and atomic absorption, *Akademii Nauk SSSR, Sibirskoe Otdelenie Institut Geokhimii, Ezhegodnik,* pp. 471-475. (Russian)

Shmakin, B. M., 1972, Pegmatites: Typomorphic characteristics of generations of minerals in muscovite pegmatites, *Tipmorfizm Mineralov i Ego Prakticheskoe Znachenie* pp. 189-194. *c*

Shmakin, B. M., 1982, Some famous pegmatite localities in the U.S.S.R., in The mineralogy of pegmatites, G. E. Brown, Jr., ed., *American Mineralogist* **67:**181.

Shumaro, O. O., 1971, Kimberlite minerals in the Dolginskian suite deposits of the junction zone of the Donets-Basin with the Azov Sea area, *Dopovidi Akademii Nauk Ukrains'koi R.S.R. Kiev.* ser. B **33:**534-536. (Russian) *c*

Sigamony, A., 1944, Magnetic properties of tourmaline and epidote, *Indian Academy of Science Proceedings* **20A:**200-203.

Sigrist, A., and Balzer, R., 1977, Untersuchungen zur Bildung von Tracks in Kristallen, *Helvetica Physica Acta* **50:**49-64.

Sillitoe, R. H., and Sawkins, F. J., 1971, Geologic, mineralogic, and fluid inclusion studies relating to the origin of copper-bearing tourmaline breccia pipes, Chile, *Economic Geology* **66:**1028-1041.

Sillitoe, R. H., Halls, C., and Grant, J. N., 1975, Porphyry tin deposits in Bolivia, *Economic Geology* **70:**913-927.

Silverman, B. D., 1968, Ultrasonic attenuation in imperfect insulating crystals, *Progress of Theoretical Physics* **39:**245-269.

Sil'vestrova, I. M., and Sil'vestrov, Yu. N., 1958, Apparatus for measurement of pyroelectric polarization of crystals, *Kristallografiya* **3:**57-63.

*Simboli, G., 1959, A tourmalinite of Alpe Valliselle (group of Cima d'Asta), *Acta Geologica Alpina,* no. 7, pp. 177-189. (Italian) *c*

Simon, K., 1908, Beiträge zur Kenntniss der Mineralfarben, *Neues Jahrbuch für Mineralogie. Beil* **26:**249-295.

*Simpson, E. S., 1931, Contributions to the mineralogy of Western Australia-Series VI, *Journal of the Royal Society of Western Australia* **17:**137-148.

Sinani, I. B., 1957, Tourmaline indicators for shock waves in liquids, *Pribory i Tekhnika Eksperimenta* **14:**85-89. (Russian)

Sinkankas, J., 1959, *Gemstones of North America* (vol. I), Van Nostrand, New York.

Sinkankas, J., 1971, Tourmaline: the electric gem of many colors and curious properties, *Rock & Gem* **1**(2):38-45.

Sinkankas, J., 1976, *Gemstones of North America* (vol. II), Van Nostrand Reinhold, New York.

*Sirotin, V. I., 1966, Tourmaline in the laterite weathering profile, *Trudy Soveshchaniya po Probleme Izucheniya Voronezh Anteklizy, 3rd, Voronezhskogo Gosudarstvennogo Universiteta,* pp. 223-225. (Russian) *c*

Takano, Y., and Takano, K., 1959, The relation between optical anomalies and crystal structure, *Journal of the Geological Society of Japan* **65:**236-247. (Japanese) *c*

*Takeshi, H., 1953, White tourmaline from Kanakura Mine, Nagana Prefecture, Japan, *Nippon Kobutsu Gakkai (Tokyo)* **1:**105-112. (Japanese)

Tarnovskii, G. N., 1960, Cookeite from East-Siberian pegmatites, *Zapiski Vostochno-Sibirskogo Otdeleniya Vsesoyuznogo Mineralogicheskogo Obschestvo* 1960, no. 2, pp. 3-23. (Russian) *c*

Tarnovskii, G. N., 1961, Crystallographic study of tourmalines from pegmatites of Siberia, *Akademiia Nauk SSSR Mineralogicheskii Muzei* (Moscow) **12:**123-144. (Russian)

Taylor, A. M., and Terrell, B. C., 1967, Synthetic tourmalines containing elements of the first transition series, *Journal of Crystal Growth* **1:**238-244.

Taylor, B. E., and Slack, J. F., 1984, Tourmalines from Appalachian-Caledonian massive sulfide deposits: Textural, chemical, and isotopic relationships, *Economic Geology* **79:**1703-1726.

Taylor, B. E., Foord, E. E., and Friedrichsen, H., 1979, Stable isotope and fluid inclusion studies of gem-bearing granite pegmatite-aplite dikes, San Diego Co., California, *Contributions to Mineralogy and Petrology* **68:**187-205.

Taylor, E. W., 1949, Correlation of the Moh's scale of hardness with the Vickers hardness numbers, *Mineralogical Magazine* **28:**718-721.

Termier, H., Owodenko, B., and Agard, J., 1950, Les gites d'étain et de tungstène de la rêgion d'Oulmes (Maroc central), *Notes et Memoires du Service Géologique du Maroc* **82:**223, 249. *ns*

Termier, P., 1907, Large tourmaline crystals from Ankaratra, *Bulletin de la Société Française de Minéralogie* **31:**138-142. (French) *c*

Tertsch, H., 1917, Trachtstudien an einem geschichteten Turmalinkristall, *Centralblatt für Mineralogie Geologie und Palaeontologie* **17:**273-289.

Teschemacher, J. E., 1842, On the occurrence of phosphate of uranium in the tourmaline locality at Chesterfield, *Boston Journal of Natural History* **4:**35-37. *ns*

Thakur, V. C., 1972, Computation of the values of the finite strains in the Molare Region, Ticino, Switzerland, using stretched tourmaline crystals, *Geological Magazine* **109:**445-449.

Theophrastus, 315 B.C., *περι λιθωυ* (On Stones). (See, for example, 1774 publication "Theophrastus' history of stones," John Hill, London; or 1956 publication E. R. Caley and J. F. C. Richards, "On Stones," Ohio State University Press, Columbus.)

Thiel, G. A., 1941, The relative resistance to abrasion of mineral grains of sand size, *Journal of Sedimentary Petrology* **10:**103-124.

Thompson, S. P., 1881, *Elementary lessons in Electricity and Magnetism,* MacMillan, London.

Thomson, T., 1836, *Outlines of Mineralogy, Geology, and Mineral Analysis* (vol. 1), London.

Thuraisingham, M. S., and Stephens, R. W. B., 1973, Effects of Gamma-Iradiation on the Ultrasonic attenuation in quartz and tourmaline at low temperatures, *International Conference on Internal Friction and Ultrasonic Attenuation in Crystalline Solids, 5th Conference* (Aachen—1973), pp. 308-313.

Tikhomirova, N. Ya., 1957, The regeneration of tourmaline in lower Cambrian

*Sjörgen, H., 1916, The chemical composition of tourmaline from Utö, *Mineralogisk-Geologiska Institut Upsala Universitet Bulletin* **15:**317-324.

Slack, J. F., 1980, Tourmaline—a prospecting guide for massive base-metal sulfide deposits in the Penobscot Bay Area, Maine, *Maine Geological Survey, Special Economic Studies Series,* no. 8.

Slack, J. F., 1982, Tourmaline in Appalachian-Caledonian massive sulphide deposits and its exploration significance, *Institute of Mining and Metallurgy. Transactions. Sect. B* **91:**B81-B90.

Slack, J. F., and Plimer, I. R., 1983, Associations of tourmaline-rich rocks with stratiform sulfide deposits, *Geological Association of Canada (Abstracts)* **8:**A62.

Slack, J. F., Herriman, N., Barnes, R. G., and Plimer, I. R., 1984, Stratiform tourmalinites in metamorphic terranes and their geologic significance, Manuscript.

Slawson, C. B., 1936, High-iron tourmaline from the Marquette Iron Range (abst.), *American Mineralogist* **21:**195.

Slawson, C. B., and Thibault, N. W., 1939, Quantitative measurement of dichroism in tourmaline, *American Mineralogist* **24:**492-498.

Slivko, M. M., 1952, Statistical study of relation of color to form frequency, *Mineralogicheskii Sbornik, L'vovskii Gosudarstvennyi Universitet, L'vov, U.S.S.R.* **6:**211-220. (Russian)

*Slivko, M. M., 1955*a,* Investigation of tourmalines of some S.S.S.R. deposits, *L'vovskoe Geologicheskoe Obshchestvo, Izdatel'stvo L'vovskogo Universiteta.* (Russian)

Slivko, M. M., 1955*b,* Tourmalinization in pegmatites, *Mineralogicheskii Sbornik, L'vovskii Gosudarstvennyi Universitet im I Franko* **9:**181-215. (Russian) *c*

*Slivko, M. M., 1957, Color changes in tourmaline and their geochemical meaning, *Mineralogicheskii Sbornik, L'vovskii Gosodarstvennyi Universitet im I. Franko* **11:**81-88. (Russian)

*Slivko, M. M., 1959*a,* On manganese tourmalines, *Mineralogicheskii Sbornik, L'vovskogo Geologicheskogo Obshchestvo* **13:**139-148. English translation appears in *International Geological Review* **3:**195-201 (1961).

Slivko, M. M., 1959*b,* Rare and dispersed elements in tourmalines, *Problemy Geokhimii, L'vovskii Gosudarstvennyi Universitet im I Franko,* **1:**261-271. (Russian)

Slivko, M. M., 1962*a,* Characteristic features of the chemical composition of tourmalines from the schorlite-dravite series, *L'vovs'kogo Derzhavnogo Universitetu. Vestnik. Seriya Geologichna,* no. 1, pp. 134-138. (Russian)

Slivko, M. M., 1962*b,* Chemical composition and isomorphic replacement in tourmalines, *Mineralogicheskii Sbornik, L'vovskii Gosudarstvennyi Universitet* **16:**113-129. (Russian)

Slivko, M. M., 1963, Crystal morphology of tourmalines from pegmatites, *Mineralogicheskii Sbornik, L'vovskii Gosudarstvennyi Universitet* **17:**45-54. (Russian)

Slivko, M. M., 1965, On inclusions of solutions in crystals of tourmaline in Yermakov, N. P. and others, *Research on the Nature of Mineral-Forming Solutions,* Pergamon, Oxford, pp. 626-634.

Slivko, M. M., 1966*a,* Sculpture on the faces of tourmaline crystals, *Mineralogicheskii Sbornik, L'vovskii Gosudarstvennyi Universitet* **20:**161-165. (Russian)

Slivko, M. M., 1966*b,* Free growth and generation of tourmaline crystals, *Materialy Nauchnoi Konferentsii Vypusk Geologicheskoi Fakul'teta L'vov Universiteta,* 1945-1965, pp. 91-92. (Russian)

Slivko, M. M., 1966*c*, Repulsion and capture of solid particles by growing tourmaline crystals, *Akademii Nauk SSSR Genezis Mineral'nykh Individov i Agregatov,* pp. 116-121. (Russian)

Slivko, M. M., 1968, The inclusions of solutions in schorlite crystals of Volhyn pegmatites, *Mineralogicheskii Sbornik, L'vovskogo Gosudarstvennogo Universitet im I. Franko* **22:**235-243. (Russian)

Slivko, M. M., 1969*a,* Fluid carbon dioxide-water inclusions in tourmalines from pegmatites of the Mamsk region pegmatites, *Mineralogicheskii Sbornik, L'vovskii Gosudarstvennyi Universitet* **23**(pt. 3):295-301. (Russian)

Slivko, M. M., 1969*b,* Thermodynamic parameters of tourmalinization in the Korets pegmatites, *Dopovidi Akademii Nauk Ukrains'koi* ser. B **31:**218-221. (Russian)

Slivko, M. M., 1973, Application of mathematical statistics to the study of tourmaline composition, *Matematicheskie Metody Geologii Tezisy Materialov Konferentsii* (1971) U.S.S.R., pp. 97-98. (Russian) *c*

Slivko, M. M., 1974, Comparative physicochemical characteristics of solutions from inclusions in tourmalines from mica versus rare metal pegmatites, *Mineralogicheskii Sbornik, L'vovskii Gosudarstvennyi Universitet* **28:**40-46. (Russian)

Slivko, M. M., and Iorysh, L. N., 1964, Relation of crystal lattice parameters of tourmalines to their composition, *Mineralogicheskii Sbornik, L'vovskii Gosudarstvennyi Universitet* **18:**433-437. (Russian)

Slivko, M. M., and Voskrensenskaya, I. E., 1966, The growth of tourmaline crystals, *Mineralogicheskii Sbornik, L'vovskii Gosudarstvennyi Universitet* **20:**407-411. (Russian)

*Slivko, M. M., Boriskov, F. F., and Kornilov, V. F., 1972, Tourmaline-asbestos: *Mineralogicheskii Sbornik, L'vovskii Gosudarstvennyi Universitet* **26:**209-212. (Russian)

Smith, D. C., 1971, A tourmaline-bearing eclogite from Sunnmore, Norway, *Norsk Geologisk Tidsskrift* **31:**141-147.

Smith, F. G., 1949, Tourmaline, (pt.) 4 of Transport and deposition of the non-sulphide vein minerals, *Economic Geology* **44:**186-192.

Smith, G., 1977, Low-temperature optical studies of metal-metal charge-transfer transitions in various minerals, *Canadian Mineralogist* **15:**500-507.

Smith, G., 1978*a,* Evidence for absorption by exchange-coupled $Fe^{2+} - Fe^{3+}$ pairs in the near infra-red spectra of minerals, *Physics and Chemistry of Minerals* **3:**375-383.

Smith, G., 1978*b,* A reassessment of the role of iron in the 5,000-30,000 cm^{-1} region of the electronic absorption spectra of tourmaline, *Physics and Chemistry of Minerals* **3:**343-373. *c*

Smith, G., and Strens, R. G. J., 1976, Intervalence-transfer absorption in some silicate, oxide and phosphate minerals, in *Physics and Chemistry of Minerals,* R. G. J. Strens, ed., Wiley, New York, pp. 583-612.

Smith, H. H., 1952, Visible tourmaline in Norwegian pyrite, *Bulletin of the Institute of Mining and Metallurgy,* no. 553, pp. 109-110. *c*

Smith, W. C., 1923, A description of the peridotites and associated rocks from St. John's Island in the Red Sea, in *Preliminary Geological Report on St. John's Island,* F. W. Moon, ed., Geological Survey Egypt 1923, appendix I, pp. 17-26. *c*

Snetsinger, K., 1966, Barium-vanadium muscovite and vanadium tourmaline from Mariposa County, California, *American Mineralogist* **51:**1623-1639.

Sobolev, V. S., Dolgov, Yu. A., Bazarov, I. T., Bakumenko, I. T., and Shcherbakov, Z. V., 1964, High-temperature xenoliths in pegmatite and granite rocks, *Doklady Akademiya Nauk SSSR* **157**:349-352. (Russian) *c*

*Sokolov, V. A., and Cheverdin, V. A., 1958, Distribution of boron silicates in a skarn deposit, *Uchenye Zapiski Kazakhskii Gosudarstvennyi Universitet* **37**:98-103. (Russian)

Solly, R. H., 1884, On the tetartohedral development of a crystal of tourmaline, *Mineralogical Magazine and Journal of the Mineralogical Society* **28**:80-82.

Spencer, E., 1925, Albite and other authigenic minerals in limestone from Bengal, *Mineralogical Magazine* **20**:371-381.

Spencer, L. J., 1907, Notes on some Bolivian minerals, *Mineralogical Magazine* **14**:308-344.

Spencer, L. J., 1934, Thirteenth list of new mineral names, *Mineralogical Magazine* **23**:624-640.

Spottswood, A., 1911, Occurrence and uses of American tourmalines, *Mining and Engineering World* **35**:1280-1281.

Staatz, M. H., Murata, K. J., and Glass, J. J., 1955, Variation of composition and physical properties of tourmaline with its position in the pegmatite, *American Mineralogist* **40**:789-804.

Stadler, H. A., 1967, Blue quartz from Calanda near Tamins, *Urner Mineralien-Freund* **5**:1-6. (German) *c*

Stamm, P., 1926, Die Absorption des sichtbaren und ultravioletten Lichtes und die Interferenz der Röntgenstrahlen beim Turmalin, *Neues Jahrbuch für Mineralogie* **54**:293-319.

Stella, A., and Tamburino, S., 1952, Radioactivity of the pegmatite of Delianova, *Rendiconti della Societa Mineralogica Italiana* **8**:185-188. (Italian) *c*

Stepnev, Yu. S., 1962, Gallium in granitic pegmatites of the Sayan Mountains, *Geokhimiya*, pp. 637-639. *c*

Sterrett, D. B., 1904, Tourmaline from San Diego County, California, *American Journal of Science* (4th ser.) **17**:459-465.

Stifter, A., 1853, *Bunte Steine und Erzählungen* (undated reprint), Winkler-Verlag, Munich.

Stoicovici, E., and von Gliszczynski, S., 1944, Druses in the granite of the Drăganului Valley Rumania, *Neues Jahrbuch für Mineralogie, Geologie und Palaeontologie, Abhandlungen* **79**:129-160. *c*

Stoicovici, E., Ghergaru, L., and Motiu, A., 1957, The beneficiation of tourmaline, *Buletinul Universitatilor "V. Babes" si "Bolyai" Cluj, Seria Stiintele Naturii* **1**:315-324. *c*

Stow, M. H., 1932, Authigenic tourmaline in the Oriskany sandstone, *American Mineralogist* **17**:150-152.

Strunz, H., 1970, *Mineralogische Tabellen*, Akademie Verlag, Geest & Portig K.-G., Leipzig.

Stuker, P., 1961, Feldergeteilter Turmalin in Pegmatit von Brissago, *Schweizerische Mineralogische und Petrographische Mitteilungen* **41**:489-492.

Switzer, G. S., 1974, Memorial of Martin L. Ehrmann, August 9, 1903-May 18, 1972, *American Mineralogist* **59**:414-415.

Sztrókay, K., 1944, Improvement of the color of gemstones, *Természettudomány Közlöny* **73**:455-459.

deposits of the Irkutsk amphitheater, *Materialy Vsesoyuznogo Nauchno-Issledovatel'skogo Geologicheskogo Instituta* **1956**(8):126-132. (Russian) *c*

Timokhina, I. V., 1976, Isomorphism and ways of valence compensation in tourmalines with a schorl composition, *Geokhimiya* **4:**537-544. (Russian) *c*

Tippe, A., and Hamilton, W. C., 1971, A neutron-diffraction study of the ferric tourmaline, buergerite, *American Mineralogist* **56:**101-113.

Titiz, S., 1977, Effects of organic compounds secreted from sugar beet roots on the solubility of tourmaline, *Bilim Kongresi Tarim Ormancilik Arastirma Grubu Tebligleri, Turkiye Bilimsel ve Teknik Arastirma Kurumu, 5th* (Istanbul) 1975, pp. 61-73. (Turkish)

Tokuda, T., and Ao, T., 1956, Abundance ratio of oxygen isotopes in siliceous rocks, *Yogyo Kyokai Shi* **64:**12-18. (Japanese, English summary)

Tomisaka, T., 1968, Syntheses of some end-members of the tourmaline group, *Mineralogical Journal* (Tokyo) **5:**355-364.

Tomlinson, H. W., 1945, Occurrence of borosilicates in diabase at Lambertville, New Jersey, *American Mineralogist* **30:**203-204.

Topkaya, M., 1950, Authigenic silicates in sedimentary rocks, *Bulletin des Laboratoires Geologie, Mineralogie, Geophysique de l'Université de Lausanne,* no. 97. *c*

Townsend, M. G., 1970, Dichroism of tourmaline, *Journal of Physics and Chemistry of Solids* **31:**2481-2488.

Traube, H., 1896, Ueber die Aetzfiguren einiger Minerale, *Neues Jahrbuch für Mineralogie, Geologie und Palaentologie Beilageband, Abteilung B* **10:**454-469.

Triche, C., Leckebush, R., and Recker, K., 1973, Quantitative Nachweise von Ti, V, Cr, Mn und Fe in Turmalinen mittels Spektro-chemischer Methoden, *Neues Jahrbuch für Mineralogie, Monatschefte* 1973, pp. 63-69.

Tsang, T., and Ghose, S., 1973, Nuclear magnetic resonance of ^{1}H, ^{7}Li, ^{11}B, ^{23}Na and ^{27}Al in tourmaline (elbaite), *American Mineralogist* **58:**224-229.

Tsang, T., and Thorpe, A. N., 1971, Magnetic susceptibility and triangular exchange coupling in tourmaline mineral group, *Journal of Physics and Chemistry of Solids* **32:**1441-1448.

Tschermak, G., 1885, *Lehrbuch der Mineralogie* (1st ed.), A. Hölder, Wien. *ns*

Tsukanov, V. A., and Esipchuk, K. E., 1971, Paragenetic associations and crystallization sequence of accessory minerals in diorites and tonalites of the western Azov area, *Geologichniy Zhurnal* **31:**122-127. (Russian) *c*

Tsvetkova-Goleva, V., 1964, Distribution of heavy minerals in the moraine sediments of the Rila Mountains, *Izvestiya na Geologicheskiya Institut, Belgarska Akademiya na Naukite,* no. 13, pp. 41-56. *c*

Ushio, M., and Sumiyoshi, Y., 1971, Synthesis of large industrial single crystals: 1. Hydrothermal synthesis of dravite-tourmaline, *Kogyo Kagaku Zasshi* **74:**2252-2259. (Japanese) *c*

Ussing, N. V., 1889, Untersuchungen der Mineralien von Fiskernäs in Grönland, *Zeitschrift für Kristallographie und Mineralogie* **15:**596-615. *ns*

Valeton, I., 1955, Veränderungen an Zirkon und Turmalin in Buntsandstein und Keuper, *Beiträge zur Mineralogie und Petrographie* **5:**100-104.

Van Horn, F. R., 1926, The occurrence of a large iron-tourmaline in Alabama pegmatite, *American Mineralogist* **10:**348-350.

Van Houten, F. B., 1971, Contact metamorphic mineral assemblages, Late Triassic Newark Group, New Jersey, *Contributions to Mineralogy and Petrology* **30:**1-14.

Vargas, J. I., and Tupinambá, G. A. C., 1962, Tourmaline as a new thermal neutron dosimeter, *Inter-American Symposium on the Peaceful Application of Nuclear Energy, 4th,* Mexico City, pp. 67-70.

Vedeneeva, N. E., and Grum-Grzhimailo, S. V., 1934, Spectro-pleochroism meter and investigation of mineral dichorism, *Comptes Rendus de l'Academie des Sciences de l'U.R.S.S.* **3:**583-585.

Vedeneeva, N. E., and Grum-Grzhimailo, S. V., 1948, Thermal discoloration of red tourmaline, *Trudy Instituta Kristallografii Akademiya Nauk SSSR* **4:**215-. *ns*

Vernadsky, W., 1913, Über die chemische Formel der Turmaline, *Zeitschrift für Kristallographie, Kristallogeometri Kristallphysik, Kristallchemie* **53:**273-288.

Vetrunov, L. N., 1955, Authigenic tourmaline in Cenozoic molasses of northern Fergana, *Doklady Akademiya Nauk Tadzhikskoi SSR* **14:**29-30. *c*

Vetrunov, L. N., 1962, Mineralogical composition of the Tertiary continental sediments in southeastern shores of the Issyk-Kul Lake, *Zapiski Kirgizskogo Otdeleniya Vsesoyuznogo Mineralogicheskogo Obshchestvo,* no. 3, pp. 81-91. *c*

Vierne, R., and Brunel, R., 1970, Spectre de réflexion infra-rouge de la tourmaline, *Comptes Rendus de l'Academie des Sciences* (Paris) **270:**488-490.

Vinokurov, V. M., and Zaripov, M. M., 1960, Magnetic properties of tourmaline, *Kristallografiya* **4:**873-877. (Russian)

Vishnevskii, V. N., and Klimovskaya, L. K., 1958, A simple method for studying absorption spectra of microcrystals, *Ukrainskyi Fizychnyi Zhurnal* **3:**239-244. (Ukrainian) *c*

*Vladykin, N. V., Antipin, V. S., Kovalenko, V. I., Afonina, G. G., Lapides, I. L., Novikov, V. M., and Gormashera, G. S., 1975, Chemical composition and genetic grouping of tourmalines from Mesozoic granitic rocks of Mongolia, *Zapiski Vsesoyuznogo Mineralogicheskogo Obshchestvo* **104:**403-412. (Russian)

Voigt, W., 1890, Bestimmungen der Elasticitäts-constanten des brasilianischen Turmalins, *Wiedemann's Annalen der Physik und Chemie* **41:**712-724.

Voigt, W., 1910, *Lehrbuch der Kristallphysik* (Copy used is reprint of 1928 revision by M. von Laue; reprint is dated 1966 and is published by Johnson Reprint Corporation, New York.)

Volborth, A., 1955, Das Bor in den finnischen Pegmatiten, *Tschermaks Mineralogische und Petrographische Mitteilungen* **5:**252-259.

von Engelhardt, W., 1942, The abrasion hardness of quartz and other solid substances in various liquids, *Nachrichten der Akademie der Wissenschaften in Göttingen, Mathematisch-Physikalische Klasse* **1942:**34p. *c*

von Leonhard, C. C., 1826, *Handbuch der Oryktognosie,* Mohr, Heidelberg.

von Worobieff, W., 1900, Krystallographische Studien über Turmalin von Ceylon und einigen anderen Vorkommen, *Zeitschrift für Kristallographie und Mineralogie* **33:**263-454.

Vorob'ev, G. G., 1955, The behavior of minerals in the flame of the electric arc, *Zapiski Vsesoyuznogo Mineralogicheskogo Obshchestvo* **84:**466-468. (Russian) *c*

Voskresenskaya, I. E., 1965, Some properties of synthetic tourmalines, *Mineralogicheskii Sbornik, L'vovskii Gosudarstvennyi Universitet* **19:**164-167. (Russian)

Voskresenskaya, I. E., 1968, Synthesis and properties of some ferruginous and non-ferruginous tourmalines, *Akademiia Nauk SSSR, Institut Kristallographii, Gidrotermal'nyi Sintez Kristallov,* pp. 175-192. (Russian)

Voskresenskaya, I. E., 1976, Phase formation in experiments on tourmaline synthesis, *Mineralogicheskii Sbornik, L'vovskii Gosudarstvennyi Universitet* **30:**14-23. (Russian)

Voskresenskaya, I. E., and Barsukova, M. L., 1971, Synthesis and properties of some iron and iron free varieties of tourmaline (English translation), in *Hydrothermal Synthesis of Crystals,* A. N. Lobachev, ed., Consultants Bureau, New York, pp. 126-138.

Voskresenskaya, I. E., and Grum-Grzhimailo, S. V., 1967, Color of synthetic iron tourmalines, *Kristallografiya* **12:**363-365. (Russian)

Voskresenskaya, I. E., and Ivanova, T. N., 1975, Study of tourmalines obtained by synthesis, *Akademiya Nauk SSSR, Mineralogicheskii Muzei* **24:**20-30. (Russian)

Voskresenskaya, I. E., and Okulov, E. N., 1974, Rare variety of tourmaline from cleavelandite pegmatites, *Mineralogicheskii Sbornik, L'vovskii Gosudarstvennyi Universitet* **28:**62-65. (Russian)

Voskresenskaya, I. E., and Plyusnina, I. I., 1975, Infrared spectra of several varieties of synthetic tourmalines, *Vestnik Moskovskogo Universiteta Geologiya* **30:**104-107. (Russian)

Voskresenskaya, I. E., and Shternberg, A. A., 1973, Synthesis of tourmaline in chloride media, I., *Kristallografiya* **18:**562-563.

Voskresenskaya, I., Korovushkin, V. V., Moiseev, B. M., and Shipko, M. N., 1979, Mössbauer spectra of irradiated elbaites, *Kristallografiya* **24:**835-837. (Russian)

Wadhawan, S. K., and Roonwal, G. S., 1977, Genetic significance of tourmaline associated with sulphide mineralization at Zawar area, Udaipur, Rajasthan, India, *Neues Jahrbuch für Mineralogie Abhandlungen Monatshefte* **5:**233-237. *ns*

Waesche, H. H., 1949, Importance and application of piezoelectric minerals, *Mining Engineering* **1:**12-16.

Wagner, A., 1973, Minerals from miarolitic holes, *Mineralienfreund* **11:**21-25. (German) *c*

Wagner, C. E., Pollard, O., Jr., Young, R. A., and Donnay, G., 1971, Texture variants in color-zoned tourmaline crystals, *American Mineralogist* **56:**114-132.

Wagner, J. F., 1818, *Notizen über die Mineralien-Sammlung, Sr. Excellenz des Herren Dr's Alexander's von Crichton,* August Semen, Moscow.

Walcott, C., Gould, J. L., and Kirschvink, J. L., 1979, Pigeons have magnets, *Science* **205:**1027-1029.

*Walenta, K., and Dunn, P. J., 1979, Ferridravite, a new mineral of the tourmaline group from Bolivia, *American Mineralogist* **64:**945-948.

Walker, T. L., 1898, Examination of triclinic minerals, *American Journal of Science* **5:**178-185.

Wallerius, J. G., 1747, *Mineralogia,* Eller Mineral-Riket, Stockholm.

*Wang, W.-S., and Hsu, H.-Y., 1966, A new variety of dravite and its significance in mineralogy, *Kexue Tongbao* (foreign language ed.) **17:**1008-1012.

Wappler, G., 1965, Messungen an Einkristallen von Mineralen, *Zeitschrift fuer Physikalische Chemie* **228:**33-38.

*Ward, G. W., 1931, A chemical and optical study of the black tourmalines, *American Mineralogist* **16:**145-190.

Warner, T. W., Jr., 1935, Spectroscopic analysis of tourmalines with correlation of color and composition, *American Mineralogist* **20:**531-536.

Wasastjerna, J. A., 1922, On light absorption in a pleochroitic uniaxial crystal (a green tourmaline from Minas Geraes in Brazil), *Oversikt av Finska Vetenskaps-Societetens Förhandlingar* **64:**1-8.

Wasserstein, B., 1951, South African granites and their boron content, *Geochimica et Cosmochimica Acta* **1:**329-338.

Watson, W., 1759, Some observations relating to the *Lyncurium* of the ancients, *Philosophical Transactions* **51:**394-398.

Weiss, C. S., 1816-1817, Krystallographische Fundamentalbestimmung des Feldspathes, *Abhandlungen der Akademie der Wissenschaften zu Berlin,* p. 318. *ns*

Weiss, D., 1969, Bestimmung von Eisen (II)-oxid in tourmalinen, *Chemicke Zvesti* **23:**671-676.

Weis, P. L., 1953, Fluid inclusions in minerals from zoned pegmatites of the Black Hills, South Dakota, *American Mineralogist* **38:**671-697.

Wells, M. K., 1946, A contribution to the study of luxullianite, *Mineralogical Magazine* **27:**186-194.

Werding, G., and Schreyer, W., 1977, Synthese von borthaltigem Kornerupin und Turmalin im System $MgO-Al_2O_3-B_2O_3-SiO_2-H_2O$, *Fortschritte der Mineralogie* (Jena) **55:**152-154.

Werner, A. G., 1780, *Cronstedt's Versuch einer Mineralogie,* Übersetzt und Vermehrt (vol. 1, pt. 1), Leipzig. *ns*

Westbrook, J. H., and Jorgensen, P. J., 1968, Effects of water desorption on indentation microhardness anisotropy in minerals, *American Mineralogist* **53:**1899-1909.

White, D. E., and Waring, G. A., 1963, Volcanic emanations, in *Data of Geochemistry,* M. Fleischer, ed., United States Geological Survey Professional Paper 440-K.

Wickersheim, K. A., and Buchanan, R. A., 1968, The near infrared spectrum of beryl: a correction, *American Mineralogist* **53:**347.

Wilcke, J. C., 1766, Historien om Tourmalin, *Königlichen Vetenskaps—Academiens* (Sweden) **27:**89-115. (Swedish)

Wild, G. O., 1931, Spectroskopische Untersuchungen an brasilianischen Turmalinen, *Zentralblatt für Mineralogie, Geologie und Palaeontologie Abteilung A* (Stuttgart), pp. 327-330.

Wilkins, R. W. T., Farrell, E. F., and Naiman, C. S., 1969, The crystal field spectra and dichroism of tourmaline, *Journal of Physics and Chemistry of Solids* **30:**43-56.

Williams, E. H., Jr., 1876, On crystals of tourmaline with enveloped orthoclase, *American Journal of Science* **11:**274-275.

Williams, W. S., Breger, L., and Johnson, M., 1975, Ceramic models for study of piezo-electricity in solids, *American Ceramic Society Journal* **58:**415-417.

Wilson, B., 1759, Experiments on the Tourmaline, *Philosophical Transactions* **51:**308-339.

Wilson, E., 1920, The measurement of magnetic susceptibilities of low order, *Royal Society of London Proceedings* (ser. A) **96:**429-455.

Wilson, W. E., 1977, What's new in minerals? *Mineralogical Record* **8:**389.

Winchell, A. N., and Winchell, H., 1951, *Elements of Optical Mineralogy . . . Part II. Descriptions of Minerals* (4th ed.), Wiley, New York.

Winchers, E., Schlect, W. G., and Gordon, C. L., 1945, Preparing refractory oxides, silicates and ceramic materials for analysis by heating with acids in sealed tubes at elevated temperatures, *National Bureau of Standards Journal of Research* **33:**451-456 (1944) (res. paper no. 1621). *c*

Woodford, A. O., Crippen, R. A., and Garner, K. B., 1940, Section across Commercial quarry, Crestmore, California, *American Mineralogist* **26:**351-381.

Wooster, W. A., 1938, *A Textbook on Crystal Physics,* Cambridge University Press, London.

Wooster, W. A., 1946, The relation between piezo-electricity and crystal structure, *Memoires de la Société Royale des Sciences de Bohème* (Praha) **10:**477-. *ns*

Wooster, W. A., 1976, Etch figures and crystal structures, *Kristall und Technik* **11:**615-623.

Wooster, W. A., and Breton, A., 1970, *Experimental Crystal Physics,* Clarendon Press, Oxford (England).

Wülfing, E. A., 1889, Berechnung der chemischen Formel der Turmaline nach den Analysen von R. B. Riggs, *Tschermaks Mineralogische und Petrographische Mitteilungen* **10:**161-173.

Wülfing, E. A., 1900, Ueber einige krystallographische Constanten des Turmalins und ihre Abhangigkeit von seiner chemischen Zussamensetzung, *Zeitschrift für Kristallographie und Mineralogie* **36:**538-543.

Wülfing, E. A., 1901, Ueber die Lichtbewegung im Turmalin, *Zentralblatt für Mineralogie, Geologie und Palaeontologie* **1:**299-302.

*Wülfing, E. A., and Becht, K., 1913, *Über neue Turmalinanalyses,* C. Winters, Heidelberg.

Wyckoff, R. W. G., 1931, *The Structure of Crystals* (2nd ed.), Chemical Catalog Company, New York.

Wyo, J. R., and Reitz, H. K., 1944, Note on the heavy residues of the recent dune sands of the eastern Cape Flats near Faure, *Annale Universitet van Stellenbosch* **22A:**135-142 *c*

*Yakovleva, M. E., and Osolodkina, G. A., 1966, Tourmaline, *Akademiya Nauk SSSR, Mineralogicheskii Muzei (Moscow)* **17:**249-252. (Russian)

Yamaguchi, S., 1964*a,* Electron diffraction of a pyroelectric tourmaline crystal, *Journal of Applied Physics* **35:**1654-1655. *c*

Yamaguchi, S., 1964*b,* Electron diffraction of a pyroelectric tourmaline crystal, *Naturwissenschaften* **51:**55. *c*

Yamaguchi, S., 1983, Tourmaline as gas detecting sensor, *Materials Chemistry and Physics* **8:**493-498. *c*

Young, B. B., and Millman, A. P., 1964, Microhardness and deformation characteristics of ore minerals, *Institute of Metallurgy (London) Transactions* **73:**437-466.

Young, R. A., Wagner, C. E., Pollard, C. O., Donnay, G., 1969, Crystalline heterogeneity, evidence from source-image distortion, *Carnegie Institute of Washington Year Book* **67:**220-221.

Zadnik, M. G., 1982, Cryogenic techniques for neon trapped in terrestrial minerals: anomalous neon in beryl and tourmalines (abst.), *Abstracts of papers presented at the 45th annual meeting of the Meteoritical Society, Meteoritics* **17**(4):302.

Zagyaneskĭi, A. L., 1958, Blast furnace production of low sulfur pig iron, *U.S.S.R. patent 74,361,* November 22, 1958. (Russian)

Zheludev, I. S., and Belov, V. F., 1967, Moessbauer effect in tourmaline and in a solid solution of $BaTi_{0.75}Sn_{0.25}O_3$, *Izvestiya Akademii Nauk SSSR Seriya Fizicheskaya* **31:**1117-1119. (Russian)

Zheludev, I. S., and Tagieva, M. M., 1963, Electric polarization of crystalline pyroelectrics under hydrostatic compression (expansion), *Kristallografiya* **7:**589-592.

Zubova, S. I., and Rezapova, N. M., 1964, Source of boron in authigenic tourmaline from the Altai volcanic-sedimentary iron ores, *Trudy Sibirskogo Nauchno-Issledovatel'skogo Instituta Geologii, Geofiziki i Mineral'nogo Syr'ya,* no. 35, pp. 155-157. (Russian)

Zwaan, P. C., 1974, Garnet, corundum, and other gem minerals from Umba, Tanzania, *Scripta Geologica,* no. 20, pp. 1-41. *c*

Index